A Guide to Monte Carlo Simulations in Statistical Physics

This book deals with all aspects of Monte Carlo simulation of complex physical systems encountered in condensed-matter physics and statistical mechanics as well as in related fields, for example polymer science and lattice gauge theory.

After briefly recalling essential background in statistical mechanics and probability theory, the authors give a succinct overview of simple sampling methods. The next several chapters develop the importance sampling method, both for lattice models and for systems in continuum space. The concepts behind the various simulation algorithms are explained in a comprehensive fashion, as are the techniques for efficient evaluation of system configurations generated by simulation (histogram extrapolation, multicanonical sampling, thermodynamic integration and so forth). The fact that simulations deal with small systems is emphasized, and the text incorporates various finite size scaling concepts to show how a careful analysis of finite size effects can be a useful tool for the analysis of simulation results. Other chapters also provide introductions to quantum Monte Carlo methods, aspects of simulations of growth phenomena and other systems far from equilibrium, and the Monte Carlo renormalization group approach to critical phenomena. Throughout the book there are many applications, examples, and exercises to help the reader in a thorough study of this book; furthermore, many up-to-date references to more specialized literature are also provided.

This book will be bought by graduate students who have to deal with computer simulations in their research, as well as by postdoctoral researchers, in both physics and physical chemistry. It can be used as a textbook for graduate courses on computer simulations in physics and related disciplines.

DAVID P. LANDAU was born on June 22, 1941 in St. Louis, MO, USA. He received a B.A. in Physics from Princeton University in 1963 and a Ph.D. in Physics from Yale University in 1967. His Ph.D. research involved experimental studies of magnetic phase transitions as did his postdoctoral research at the CNRS in Grenoble, France. After teaching at Yale for a year he moved to the University of Georgia and initiated a research program of Monte Carlo studies in statistical physics. He is currently the Research Professor of Physics and founding Director of the Center for Simulational Physics at the University of Georgia. He has been teaching graduate courses in computer simulations since 1982. David Landau has authored/co-authored almost 300 research publications and is editor/co-editor of more than a dozen books. The University of Georgia awarded him a Creative Research Medal in 1981 and named him a Senior Teaching Fellow in 1993. In 1998 he also became an Adjunct Professor at the Helsinki University of Technology. He is a Fellow of the American Physical Society and a past Chair of the Division of Computational Physics of the APS. He received the Jesse W. Beams award from SESAPS in 1987, and a Humboldt Fellowship and Humboldt Senior U.S. Scientist award in 1975 and 1988 respectively. In 1999 he was named a Fellow of the Japan Society for the Promotion of Science. He is currently a Principal Editor for the journal *Computer Physics Communications*.

KURT BINDER was born on February 10, 1944 in Korneuburg, Austria and then lived in Vienna, where he received his Ph.D. in 1969 at the Technical University of Vienna. Even then his thesis dealt with Monte Carlo simulations of Ising and Heisenberg magnets, and since then he has pioneered the development of Monte Carlo simulation methods in statistical physics. From 1969 until 1974 Kurt Binder worked at the Technical University in Munich, where he defended his Habilitation thesis in 1973 after a stay as IBM postdoctoral fellow in Zurich in 1972/73. Further key times in his career were spent at Bell Laboratories, Murray Hill, NJ (1974) and a first appointment as Professor of Theoretical Physics at the University of Saarbrücken back in Germany (1974–1977), followed by a joint appointment as full professor at the University of Cologne and the position as one of the directors of the Institute of Solid State Research at Jülich (1977–1983). He has held his present position as Professor of Theoretical Physics at the University of Mainz, Germany since 1983, and since 1989 he has also been an external member of the Max-Planck-Institut for Polymer Research at Mainz. Kurt Binder has written more than 600 research publications and edited five books dealing with computer simulation. His book (with Dieter W. Hermann) *Monte Carlo Simulation in Statistical Physics: An Introduction*, first published in 1988, is in its 3rd edition. Kurt Binder has been a corresponding member of the Austrian Academy of Sciences in Vienna since 1992 and received the Max Planck Medal of the German Physical Society in 1993. He also acts as Editorial Board member of several journals and presently serves as chairman of the IUPAP Commission on Statistical Physics.

A Guide to
Monte Carlo Simulations in
Statistical Physics

David P. Landau
Center for Simulational Physics, The University of Georgia

Kurt Binder
Institut für Physik, Johannes-Gutenberg-Universität Mainz

CAMBRIDGE
UNIVERSITY PRESS

PUBLISHED BY THE PRESS SYNDICATE OF THE UNIVERSITY OF CAMBRIDGE
The Pitt Building, Trumpington Street, Cambridge, United Kingdom

CAMBRIDGE UNIVERSITY PRESS
The Edinburgh Building, Cambridge CB2 2RU, UK   http://www.cup.cam.ac.uk
40 West 20th Street, New York, NY 10011-4211, USA   http://www.cup.org
10 Stamford Road, Oakleigh, Melbourne 3166, Australia
Ruiz de Alarcón 13, 28014 Madrid, Spain

First published 2000

Printed in the United Kingdom at the University Press, Cambridge

Typeface Ehrhardt 10.25/12.5pt   *System* Advent 3B2   [KT]

A catalogue record for this book is available from the British Library

Library of Congress Cataloging in Publication data
Landau, David P.
    A guide to Monte Carlo simulations in statistical physics / David P. Landau, Kurt Binder.
        p.   cm.
    Includes index.
    ISBN 0 521 65314 2 (hardbound)
    1. Monte Carlo method. 2. Statistical physics. I. Binder, K. (Kurt), 1944– . II. Title.
QC174.85.M64L36 2000
530.13–dc21   99-38308CIP

ISBN 0 521 65314 2 hardback
ISBN 0 521 65366 5 paperback

Contents

Preface

Historically physics was first known as 'natural philosophy' and research was carried out by purely theoretical (or philosophical) investigation. True progress was obviously limited by the lack of real knowledge of whether or not a given theory really applied to nature. Eventually experimental investigation became an accepted form of research although it was always limited by the physicist's ability to prepare a sample for study or to devise techniques to probe for the desired properties. With the advent of computers it became possible to carry out simulations of models which were intractable using 'classical' theoretical techniques. In many cases computers have, for the first time in history, enabled physicists not only to invent new models for various aspects of nature but also to solve those same models without substantial simplification. In recent years computer power has increased quite dramatically, with access to computers becoming both easier and more common (e.g. with personal computers and workstations), and computer simulation methods have also been steadily refined. As a result computer simulations have become another way of doing physics research. They provide another perspective; in some cases simulations provide a theoretical basis for understanding experimental results, and in other instances simulations provide 'experimental' data with which theory may be compared. There are numerous situations in which direct comparison between analytical theory and experiment is inconclusive. For example, the theory of phase transitions in condensed matter must begin with the choice of a Hamiltonian, and it is seldom clear to what extent a particular model actually represents a real material on which experiments are done. Since analytical treatments also usually require mathematical approximations whose accuracy is difficult to assess or control, one does not know whether discrepancies between theory and experiment should be attributed to shortcomings of the model, the approximations, or both. The goal of this text is to provide a basic understanding of the methods and philosophy of computer simulations research with an emphasis on problems in statistical thermodynamics as applied to condensed matter physics or materials science. There exist many other simulational problems in physics (e.g. simulating the spectral intensity reaching a detector in a scattering experiment) which are more straightforward and which will only occasionally be mentioned. We shall use many specific examples and, in some cases, give explicit computer programs, but we wish to emphasize that these methods are applicable to a wide variety of systems including those which are not treated here at all. As computer architecture changes the methods presented here will in some cases require relatively minor reprogramming and in other instances will require new algorithm development in order to be truly efficient. We hope that this material will prepare the reader for studying new and different problems using both existing as well as new computers.

At this juncture we wish to emphasize that it is important that the simulation algorithm and conditions be chosen with the physics problem at hand in mind. The *interpretation* of the resultant output is critical to the success of any simulational project, and we thus include substantial information about various aspects of thermodynamics and statistical physics to help strengthen this connection. We also wish to draw the reader's attention to the rapid development of scientific visualization and the important role that it can play in producing *understanding* of the results of some simulations.

This book is intended to serve as an introduction to Monte Carlo methods for graduate students, and advanced undergraduates, as well as more senior researchers who are not yet experienced in computer simulations. The book is divided up in such a way that it will be useful for courses which only wish to deal with a restricted number of topics. Some of the later chapters may simply be skipped without affecting the understanding of the chapters which follow. Because of the immensity of the subject, as well as the existence of a number of very good monographs and articles on advanced topics which have become quite technical, we will limit our discussion in certain areas, e.g. polymers, to an introductory level. The examples which are given are in FORTRAN, not because it is necessarily the best scientific computer language, but because it is certainly the most widespread. Many existing Monte Carlo programs and related subprograms are in FORTRAN and will be available to the student from libraries, journals, etc. A number of sample problems are suggested in the various chapters; these may be assigned by course instructors or worked out by students on their own. Our experience in assigning problems to students taking a graduate course in simulations at the University of Georgia over a 15 year period suggests that for maximum pedagogical benefit, students should be required to prepare cogent reports after completing each assigned simulational problem. Students were required to complete seven 'projects' in the course of the quarter for which they needed to write and debug programs, take and analyze data, and prepare a report. Each report should briefly describe the algorithm used, provide sample data and data analysis, draw conclusions and add comments. (A sample program/output should be included.) In this way, the students obtain practice in the summary and presentation of simulational results, a skill which will prove to be valuable later in their careers. For convenience, the case studies that are described have been simply taken from the research of the authors of this book – the reader should be aware that this is by no means meant as a negative statement on the quality of the research of numerous other groups in the field. Similarly, selected references are given to aid the reader in finding more detailed information, but because of length restrictions it is simply not possible to provide a complete list of relevant literature. Many coworkers have been involved in the work which is mentioned here, and it is a pleasure to thank them for their fruitful collaboration. We have also benefited from the stimulating comments of many of our colleagues and we wish to express our thanks to them as well.

1 Introduction

1.1 WHAT IS A MONTE CARLO SIMULATION?

In a Monte Carlo simulation we attempt to follow the 'time dependence' of a model for which change, or growth, does not proceed in some rigorously predefined fashion (e.g. according to Newton's equations of motion) but rather in a stochastic manner which depends on a sequence of random numbers which is generated during the simulation. With a second, different sequence of random numbers the simulation will not give identical results but will yield values which agree with those obtained from the first sequence to within some 'statistical error'. A very large number of different problems fall into this category: in percolation an empty lattice is gradually filled with particles by placing a particle on the lattice randomly with each 'tick of the clock'. Lots of questions may then be asked about the resulting 'clusters' which are formed of neighboring occupied sites. Particular attention has been paid to the determination of the 'percolation threshold', i.e. the critical concentration of occupied sites for which an 'infinite percolating cluster' first appears. A percolating cluster is one which reaches from one boundary of a (macroscopic) system to the opposite one. The properties of such objects are of interest in the context of diverse physical problems such as conductivity of random mixtures, flow through porous rocks, behavior of dilute magnets, etc. Another example is diffusion limited aggregation (DLA) where a particle executes a random walk in space, taking one step at each time interval, until it encounters a 'seed' mass and sticks to it. The growth of this mass may then be studied as many random walkers are turned loose. The 'fractal' properties of the resulting object are of real interest, and while there is no accepted analytical theory of DLA to date, computer simulation is the method of choice. In fact, the phenomenon of DLA was first discovered by Monte Carlo simulation!

Considering problems of statistical mechanics, we may be attempting to sample a region of phase space in order to estimate certain properties of the model, although we may not be moving in phase space along the same path which an exact solution to the time dependence of the model would yield. Remember that the task of equilibrium statistical mechanics is to calculate thermal averages of (interacting) many-particle systems: Monte Carlo simulations can do that, taking proper account of statistical fluctuations and their effects in such systems. Many of these models will be discussed in more detail in later chapters so we shall not provide further details here. Since the accuracy of a Monte Carlo estimate depends upon the thoroughness with which phase space is probed, improvement may be obtained by simply running the calculation a little longer to increase the number of samples. Unlike in the application of many analytic techniques (e.g. perturbation theory for which the extension to higher order may be prohibitively difficult), the improvement of the accuracy of Monte Carlo results is possible not just in principle but also in practice!

1.2. WHAT PROBLEMS CAN WE SOLVE WITH IT?

The range of different physical phenomena which can be explored using Monte Carlo methods is exceedingly broad. Models which either naturally or through approximation can be discretized can be considered. The motion of individual atoms may be examined directly; e.g. in a binary (AB) metallic alloy where one is interested in interdiffusion or unmixing kinetics (if the alloy was prepared in a thermodynamically unstable state) the random hopping of atoms to neighboring sites can be modeled directly. This problem is complicated because the jump rates of the different atoms depend on the locally differing environment. Of course, in this description the quantum mechanics of atoms with potential barriers in the eV range is not explicitly considered, and the sole effect of phonons (lattice vibrations) is to provide a 'heat bath' which provides the excitation energy for the jump events. Because of a separation of time scales (the characteristic times between jumps are orders of magnitude larger than atomic vibration periods) this approach provides very good approximation. The same kind of arguments hold true for growth phenomena involving macroscopic objects, such as DLA growth of colloidal particles; since their masses are orders of magnitude larger than atomic masses, the motion of colloidal particles in fluids is well described by classical, random Brownian motion. These systems are hence well suited to study by Monte Carlo simulations which use random numbers to realize random walks. The motion of a fluid may be studied by considering 'blocks' of fluid as individual particles, but these blocks will be far larger than individual molecules. As an example, we consider 'micelle formation' in lattice models of microemulsions (water–oil–surfactant fluid mixtures) in which each surfactant molecule may be modeled by two 'dimers' on the lattice (two occupied nearest neighbor sites on the lattice). Different effective interactions allow one dimer to mimic the hydrophilic group and the other dimer the hydrophobic group of the surfactant molecule. This model then allows the study of the size and shape of the aggregates of surfactant molecules (the micelles) as well as the kinetic aspects of their formation. In reality, this process is quite slow so that a deterministic molecular dynamics simulation (i.e. numerical integration of Newton's second law) is not feasible. This example shows that part of the 'art' of simulation is the appropriate choice (or invention!) of a suitable (coarse-grained) model. Large collections of interacting classical particles are directly amenable to Monte Carlo simulation, and the behavior of interacting quantized particles is being studied either by transforming the system into a pseudo-classical model or by considering permutation properties directly. These considerations will be discussed in more detail in later chapters. Equilibrium properties of systems of interacting atoms have been extensively studied as have a wide range of models for simple and complex fluids, magnetic materials, metallic alloys, adsorbed surface layers, etc. More recently polymer models have been studied with increasing frequency; note that the simplest model of a flexible polymer is a random walk, an object which is well suited for Monte Carlo simulation. Furthermore, some of the most significant advances in understanding the theory of elementary particles have been made using Monte Carlo simulations of lattice gauge models.

1.3 WHAT DIFFICULTIES WILL WE ENCOUNTER?

1.3.1 Limited computer time and memory

Because of limits on computer speed there are some problems which are inherently not suited to computer simulation, at this time. A simulation which requires years of cpu time on whatever machine is available is simply impractical. Similarly a calculation which requires memory which far exceeds that which is available can be carried out only by using very sophisticated programming techniques which slow down running speeds and greatly increase the probability of errors. It is therefore important that the user first consider the requirements of both memory and cpu time *before* embarking on a project to ascertain whether or not there is a realistic possibility of obtaining the resources to simulate a problem properly. Of course, with the rapid advances being made by the computer industry, it may be necessary to wait only a few years for computer facilities to catch up to your needs. Sometimes the tractability of a problem may require the invention of a new, more efficient simulation algorithm. Of course, developing new strategies to overcome such difficulties constitutes an exciting field of research by itself.

1.3.2 Statistical and other errors

Assuming that the project can be done, there are still potential sources of error which must be considered. These difficulties will arise in many different situations with different algorithms so we wish to mention them briefly at this time without reference to any specific simulation approach. All computers operate with limited word length and hence limited precision for numerical values of any variable. Truncation and round-off errors may in some cases lead to serious problems. In addition there are statistical errors which arise as an inherent feature of the simulation algorithm due to the finite number of members in the 'statistical sample' which is generated. These errors must be estimated and then a 'policy' decision must be made, i.e. should more cpu time be used to reduce the statistical errors or should the cpu time available be used to study the properties of the system under other conditions. Lastly there may be systematic errors. In this text we shall not concern ourselves with tracking down errors in computer programming – although the practitioner must make a special effort to eliminate any such errors! – but with more fundamental problems. An algorithm may fail to treat a particular situation properly, e.g. due to the finite number of particles which are simulated, etc. These various sources of error will be discussed in more detail in later chapters.

1.4 WHAT STRATEGY SHOULD WE FOLLOW IN APPROACHING A PROBLEM?

Most new simulations face hidden pitfalls and difficulties which may not be apparent in early phases of the work. It is therefore often advisable to begin with a relatively simple program and use relatively small system sizes and modest running times. Sometimes there are special values of parameters for which the answers are already known (either from analytic solutions or from previous, high quality simulations) and these cases can be used to test a new simulation program. By proceeding in this manner one is able to uncover which are the parameter ranges of interest and what unexpected difficulties are present. It is then possible to refine the program and then to increase running times. Thus *both* cpu time and human time can be used most effectively. It makes little sense of course to spend a month to rewrite a computer program which may result in a total saving of only a few minutes of cpu time. If it happens that the outcome of such test runs shows that a new problem is not tractable with reasonable effort, it may be desirable to attempt to improve the situation by redefining the model or redirect the focus of the study. For example, in polymer physics the study of short chains (oligomers) by a given algorithm may still be feasible even though consideration of huge macromolecules may be impossible.

1.5 HOW DO SIMULATIONS RELATE TO THEORY AND EXPERIMENT?

In many cases theoretical treatments are available for models for which there is no perfect physical realization (at least at the present time). In this situation the only possible test for an approximate theoretical solution is to compare with 'data' generated from a computer simulation. As an example we wish to mention recent activity in growth models, such as diffusion limited aggrega-

Fig. 1.1 Schematic
view of the
relationship between
theory, experiment,
and computer
simulation.

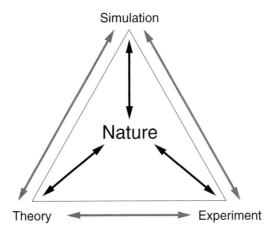

tion, for which a very large body of simulation results already exists but for which extensive experimental information is just now becoming available. It is not an exaggeration to say that interest in this field was created by simulations. Even more dramatic examples are those of reactor meltdown or large scale nuclear war: although we want to know what the results of such events would be we do not want to carry out experiments! There are also real physical systems which are sufficiently complex that they are not presently amenable to theoretical treatment. An example is the problem of understanding the specific behavior of a system with many competing interactions and which is undergoing a phase transition. A model Hamiltonian which is believed to contain all the essential features of the physics may be proposed, and its properties may then be determined from simulations. If the simulation (which now plays the role of theory) disagrees with experiment, then a new Hamiltonian must be sought. An important advantage of the simulations is that different physical effects which are simultaneously present in real systems may be isolated and through separate consideration by simulation may provide a much better understanding. Consider, for example, the phase behavior of polymer blends – materials which have ubiquitous applications in the plastics industry. The miscibility of different macromolecules is a challenging problem in statistical physics in which there is a subtle interplay between complicated enthalpic contributions (strong covalent bonds compete with weak van der Waals forces, and Coulombic interactions and hydrogen bonds may be present as well) and entropic effects (configurational entropy of flexible macromolecules, entropy of mixing, etc.). Real materials are very difficult to understand because of various asymmetries between the constituents of such mixtures (e.g. in shape and size, degree of polymerization, flexibility, etc.). Simulations of simplified models can 'switch off' or 'switch on' these effects and thus determine the particular consequences of each contributing factor. We wish to emphasize that the aim of simulations is not to provide better 'curve fitting' to experimental data than does analytic theory. The goal is to create an understanding of physical properties and processes which is as complete as possible, making use of the perfect control of 'experimental' conditions in the 'computer experiment' and of the possibility to examine every aspect of system configurations in detail. The desired result is then the elucidation of the physical mechanisms that are responsible for the observed phenomena. We therefore view the relationship between theory, experiment, and simulation to be similar to those of the vertices of a triangle, as shown in Fig. 1.1: each is distinct, but each is strongly connected to the other two.

With the rapidly increasing growth of computer power which we are now seeing, coupled with the steady drop in price, it is clear that computer simulations will be able to increase rapidly in sophistication to allow more subtle comparisons to be made.

2 Some necessary background

2.1 THERMODYNAMICS AND STATISTICAL MECHANICS: A QUICK REMINDER

2.1.1 Basic notions

In this chapter we shall review some of the basic features of thermodynamics and statistical mechanics which will be used later in this book when devising simulation methods and interpreting results. Many good books on this subject exist and we shall not attempt to present a complete treatment. This chapter is hence *not* intended to *replace* any textbook for this important field of physics but rather to 'refresh' the reader's knowledge and to draw attention to notions in thermodynamics and statistical mechanics which will henceforth be assumed to be known throughout this book.

2.1.1.1 Partition function

Equilibrium statistical mechanics is based upon the idea of a partition function which contains all of the essential information about the system under consideration. The general form for the partition function for a classical system is

$$Z = \sum_{\text{all states}} e^{-\mathcal{H}/k_{\text{B}}T}, \tag{2.1}$$

where $\mathcal{H}$ is the Hamiltonian for the system, T is the temperature, and k_{B} is the Boltzmann constant. The sum in Eqn. (2.1) is over all possible states of the system and thus depends upon the size of the system and the number of degrees of freedom for each particle. For systems consisting of only a few interacting particles the partition function can be written down exactly with the consequence that the properties of the system can be calculated in closed form. In a few other cases the interactions between particles are so simple that evaluating the partition function is possible.

Example

Let us consider a system with N particles each of which has only two states, e.g. a non-interacting Ising model in an external magnetic field H, and which has the Hamiltonian

$$\mathcal{H} = -H \sum_i \sigma_i, \qquad (2.2)$$

where $\sigma_i = \pm 1$. The partition function for this system is simply

$$Z = \left(e^{-H/k_B T} + e^{+H/k_B T} \right)^N, \qquad (2.3)$$

where for a single spin the sum in Eqn. (2.1) is only over two states. The energies of the states and the resultant temperature dependence of the internal energy appropriate to this situation are pictured in Fig. 2.1.

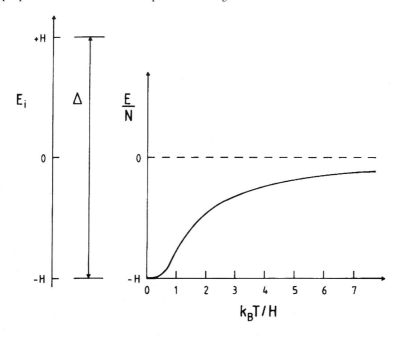

Fig. 2.1 (left) Energy levels for the two level system in Eqn. (2.2); (right) internal energy for a two level system as a function of temperature.

Problem 2.1 Work out the average magnetization per spin, using Eqn. (2.3), for a system of N non-interacting Ising spins in an external magnetic field. [Solution $M = -(1/N)\partial F/\partial H$, $F = -k_B T \ln Z \Rightarrow M = \tanh(H/k_B T)$]

There are also a few examples where it is possible to extract exact results for very large systems of interacting particles, but in general the partition function cannot be evaluated exactly. Even enumerating the terms in the partition function on a computer can be a daunting task. Even if we have only 10 000 interacting particles, a very small fraction of Avogadro's number, with only two possible states per particle, the partition function would contain $2^{10\,000}$

terms! The probability of any particular state of the system is also determined by the partition function. Thus, the probability that the system is in state μ is given by

$$P_\mu = e^{-\mathcal{H}(\mu)/k_B T}/Z, \tag{2.4}$$

where $\mathcal{H}(\mu)$ is the Hamiltonian when the system is in the μth state. As we shall show in succeeding chapters, the Monte Carlo method is an excellent technique for estimating probabilities, and we can take advantage of this property in evaluating the results.

2.1.1.2 Free energy, internal energy, and entropy

It is possible to make a direct connection between the partition function and thermodynamic quantities and we shall now briefly review these relationships. The free energy of a system can be determined from the partition function (Callen, 1985) from

$$F = -k_B T \ln Z \tag{2.5}$$

and all other thermodynamic quantities can be calculated by appropriate differentiation of Eqn. (2.5). This relation then provides the connection between statistical mechanics and thermodynamics. The internal energy of a system can be obtained from the free energy via

$$U = -T^2 \partial(F/T)/\partial T. \tag{2.6}$$

By the use of a partial derivative we imply here that F will depend upon other variables as well, e.g. the magnetic field H in the above example, which are held constant in Eqn. (2.6). This also means that if the internal energy of a system can be measured, the free energy can be extracted by appropriate integration, assuming, of course, that the free energy is known at some reference temperature. We shall see that this fact is important for simulations which do not yield the free energy directly but produce instead values for the internal energy. Free energy differences may then be estimated by integration, i.e. from $\Delta(F/T) = \int d(1/T)U$.

Using Eqn. (2.6) one can easily determine the temperature dependence of the internal energy for the non–interacting Ising model, and this is also shown in Fig. 2.1. Another important quantity, the entropy, measures the amount of disorder in the system. The entropy is defined in statistical mechanics by

$$S = -k_B \ln P, \tag{2.7}$$

where P is the probability of occurrence of a state. The entropy can be determined from the free energy from

$$S = -(\partial F/\partial T)_{V,N}. \tag{2.8}$$

2.1.1.3 Thermodynamic potentials and corresponding ensembles

The internal energy is expressed as a function of the extensive variables, S, V, N, etc. There are situations when it is appropriate to replace some of these variables by their conjugate intensive variables, and for this purpose additional thermodynamic potentials can be defined by suitable Legendre transforms of the internal energy; in terms of liquid–gas variables such relations are given by:

$$F = U - TS, \tag{2.9a}$$

$$H = U + pV, \tag{2.9b}$$

$$G = U - TS + pV, \tag{2.9c}$$

where F is the Helmholtz free energy, H is the enthalpy, and G is the Gibbs free energy. Similar expressions can be derived using other thermodynamic variables, e.g. magnetic variables. The free energy is important since it is a minimum in equilibrium when T and V are held constant, while G is a minimum when T and p are held fixed. Moreover, the difference in free energy between any two states does not depend on the path between the states. Thus, in Fig. 2.2 we consider two points in the p–T plane. Two different paths which connect points 1 and 2 are shown; the difference in free energy between these two points is identical for both paths, i.e.

$$F_2 - F_1 = \int_{\text{path I}} dF = \int_{\text{path II}} dF. \tag{2.10}$$

The multidimensional space in which each point specifies the complete microstate (specified by the degrees of freedom of all the particles) of a system is termed 'phase space'. Averages over phase space may be constructed by considering a large number of identical systems which are held at the same fixed conditions. These are called 'ensembles'. Different ensembles are relevant for different constraints. If the temperature is held fixed the set of systems is said to belong to the 'canonical ensemble' and there will be some distribution of energies among the different systems. If instead the energy is fixed, the ensemble is termed the 'microcanonical' ensemble. In

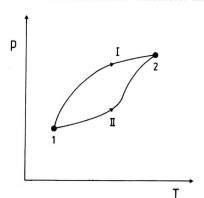

Fig. 2.2 Schematic view of different paths between two different points in thermodynamic p–T space.

the first two cases the number of particles is held constant; if the number of particles is allowed to fluctuate the ensemble is the 'grand canonical' ensemble.

Systems are often held at fixed values of intensive variables, such as temperature, pressure, etc. The conjugate extensive variables, energy, volume, etc. will fluctuate with time; indeed these fluctuations will actually be observed during Monte Carlo simulations.

Problem 2.2 Consider a two level system composed of N non-interacting particles where the groundstate of each particle is doubly degenerate and separated from the upper level by an energy ΔE. What is the partition function for this system? What is the entropy as a function of temperature?

2.1.1.4 Fluctuations

Equations (2.4) and (2.5) imply that the probability that a given 'microstate' μ occurs is $P_\mu = \exp\{[F - \mathcal{H}(\mu))]/k_{\mathrm{B}}T\} = \exp\{-S/k_{\mathrm{B}}\}$. Since the number of different microstates is so huge, we are not only interested in probabilities of individual microstates but also in probabilities of macroscopic variables, such as the internal energy U. We first form the moments (where $\beta \equiv 1/k_{\mathrm{B}}T$; the average energy is denoted $\overline{U}$ and U is a fluctuating quantity),

$$\overline{U}(\beta) = \langle \mathcal{H}(\mu) \rangle \equiv \sum_\mu P_\mu \mathcal{H}(\mu) = \sum_\mu \mathcal{H}(\mu) e^{-\beta \mathcal{H}(\mu)} \Big/ \sum_\mu e^{-\beta \mathcal{H}(\mu)},$$

$$\langle \mathcal{H}^2 \rangle = \sum_\mu \mathcal{H}^2 e^{-\beta \mathcal{H}(\mu)} \Big/ \sum_\mu e^{-\beta \mathcal{H}(\mu)}, \qquad (2.11)$$

and note the relation $-(\partial U(\beta)/\partial \beta)_V = \langle \mathcal{H}^2 \rangle - \langle \mathcal{H} \rangle^2$. Since $(\partial U/\partial T)_V = C_V$, the specific heat thus yields a fluctuation relation

$$k_{\mathrm{B}} T^2 C_V = \langle \mathcal{H}^2 \rangle - \langle \mathcal{H} \rangle^2 = \langle (\Delta U)^2 \rangle_{NVT}, \qquad \Delta U \equiv \mathcal{H} - \langle \mathcal{H} \rangle. \quad (2.12)$$

Now for a macroscopic system ($N \gg 1$) away from a critical point, $U \propto N$ and the energy and specific heat are extensive quantities. However, since both $\langle \mathcal{H}^2 \rangle$ and $\langle \mathcal{H} \rangle^2$ are clearly proportional to N^2, we see that the relative fluctuation of the energy is very small, of order $1/N$. While in real experiments (where often $N \approx 10^{22}$) such fluctuations may be too small to be detectable, in simulations these thermal fluctuations are readily observable, and relations such as Eqn. (2.12) are useful for the actual estimation of the specific heat from energy fluctuations. Similar fluctuation relations exist for many other quantities, for example the isothermal susceptibility $\chi = (\partial \langle M \rangle / \partial H)_T$ is related to fluctuations of the magnetization $M = \sum_i \sigma_i$, as

$$k_{\mathrm{B}} T \chi = \langle M^2 \rangle - \langle M \rangle^2 = \sum_{i,j} (\langle \sigma_i \sigma_j \rangle - \langle \sigma_i \rangle \langle \sigma_j \rangle). \qquad (2.13)$$

Writing the Hamiltonian of a system in the presence of a magnetic field H as $\mathcal{H} = \mathcal{H}_0 - HM$, we can easily derive Eqn. (2.13) from $\langle M \rangle = \sum_\mu M \exp$

$[-\beta\mathcal{H}(\mu)]/\sum_{\mu}\exp[-\beta\mathcal{H}(\mu)]$ in a similar fashion as above. The relative fluctuation of the magnetization is also small, of order $1/N$.

It is not only of interest to consider for quantities such as the energy or magnetization the lowest order moments but to discuss the full probability distribution $P(U)$ or $P(M)$, respectively. For a system in a pure phase the probability is given by a simple Gaussian distribution

$$P(U) = (2\pi k_B C_V T^2)^{-1/2}\exp\left[-(\Delta U)^2/2k_B Tx\right] \qquad (2.14)$$

while the distribution of the magnetization for the paramagnetic system becomes

$$P(M) = (2\pi k_B T\chi)^{-1/2}\exp\left[-(M - \langle M\rangle)^2/2k_B T\chi\right]. \qquad (2.15)$$

It is straightforward to verify that Eqns. (2.14), (2.15) are fully consistent with the fluctuation relations (2.12), (2.13). Since Gaussian distributions are completely specified by the first two moments, higher moments $\langle H^k\rangle$, $\langle M^k\rangle$, which could be obtained analogously to Eqn. (2.11), are not required. Note that on the scale of $\overline{U}/N$ and $\langle M\rangle/N$ the distributions $P(U)$, $P(M)$ are extremely narrow, and ultimately tend to δ-functions in the thermodynamic limit. Thus these fluctuations are usually neglected altogether when dealing with relations between thermodynamic variables.

An important consideration is that the thermodynamic state variables do not depend on the ensemble chosen (in pure phases) while the fluctuations do. Therefore, one obtains the same average internal energy $\overline{U}(N, V, T)$ in the canonical ensemble as in the NpT ensemble while the specific heats and the energy fluctuations differ (see Landau and Lifshitz, 1980):

$$\left\langle(\Delta U)^2\right\rangle_{NpT} = k_B T^2 C_V - \left[T\left(\frac{\partial p}{\partial T}\right)_V - p\right]^2 T\left(\frac{\partial V}{\partial p}\right)_T. \qquad (2.16)$$

It is also interesting to consider fluctuations of several thermodynamic variables together. Then one can ask whether these quantities are correlated, or whether the fluctuations of these quantities are independent of each other. Consider the NVT ensemble where entropy S and the pressure p (an intensive variable) are the (fluctuating) conjugate variables $\{p = -(\partial F/\partial V)_{NT}$, $S = -(\partial F/\partial T)_{NV}\}$. What are the fluctuations of S and p, and are they correlated? The answer to these questions is given by

$$\left\langle(\Delta S)^2\right\rangle_{NVT} = C_p, \qquad (2.17a)$$

$$\left\langle(\Delta p)^2\right\rangle_{NVT} = -T(\partial p/\partial V)_S, \qquad (2.17b)$$

$$\left\langle(\Delta S)(\Delta p)\right\rangle_{NVT} = 0. \qquad (2.17c)$$

One can also see here an illustration of the general principle that fluctuations of extensive variables (like S) scale with the volume, while fluctuations of intensive variables (like p) scale with the inverse volume.

2.1.2 Phase transitions

The emphasis in the standard texts on statistical mechanics clearly is on those problems that can be dealt with analytically, e.g. ideal classical and quantum gases, dilute solutions, etc. The main utility of Monte Carlo methods is for problems which evade exact solution such as phase transitions, calculations of phase diagrams, etc. For this reason we shall emphasize this topic here. The study of phase transitions has long been a topic of great interest in a variety of related scientific disciplines and plays a central role in research in many fields of physics. Although very simple approaches, such as mean field theory, provide a very simple, intuitive picture of phase transitions, they generally fail to provide a quantitative framework for explaining the wide variety of phenomena which occur under a range of different conditions and often do not really capture the conceptual features of the important processes which occur at a phase transition. The last half century has seen the development of a mature framework for the understanding and classification of phase transitions using a combination of (rare) exact solutions as well as theoretical and numerical approaches.

2.1.2.1 Order parameter

The distinguishing feature of most phase transitions is the appearance of a non–zero value of an 'order parameter', i.e. of some property of the system which is non-zero in the ordered phase but identically zero in the disordered phase. The order parameter is defined differently in different kinds of physical systems. In a ferromagnet it is simply the spontaneous magnetization. In a liquid–gas system it will be the difference in the density between the liquid and gas phases at the transition; for liquid crystals the degree of orientational order is telling. An order parameter may be a scalar quantity or may be a multicomponent (or even complex) quantity. Depending on the physical system, an order parameter may be measured by a variety of experimental methods such as neutron scattering, where Bragg peaks of superstructures in antiferromagnets allow the estimation of the order parameter from the integrated intensity, oscillating magnetometer measurement directly determines the spontaneous magnetization of a ferromagnet, while NMR is suitable for the measurement of local orientational order.

2.1.2.2 Correlation function

Even if a system is not ordered, there will in general be microscopic regions in the material in which the characteristics of the material are correlated. Correlations are generally measured through the determination of a two-point correlation function

$$\Gamma(r) = \langle \rho(0)\rho(r)\rangle, \tag{2.18}$$

where r is the spatial distance and ρ is the quantity whose correlation is being measured. (The behavior of this correlation function will be discussed shortly.) It is also possible to consider correlations that are both space-dependent and time-dependent, but at the moment we only consider equal time correlations that are time-independent. As a function of distance they will decay (although not always monotonically), and if the correlation for the appropriate quantity decays to zero as the distance goes to infinity, then the order parameter is zero.

2.1.2.3 First order vs. second order

These remarks will concentrate on systems which are in thermal equilibrium and which undergo a phase transition between a disordered state and one which shows order which can be described by an appropriately defined order parameter. If the first derivatives of the free energy are discontinuous at the transition temperature T_c, the transition is termed first order. The magnitude of the discontinuity is unimportant in terms of the classification of the phase transition, but there are diverse systems with either very large or rather small 'jumps'. For second order phase transitions first derivatives are continuous; transitions at some temperature T_c and 'field' H are characterized by singularities in the second derivatives of the free energy, and properties of rather disparate systems can be related by considering not the absolute temperature but rather the reduced distance from the transition $\varepsilon = |1 - T/T_c|$. (Note that in the 1960s and early 1970s the symbol ε was used to denote the reduced distance from the critical point. As renormalization group theory came on the scene, and in particular ε-expansion techniques became popular, the notation changed to use the symbol t instead. In this book, however, we shall often use the symbol t to stand for time, so to avoid ambiguity we have returned to the original notation.) In Fig. 2.3 we show characteristic behavior for both kinds of phase transitions. At a first order phase transition the free energy curves for ordered and disordered states cross with a finite difference in slope and both stable and metastable states exist for some region of temperature. In contrast, at a second order transition the two free energy curves meet tangentially.

2.1.2.4 Phase diagrams

Phase transitions occur as one of several different thermodynamic fields is varied. Thus, the loci of all points at which phase transitions occur form phase boundaries in a multidimensional space of thermodynamic fields. The classic example of a phase diagram is that of water, shown in pressure–temperature space in Fig. 2.4, in which lines of first order transitions separate ice–water, water–steam, and ice–steam. The three first order transitions join at a 'triple point', and the water–steam phase line ends at a 'critical point' where a second order phase transition occurs. (Ice actually has multiple inequivalent phases and we have ignored this complexity in this figure.) Predicting the phase diagram of simple atomic or molecular systems, as well as of mixtures, given the knowledge of the microscopic interactions, is

Fig. 2.3 (left)
Schematic temperature
dependence of the free
energy and the
internal energy for a
system undergoing a
first order transition;
(right) schematic
temperature
dependence of the free
energy and the
internal energy for a
system undergoing a
second order
transition.

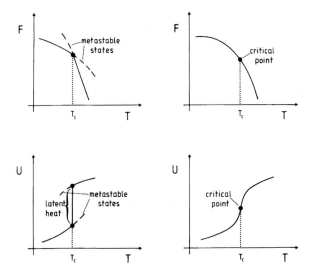

an important task of statistical mechanics which relies on simulation methods quite strongly, as we shall see in later chapters. A much simpler phase diagram than for water occurs for the two-dimensional Ising ferromagnet with Hamiltonian

$$\mathcal{H} = -\mathcal{J}_{nn} \sum_{nn} \sigma_i \sigma_j - H \sum_i \sigma_i, \qquad (2.19)$$

where $\sigma_i = \pm 1$ represents a 'spin' at lattice site i which interacts with nearest neighbors on the lattice with interaction constant $\mathcal{J}_{nn} > 0$. In many respects this model has served as a 'fruit fly' system for studies in statistical mechanics. At low temperatures a first order transition occurs as H is swept through zero, and the phase boundary terminates at the critical temperature T_c as shown in Fig. 2.4. In this model it is easy to see, by invoking the symmetry involving reversal of all the spins and the sign of H, that the phase boundary must occur at $H = 0$ so that the only remaining 'interesting' question is the location of the critical point. Of course, many physical systems do not possess this symmetry. As a third example, in Fig. 2.4 we also show the phase boundary for an Ising antiferromagnet for which $\mathcal{J} < 0$. Here the antiferromagnetic phase remains stable in non-zero field, although the critical temperature is depressed. As in the case of the ferromagnet, the phase diagram is symmetric about $H = 0$. We shall return to the question of phase diagrams for the antiferromagnet later in this section when we discuss 'multicritical points'.

2.1.2.5 Critical behavior and exponents

We shall attempt to explain thermodynamic singularities in terms of the reduced distance from the critical temperature. Extensive experimental research has long provided a testing ground for developing theories (Kadanoff *et al.*, 1967) and more recently, of course, computer simulations

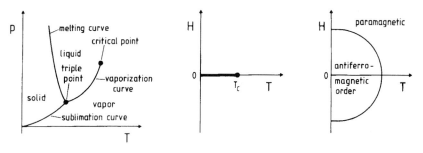

Fig. 2.4 (left) Simplified pressure–temperature phase diagram for water; (center) magnetic field–temperature phase diagram for an Ising ferromagnet; (right) magnetic field–temperature phase diagram for an Ising antiferromagnet.

have been playing an increasingly important role. Of course, experiment is limited not only by instrumental resolution but also by unavoidable sample imperfections. Thus, the beautiful specific heat peak for $RbMnF_3$, shown in Fig. 2.5, is quite difficult to characterize for $\varepsilon \le 10^{-4}$. Data from multiple experiments as well as results for a number of exactly soluble models show that the thermodynamic properties can be described by a set of simple power laws in the vicinity of the critical point T_c, e.g. for a magnet the order parameter m, the specific heat C, the susceptibility χ and the correlation length ξ vary as (Stanley, 1971; Fisher, 1974)

$$m = m_0 \varepsilon^{\beta}, \tag{2.20a}$$

$$\chi = \chi_0 \varepsilon^{-\gamma}, \tag{2.20b}$$

$$C = C_0 \varepsilon^{-\alpha}, \tag{2.20c}$$

$$\xi = \xi_0 \varepsilon^{-\nu}, \tag{2.20d}$$

where $\epsilon = |1 - T/T_c|$ and the powers (Greek characters) are termed 'critical exponents'. Note that Eqns. (2.20a–d) represent asymptotic expressions which are valid only as $\varepsilon \to 0$ and more complete forms would include additional 'corrections to scaling' terms which describe the deviations from the asymptotic behavior. Although the critical exponents for a given quantity are believed to be identical when T_c is approached from above or below, the prefactors, or 'critical amplitudes' are not usually the same. The determination of particular amplitude ratios does indeed form the basis for rather extended studies (Privman *et al.*, 1991). Along the critical isotherm, i.e. at $T = T_c$ we can define another exponent (for a ferromagnet) by

$$m = DH^{1/\delta}, \tag{2.21}$$

where H is an applied, uniform magnetic field. (Here too, an analogous expression would apply for a liquid–gas system at the critical temperature as a function of the deviation from the critical pressure.) For a system in d-dimensions the two-body correlation function $\Gamma(r)$, which well above the critical temperature has the Ornstein–Zernike form (note that for a ferromagnet in zero field $\rho(r)$ in Eqn. (2.18) corresponds to the magnetization density at r while for a fluid $\rho(r)$ means the local deviation from the average density)

Fig. 2.5 (top)
Experimental data and
(bottom) analysis of
the critical behavior of
the specific heat of the
Heisenberg-like
antiferromagnet
RbMnF$_3$. The critical
temperature is T_c.
After Kornblit and
Ahlers (1973).

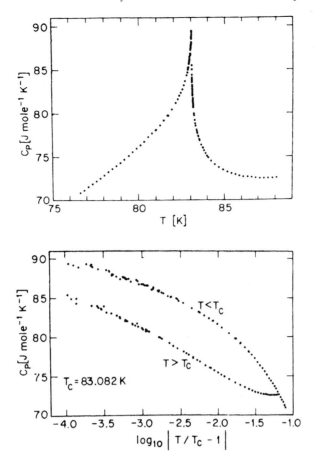

$$\Gamma(r) \propto r^{-(d-1)/2} \exp(-r/\xi), \qquad r \to \infty, \qquad (2.22)$$

also shows a power law decay at T_c,

$$\Gamma(r) = \Gamma_0 r^{-(d-2+\eta)}, \qquad r \to \infty, \qquad (2.23)$$

where η is another critical exponent. These critical exponents are known
exactly for only a small number of models, most notably the two-dimensional
Ising square lattice (Onsager, 1944) (cf. Eqn. (2.5)), whose exact solution
shows that $\alpha = 0$, $\beta = 1/8$, and $\gamma = 7/4$. Here, $\alpha = 0$ corresponds to a
logarithmic divergence of the specific heat. We see in Fig. 2.5, however,
that the experimental data for the specific heat of RbMnF$_3$ increases even
more slowly than a logarithm as $\varepsilon \to 0$, implying that $\alpha < 0$, i.e. the specific
heat is non-divergent. In fact, a suitable model for RbMnF$_3$ is not the Ising
model but a three-dimensional Heisenberg model with classical spins of unit
length and nearest neighbor interactions

$$\mathcal{H} = -\mathcal{J} \sum_{nn} (S_{ix}S_{jx} + S_{iy}S_{jy} + S_{iz}S_{jz}), \qquad (2.24)$$

which has different critical exponents than does the Ising model. (Although no exact solutions are available, quite accurate values of the exponents have been known for some time due to application of the field theoretic renormalization group (Zinn-Justin and LeGuillou, 1980), and extensive Monte Carlo simulations have yielded some rather precise results, at least for classical Heisenberg models (Chen *et al.*, 1993).)

The above picture is not complete because there are also special cases which do not fit into the above scheme. Most notable are two-dimensional XY-models with Hamiltonian

$$\mathcal{H} = -\mathcal{J}\sum_{nn}(S_{ix}S_{jx} + S_{iy}S_{jy}), \tag{2.25}$$

where S_i is a unit vector which may have either two components (plane rotator model) or three components (XY-model). These models develop no long range order at low temperature but have topological excitations, termed vortex–antivortex pairs, which unbind at the transition temperature K_{KT} (Kosterlitz and Thouless, 1973). The correlation length and susceptibility for this model diverge exponentially fast as the transition temperature is approached from above, e.g.

$$\xi \propto \exp(a\varepsilon^{-\nu}), \tag{2.26}$$

and every temperature below T_{KT} is a critical point. Other classical models with suitable competing interactions or lattice structures may also show 'unusual' transitions (Landau, 1994) which in some cases include different behavior of multiple order parameters at T_c and which are generally amenable to study by computer simulation.

The above discussion was confined to static aspects of phase transitions and critical phenomena. The entire question of dynamic behavior will be treated in a later section using extensions of the current formulation.

2.1.2.6 Universality and scaling

Homogeneity arguments also provide a way of simplifying expressions which contain thermodynamic singularities. For example, for a simple Ising ferromagnet in a small magnetic field H and at a temperature T which is near the critical point, the singular portion of the free energy $F(T, H)$ can be written as

$$F_s = \varepsilon^{2-\alpha}\mathcal{F}^{\pm}(H/\varepsilon^{\Delta}), \tag{2.27}$$

where the 'gap exponent' Δ is equal to $\frac{1}{2}(2 - \alpha + \gamma)$ and $\mathcal{F}^{\pm}$ is a function of the 'scaled' variable (H/ε^{Δ}), i.e. does not depend upon ε independently. This formula has the consequence, of course, that all other expressions for thermodynamic quantities, such as specific heat, susceptibility, etc. can be written in scaling forms as well. Similarly, the correlation function can be expressed as a scaling function of two variables

$$\Gamma(r, \xi, \varepsilon) = r^{-(d-2+\eta)}\mathcal{G}(r/\xi, H/\varepsilon^{\Delta}), \tag{2.28}$$

where $\mathcal{G}(x, y)$ is now a scaling function of two variables.

Not all of the six critical exponents defined in the previous section are independent, and using a number of thermodynamic arguments one can derive a series of exponent relations called scaling laws which show that only two exponents are generally independent. For example, taking the derivative of the free energy expressed above in a scaling form yields

$$-\partial F_s/\partial H = M = \varepsilon^{2-\alpha-\Delta}\mathcal{F}'(H/\varepsilon^{\Delta}), \quad (2.29)$$

where $\mathcal{F}'$ is the derivative of $\mathcal{F}$, but this equation can be compared directly with the expression for the decay of the order parameter to show that $\beta = 2 - \alpha - \Delta$. Furthermore, using a scaling expression for the magnetic susceptibility

$$\chi = \varepsilon^{-\gamma}\mathcal{C}(H/\varepsilon^{\Delta}) \quad (2.30)$$

one can integrate to obtain the magnetization, which for $H = 0$ becomes

$$m \propto \varepsilon^{\Delta-\gamma}. \quad (2.31)$$

Combining these simple relations one obtains the so-called Rushbrooke equality

$$\alpha + 2\beta + \gamma = 2 \quad (2.32)$$

which should be valid regardless of the individual exponent values. Another scaling law which has important consequences is the 'hyperscaling' expression which involves the lattice dimensionality d

$$d\nu = 2 - \alpha. \quad (2.33)$$

Of course, here we are neither concerned with a discussion of the physical justification of the homogeneity assumption given in Eqn. (2.27), nor with this additional scaling relation, Eqn. (2.32), see e.g. Yeomans (1992). However, these scaling relations are a prerequisite for the understanding of finite size scaling which is a basic tool in the analysis of simulational data near phase transitions, and we shall thus summarize them here. Hyperscaling may be violated in some cases, e.g. the upper critical (spatial) dimension for the Ising model is $d = 4$ beyond which mean-field (Landau theory) exponents apply and hyperscaling is no longer obeyed. Integration of the correlation function over all spatial displacement yields the susceptibility

$$\chi = \varepsilon^{-\nu(2-\eta)}, \quad (2.34)$$

and by comparing this expression with the 'definition', cf. Eqn. (2.20b), of the critical behavior of the susceptibility we have

$$\gamma = \nu(2 - \eta). \quad (2.35)$$

Those systems which have the same set of critical exponents are said to belong to the same universality class (Fisher, 1974). Relevant properties which play a role in the determination of the universality class are known to include spatial dimensionality, spin dimensionality, symmetry of the ordered state, the presence of symmetry breaking fields, and the range of interaction. Thus, nearest neighbor Ising ferromagnets (see Eqn. (2.19)) on the square and triangular lattices have identical critical exponents and belong to the same universality class. Further, in those cases where lattice models and similar continuous models with the same symmetry can be compared, they generally belong to the same universality class. A simple, nearest neighbor Ising antiferromagnet in a field has the same exponents for all field values below the zero temperature critical field. This remarkable behavior will become clearer when we consider the problem in the context of renormalization group theory (Wilson, 1971) in Chapter 9. At the same time there are some simple symmetries which can be broken quite easily. For example, an isotropic ferromagnet changes from the Heisenberg universality class to the Ising class as soon as a uniaxial anisotropy is applied to the system:

$$\mathcal{H} = -\mathcal{J}\sum[(1 - \Delta)(S_{ix}S_{jx} + S_{iy}S_{jy}) + S_{iz}S_{jz}], \qquad (2.36)$$

where $\Delta > 0$. The variation of the critical temperature is then given by

$$T_c(\Delta) - T_c(\Delta = 0) \propto \Delta^{1/\phi}, \qquad (2.37)$$

where ϕ is termed the 'crossover exponent' (Riedel and Wegner, 1972). There are systems for which the lattice structure and/or the presence of competing interactions give rise to behavior which is in a different universality class than one might at first believe from a cursory examination of the Hamiltonian. From an analysis of the symmetry of different possible adlayer structures for adsorbed films on crystalline substrates Domany *et al.* (1980) predict the universality classes for a number of two-dimensional Ising-lattice gas models. Among the most interesting and unusual results of this symmetry analysis is the phase diagram for the triangular lattice gas (Ising) model with nearest neighbor repulsive interaction and next-nearest neighbor attractive coupling (Landau, 1983). In the presence of non-zero chemical potential, the groundstate is a three-fold degenerate state with 1/3 or 2/3 filling (the triangular lattice splits into three sublattices and one is full and the other two are empty, or vice versa, respectively) and is predicted to be in the universality class of the 3-state Potts model (Potts, 1952; Wu, 1982)

$$\mathcal{H} = -\mathcal{J}\sum\delta_{\sigma_i\sigma_j}, \qquad (2.38)$$

where $\sigma_i = 1, 2,$ or 3. In zero chemical potential all six states become degenerate and a symmetry analysis predicts that the system is then in the universality class of the XY-model with sixth order anisotropy

$$\mathcal{H} = -\mathcal{J}\sum(S_{ix}S_{jx} + S_{iy}S_{jy}) + \Delta\sum\cos(6\theta_i), \qquad (2.39)$$

where θ_i is the angle which a spin makes with the *x*-axis. Monte Carlo results (Landau, 1983), shown in Fig. 2.6, confirm these expectations: in non-zero chemical potential there is a Potts-like phase boundary, complete with a 3-state Potts tricritical point. (Tricritical points will be discussed in the following sub-section.) In zero field, there are two Kosterlitz–Thouless transitions with an XY-like phase separating a low temperature ordered phase from a

Fig. 2.6 Phase diagram for the triangular Ising (lattice gas) model with antiferromagnetic nearest neighbor and ferromagnetic next-nearest neighbor interactions. T_1 and T_2 denote Kosterlitz–Thouless phase transitions and the + sign on the non-zero field phase boundary is a tricritical point. The arrangement of open and closed circles shows examples of the two different kinds of ground states using lattice gas language. From Landau (1983).

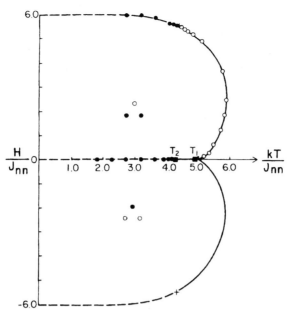

high temperature disordered state. Between the upper and lower transitions 'vortex-like' excitations can be identified and followed. Thus, even though the Hamiltonian is that of an Ising model, there is no Ising behavior to be seen and instead a very rich scenario, complete with properties expected only for continuous spin models is found! At the same time, Fig. 2.6 is an example of a phase diagram containing both continuous and first order phase transitions which cannot yet be found with *any* other technique with an accuracy which is competitive to that obtainable by the Monte Carlo methods which will be described in this book.

2.1.2.7 Multicritical phenomena

Under certain circumstances the order of a phase transition changes as some thermodynamic parameter is altered. Although such behavior appears to violate the principles of universality which we have just discussed, examination of the system in a larger thermodynamic space makes such behavior easy to understand. The intersection point of multiple curves of second order phase transitions is known as a multicritical point. Examples include the tricritical point (Griffiths, 1970; Stryjewski and Giordano, 1977; Lawrie and Sarbach, 1984) which occurs in He^3–He^4 mixtures, strongly anisotropic ferromagnets, and ternary liquid mixtures, as well as the bicritical point (Nelson *et al.*, 1974) which appears on the phase boundary of a moderately anisotropic Heisenberg antiferromagnet in a uniform magnetic field. The characteristic phase diagram for a tricritical point is shown in Fig. 2.7 in which one can see that the three second order boundaries to first order surfaces of phase transitions meet at a tricritical point. One of the simplest

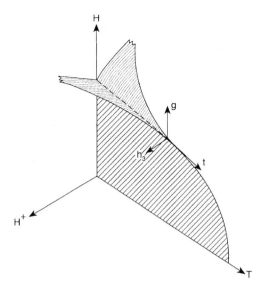

Fig. 2.7 Phase diagram for a system with a tricritical point in the three-dimensional thermodynamic field space which includes both ordering and non-ordering fields. Tricritical scaling axes are labeled t, g, and h_3.

models which exhibits such behavior is the Ising antiferromagnet with nearest and next-nearest neighbor coupling

$$\mathcal{H} = -\mathcal{J}_{nn} \sum_{nn} \sigma_i \sigma_j - \mathcal{J}_{nnn} \sum_{nnn} \sigma_i \sigma_j - H \sum_i \sigma_i - H^+ \sum_i \sigma_i, \qquad (2.40)$$

where $\sigma_i = \pm 1$, H is the uniform magnetic field, and H^+ is the staggered magnetic field which couples to the order parameter. The presence of a multicritical point introduces a new 'relevant' field g, which as shown in Fig. 2.7 makes a non-zero angle with the phase boundary, and a second scaling field t, which is tangential to the phase boundary at the tricritical point. In the vicinity of a multicritical point a 'crossover' scaling law is valid

$$F(\varepsilon, H^+, g) = |g|^{2-\alpha_\varepsilon} \mathcal{F}(H^+/|g|^{\Delta_\varepsilon}, \varepsilon/|g|^{\phi_\varepsilon}), \qquad (2.41)$$

where α_ε is the specific heat exponent appropriate for a tricritical point, Δ_ε the corresponding 'gap exponent', and ϕ_ε a new 'crossover' exponent. In addition, there are power law relations which describe the vanishing of discontinuities as the tricritical point is approached from below. For example, the discontinuity in the magnetization from M^+ to M^- as the first order phase boundary for $T < T_t$ is crossed decreases as

$$\Delta M = M^+ - M^- \propto |1 - T/T_t|^{\beta_u}. \qquad (2.42)$$

The 'u-subscripted' exponents are related to the 'ε-subscripted' ones by a crossover exponent,

$$\beta_u = (1 - \alpha_\varepsilon)/\phi_\varepsilon. \qquad (2.43)$$

As will be discussed below, the mean field values of the tricritical exponents are $\alpha_\varepsilon = 1/2$, $\Delta_\varepsilon = 5/2$, $\phi_\varepsilon = 1/2$, and hence $\beta_u = 1$. Tricritical points have been explored using both computer simulations of model systems as well as by experimental investigation of physical systems, and their theoretical aspects have been studied in detail (Lawrie and Sarbach, 1984).

2.1.2.8 Landau theory

One of the simplest theories with which simulations are often compared is the Landau theory which begins with the assumption that the free energy of a system can be expanded about the phase transition in terms of the order parameter. The free energy of a d-dimensional system near a phase transition is expanded in terms of a simple one-component order parameter $m(x)$

$$
F = F_{\mathrm{o}} + \int d^d x \left\{ \frac{1}{2} rm^2(x) + \frac{1}{4} um^4(x) + \frac{1}{6} vm^6(x) - \frac{H}{k_{\mathrm{B}}T} m(x) \right.
$$
$$
\left. + \frac{1}{2d} [R\nabla m(x)]^2 + \cdots \right\}. \tag{2.44}
$$

Here a factor of $(k_{\mathrm{B}}T)^{-1}$ has been absorbed into F and F_{o} and the coefficients r, u, and v are dimensionless. Note that the coefficient R can be interpreted as the interaction range of the model. This equation is in the form of a Taylor series in which symmetry has already been used to eliminate all odd order terms for $H = 0$. For more complex systems it is possible that additional terms, e.g. cubic products of components of a multicomponent order parameter might appear, but such situations are generally beyond the scope of our present treatment. In the simplest possible case of a homogeneous system this equation becomes

$$
F = F_{\mathrm{o}} + V\left(\tfrac{1}{2} rm^2 + \tfrac{1}{4} um^4 + \tfrac{1}{6} vm^6 - mH/k_{\mathrm{B}}T + \cdots \right). \tag{2.45}
$$

In equilibrium the free energy must be a minimum; and if $u > 0$ we can truncate the above equation and the minimization criterion $\partial F/\partial m = 0$ yields three possible solutions:

$$
m_1 = 0, \tag{2.46a}
$$
$$
m_{2,3} = \pm\sqrt{-r/u}. \tag{2.46b}
$$

Expanding r in the vicinity of T_{c} so that $r = r'(T - T_{\mathrm{c}})$, we find then for $r < 0$ (i.e. $T < T_{\mathrm{c}}$)

$$
m_{2,3} = \pm\left[(r'T_{\mathrm{c}}/u)(1 - T/T_{\mathrm{c}}) \right]^{1/2}. \tag{2.47}
$$

Thus, m_1 corresponds to the solution above T_{c} where there is no long range order, and $m_{2,3}$ correspond to solutions below T_{c} where the order parameter approaches zero with a characteristic power law (see Eqn. (2.20a)) with exponent $\beta = 1/2$. A similar analysis of the susceptibility produces $\gamma = 1$, $\delta = 3$. (Although straightforward to apply, Landau theory does not correctly describe the behavior of many physical systems. For liquid–gas critical points and most magnetic systems $\beta \approx 1/3$ (Kadanoff et al., 1967) instead of the Landau value of $\beta = 1/2$.) The appearance of tricritical points can be easily understood from the Landau theory. If the term in m^4 is negative it becomes necessary to keep the sixth order term and the minimization process yields five solutions:

$$m_1 = 0, \tag{2.48a}$$

$$m_{2,3} = \pm\left[\frac{1}{2v}\left(-u + \sqrt{u^2 - 4rv}\right)\right]^{1/2}, \tag{2.48b}$$

$$m_{4,5} = \pm\left[\frac{1}{2v}\left(-u - \sqrt{u^2 - 4rv}\right)\right]^{1/2}. \tag{2.48c}$$

If v is positive, there are multiple solutions and the transition is first order. A tricritical point thus appears when $r = u = 0$, and the tricritical exponents which result from this analysis are

$$\alpha_t = \tfrac{1}{2}, \tag{2.49a}$$
$$\beta_t = \tfrac{1}{4}, \tag{2.49b}$$
$$\gamma_t = 1, \tag{2.49c}$$
$$\delta_t = 5. \tag{2.49d}$$

Note that these critical exponents are different from the values predicted for the critical point. The crossover exponent is predicted by Landau theory to be $\phi = \tfrac{1}{2}$.

2.1.3 Ergodicity and broken symmetry

The principle of ergodicity states that all possible configurations of the system should be attainable. As indicated in Eqn. (2.4) the different states will not all have the same probability, but it must nonetheless be possible to reach each state with non-zero probability. Below a phase transition multiple different ordered states may appear, well separated in phase space. If the phase transition from the disordered phase to the ordered phase is associated with 'symmetry breaking', the separate ordered states are related by a symmetry operation acting on the order parameter (e.g. a reversal of the sign of the order parameter for an Ising ferromagnet). In the context of a discussion of dynamical behavior of such systems, symmetry breaking usually means ergodicity breaking, i.e. the system stays in one separate region in phase space. The question of non-ergodic behavior in the context of simulations is complex. For example, in the simulation of an Ising system which may have all spins up or all spins down, we may wish to keep the system from exploring all of phase space so that only positive values of the order parameter are observed. If instead the simulation algorithm is fully ergodic, then both positive and negative values of order parameter will appear and the average will be zero. A danger for simulations is that specialized algorithms may be unintentionally non-ergodic, thus yielding incorrect results.

2.1.4 Fluctuations and the Ginzburg criterion

As mentioned earlier, the thermodynamic properties of a system are not perfectly constant but fluctuate with time as the system explores different regions of phase space. In the discussion of fluctuations in Section 2.1.1.4 we have seen that relative fluctuations of extensive thermodynamic variables scale inversely with V or N, and hence such global fluctuations vanish in the thermodynamic limit. One should not conclude, however, that fluctuations are generally unimportant; indeed local fluctuations can have dramatic consequences and require a separate discussion.

What is the importance of local fluctuations? As long as they do not play a major role, we can expect that Landau theory will yield correct predictions. Let us compare the fluctuations in $m(x)$ for a d-dimensional system over the 'correlation volume' ξ^d with its mean value m_o. If Landau theory is valid and fluctuations can be ignored, then

$$\frac{\langle [m(x) - m_o]^2 \rangle}{m_o^2} \ll 1. \tag{2.50}$$

This condition, termed the Ginzburg criterion, leads to the expression

$$\xi^d m_o^2 \chi^{-1} \gg \text{const.}, \tag{2.51}$$

and following insertion of the critical behavior power laws we obtain

$$\varepsilon^{-\nu d + 2\beta + \gamma} \gg \text{const.} \tag{2.52}$$

Inserting Landau exponents into this expression we find

$$\varepsilon^{(d-4)/2} \ll \text{const.}, \tag{2.53}$$

i.e. for Landau theory to be valid the lattice dimensionality must be greater than or equal to the upper critical dimension $d_u = 4$. In addition, below some lower critical dimensionality d_l fluctuations dominate completely and no transition occurs. In order to consider the tricritical point scenario depicted in Fig. 2.7, it becomes necessary to retain the next order term $\sim v m^6$ in the Landau free energy. The shape of the resultant free energy is shown in Fig. 2.8 below, at and above the tricritical point. It turns out that mean field (i.e. Landau) theory is valid for tricritical behavior above an upper critical dimension; for the Ising model with competing interactions $d_u = 3$, but for $d = 3$ there are logarithmic corrections (Wegner and Riedel, 1973).

2.1.5 A standard exercise: the ferromagnetic Ising model

The Ising model of magnetism, defined in Eqn. (2.19), is extremely well suited to Monte Carlo simulation. The same model is equivalent to simple lattice gas models for liquid–gas transitions or binary alloy models. The transformation to a lattice gas model is straightforward. We first define site occupation variables c_i which are equal to 1 if the site is occupied and 0 if the site is empty. These variables are simple related to the Ising variables by

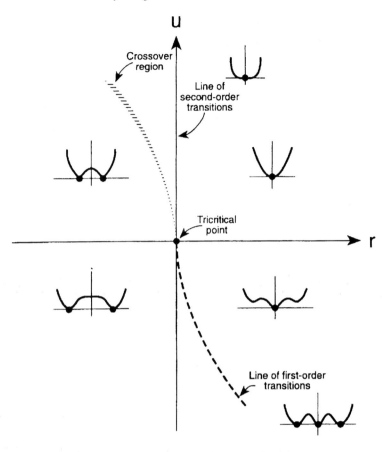

Fig. 2.8 Landau free
energy and phase
boundaries for the m^6
model in the r–u
plane. The heavy solid
line shows the second
order phase boundary
and the dashed line
represents the first
order portion of the
phase boundary. The
heavy dots show the
location(s) of the
minimum free energy.

$$c_i = (1 + \sigma_i)/2. \tag{2.54}$$

If we now substitute these into the Ising Hamiltonian we find

$$\mathcal{H}_{lg} = -\phi \sum c_i c_j - \mu \sum c_i + \text{const.} \tag{2.55}$$

where $\phi = 4\mathcal{J}$ and $\mu = 2(H + 4z\mathcal{J})$ if there are z interacting neighbors. Note that if the Ising model is studied in the canonical ensemble, any spin-flips change the number of particles in the lattice gas language and the system is effectively being studied in the grand canonical ensemble. A Monte Carlo program follows a stochastic path through phase space, a procedure which will be discussed in detail in the following chapters, yielding a sequence of states from which mean values of system properties may be determined. In the following example we show what a sample output from a Monte Carlo run might look like. A complete description of the simulation algorithm, methods of analysis, and error determination will be discussed in Chapter 4.

Example

Sample output from a Monte Carlo program simulating the two-dimensional Ising model ($\mathcal{J} = 1$) at $k_B T = 1.5$ for $L = 6$, with periodic boundary conditions.

1000 MCS discarded for equilibration
5000 MCS retained for averages
1000 MCS per bin

bin	$E(t)$	$M(t)$
1	−1.9512	0.9866
2	−1.9540	0.9873
3	−1.9529	0.9867
4	−1.9557	0.9878
5	−1.9460	0.9850

Averages: $\langle E \rangle = -1.952 \pm 0.026$
$\langle M \rangle = 0.987 \pm 0.014$

specific heat $= 0.202$
susceptibility $= 0.027$

final state :

```
+  +  +  +  +  +
+  −  +  +  +  +
+  +  +  +  +  +
+  +  +  +  −  +
+  +  −  +  +  +
+  +  +  +  +  +
```

Problem 2.3 Use the fluctuation relation for the magnetization together with Eqn. (2.49) to derive a fluctuation relation for the particle number in the grand canonical ensemble of the lattice gas.

2.2 PROBABILITY THEORY

2.2.1 Basic notions

It will soon become obvious that the notions of probability and statistics are essential to statistical mechanics and, in particular, to Monte Carlo simulations in statistical physics. In this section we want to remind the reader about some fundamentals of probability theory. We shall restrict ourselves to the basics; far more detailed descriptions may be found elsewhere, for example in the books by Feller (1968) or Kalos and Whitlock (1986). We begin by considering an elementary event with a countable set of random outcomes, $A_1, A_2, \ldots, A_k$ (e.g. rolling a die). Suppose this event occurs repeatedly, say N times, with $N \gg 1$, and we count how often the outcome A_k is observed (N_k). Then it makes sense to define probabilities p_k for the outcome A_k or (we assume that all possible events have been enumerated)

$$p_k = \lim_{N \to \infty} (N_k/N), \qquad \sum_k p_k = 1. \tag{2.56}$$

Obviously we have $0 \le p_k \le 1$ (if A_k never occurs, $p_k = 0$; if it is certain to occur, $p_k = 1$). An equivalent notation, convenient for our purposes, is $P(A_k) \equiv p_k$. From its definition, we conclude that $P(A_i$ and/or $A_j)$ $\le [P(A_i) + P(A_j)]$. We call A_i and A_j 'mutually exclusive' events, if and only if the occurrence of A_i implies that A_j does not occur and vice versa. Then

$$P(A_i \text{ and } A_j) = 0, \qquad P(A_i \text{ or } A_j) = P(A_i) + P(A_j). \tag{2.57}$$

Let us now consider two events, one with outcomes $\{A_i\}$ and probabilities p_{1i}; the second with outcomes $\{B_j\}$ and probabilities p_{2j}, respectively. We consider now the outcome (A_i, B_j) and define p_{ij} as the joint probability that both A_i and B_j occur. If the events are independent, we have

$$p_{ij} = p_{1i} \times p_{2j}. \tag{2.58}$$

If they are not independent, it makes sense to define the conditional probability $p(j|i)$ that B_j occurs, given that A_i occurs

$$p(j|i) = \frac{p_{ij}}{\sum_k p_{ik}} = \frac{p_{ij}}{p_{1i}}. \tag{2.59}$$

Of course we have $\sum_j p(j|i) = 1$ since some B_j must occur.

The outcome of such random events may be logical variables (True or False) or real numbers x_i. We call these numbers random variables. We now define the expectation value of this random variable as follows:

$$\langle x \rangle \equiv E(x) \equiv \sum_i p_i x_i. \tag{2.60}$$

Similarly, any (real) function $g(x_i)$ then has the expectation value

$$\langle g(x) \rangle \equiv E(g, x)) = \sum_i p_i g(x_i). \tag{2.61}$$

In particular, if we begin with two functions $g_1(x)$, $g_2(x)$ and consider the linear combination (λ_1, λ_2 being constants), we have $\langle \lambda_1 g_1(x) + \lambda_2 g_2(x) \rangle = \lambda_1 \langle g_1 \rangle + \lambda_2 \langle g_2 \rangle$. Of particular interest are the powers of x. Defining the nth moment as

$$\langle x^n \rangle = \sum_i p_i x_i^n \tag{2.62}$$

we then consider the so-called cumulants

$$\langle (x - \langle x \rangle)^n \rangle = \sum_i p_i (x_i - \langle x \rangle)^n. \tag{2.63}$$

Of greatest importance is the case $n = 2$, which is called the 'variance',

$$\text{var}(x) = \langle (x - \langle x \rangle)^2 \rangle = \langle x^2 \rangle - \langle x \rangle^2. \tag{2.64}$$

If we generalize these definitions to two random variables (x_i and y_j), the analogue of Eqn. (2.60) is

$$\langle xy \rangle = \sum_{i,j} p_{ij} x_i y_j. \tag{2.65}$$

If x and y are independent, then $p_{ij} = p_{1i} p_{2j}$ and hence

$$\langle xy \rangle = \sum_i p_{1i} x_i \sum_j p_{2j} y_j = \langle x \rangle \langle y \rangle. \tag{2.66}$$

As a measure of the degree of independence of the two random variables it is hence natural to take their covariance

$$\mathrm{cov}(x, y) = \langle xy \rangle - \langle x \rangle \langle y \rangle. \tag{2.67}$$

2.2.2 Special probability distributions and the central limit theorem

Do we find any special behavior which arises when we consider a very large number of events? Consider two events A_0 and A_1 that are mutually exclusive and exhaustive:

$$P(A_1) = p, \quad x = 1; \quad P(A_0) = 1 - p, \quad x = 0. \tag{2.68}$$

Suppose now that N independent samples of these events occur. Each outcome is either 0 or 1, and we denote the sum X of these outcomes, $X = \sum_{r=1} x_r$. Now the probability that $X = n$ is the probability that n of the X_r were 1 and $(N - n)$ were 0. This is called the binomial distribution,

$$P(X > n) = \binom{N}{n} p^n (1 - p)^{N-n}, \tag{2.69}$$

$\binom{N}{n}$ being the binomial coefficients. It is easy to show from Eqn. (2.69) that

$$\langle X \rangle = Np, \quad \langle (X - \langle X \rangle)^2 \rangle = Np(1 - p). \tag{2.70}$$

Suppose now we still have two outcomes $(1, 0)$ of an experiment: if the outcome is 0, the experiment is repeated, otherwise we stop. Now the random variable of interest is the number n of experiments until we get the outcome 1:

$$P(x = n) = (1 - p)^{n-1} p, \quad n = 1, 2, 3, \ldots. \tag{2.71}$$

This is called the geometrical distribution. In the case that the probability of 'success' is very small, the Poisson distribution

$$P(x = n) = \frac{\lambda^n}{n!} \exp(-\lambda), \quad n = 0, 1, \ldots \tag{2.72}$$

represents an approximation to the binomial distribution. The most important distribution that we will encounter in statistical analysis of data is the Gaussian distribution

$$p_G(x) = \frac{1}{\sqrt{2\pi\sigma^2}} \exp\left[-\frac{(x - \langle x \rangle)^2}{2\sigma^2}\right] \tag{2.73}$$

which is an approximation to the binomial distribution in the case of a very large number of possible outcomes and a very large number of samples. If random variables $x_1, x_2, \ldots, x_n$ are all independent of each other and drawn from the same distribution, the average value $\overline{X}_N = \sum_{i=1}^{N} x_i/N$ in the limit $N \to \infty$ will always be distributed according to Eqn. (2.73), irrespective of the distribution from which the x_i were drawn. This behavior is known as the 'central limit theorem' and plays a *very* important role in the sampling of states of a system One also can show that the variance of $\overline{X}_N$ is the quantity σ^2 that appears in Eqn. (2.73), and that $\sigma^2 \propto 1/N$.

Of course, at this point it should be clear to those unfamiliar with probability theory that there is no way to fully understand this subject from this 'crash course' of only a few pages which we are presenting here. For the uninitiated, our goal is only to 'whet the appetite' about this subject since it is central to the estimation of errors in the simulation results. (This discussion may then also serve to present a guide to the most pertinent literature.)

Problem 2.4 Compute the average value and the variance for the exponential distribution and for the Poisson distribution.

2.2.3 Statistical errors

Suppose the quantity A is distributed according to a Gaussian with mean value $\langle A \rangle$ and width σ. We consider n statistically independent observations $\{A_i\}$ of this quantity A. An unbiased estimator of the mean $\langle A \rangle$ of this distribution is

$$\overline{A} = \frac{1}{n}\sum_{i=1}^{n} A_i \tag{2.74}$$

and the standard error of this estimate is

$$\text{error} = \sigma/\sqrt{n}. \tag{2.75}$$

In order to estimate the variance σ itself from the observations, consider deviations $\delta A_i = A_i - \overline{A}$. Trivially we have $\overline{\delta A_i} = 0$ and $\langle \delta A \rangle = 0$. Thus we are interested in the mean square deviation

$$\overline{\delta A^2} = \frac{1}{n}\sum_{i=1}^{n}(\delta A_i)^2 = \overline{A^2} - \left(\overline{A}\right)^2. \tag{2.76}$$

The expectation value of this quantity is easily related to $\sigma^2 = \langle A^2 \rangle - \langle A \rangle^2$ as

$$\langle \overline{\delta A^2} \rangle = \sigma^2(1 - 1/n). \tag{2.77}$$

Combining Eqns. (2.75) and (2.77) we recognize the usual formula for the computation of errors of averages from uncorrelated estimates,

$$\text{error} = \sqrt{\overline{\langle \delta A^2 \rangle}/(n-1)} = \sqrt{\sum_{i=1}^{n}(\delta A_i)^2/[n(n-1)]}. \quad (2.78)$$

Equation (2.78) is immediately applicable to simple sampling Monte Carlo methods. However, as we shall see later, the usual form of Monte Carlo sampling, namely importance sampling Monte Carlo, leads to 'dynamic' correlations between subsequently generated observations $\{A_i\}$. Then Eqn. (2.78) is replaced by

$$(\text{error})^2 = \frac{\sigma^2}{n}(1 + 2\tau_A/\delta t), \quad (2.79)$$

where δt is the 'time interval' between subsequently generated states A_i, A_{i+1} and τ_A is the 'correlation time' (measured in the same units as δt).

2.2.4 Markov chains and master equations

The concept of Markov chains is so central to Monte Carlo simulations that we wish to present at least a brief discussion of the basic ideas. We define a stochastic process at discrete times labeled consecutively $t_1, t_2, t_3, \ldots$, for a system with a finite set of possible states $S_1, S_2, S_3, \ldots$, and we denote by X_t the state the system is in at time t. We consider the conditional probability that $X_{t_n} = S_{i_n}$,

$$P(X_{t_n} = S_{i_n}|X_{t_{n-1}} = S_{i_{n-1}}, X_{t_{n-2}} = S_{i_{n-2}}, \ldots, X_{t_1} = S_{i_1}), \quad (2.80)$$

given that at the preceding time the system state $X_{t_{n-1}}$ was in state $S_{i_{n-1}}$, etc. Such a process is called a Markov process, if this conditional probability is in fact independent of all states but the immediate predecessor, i.e. $P = P(X_{t_n} = S_{i_n}|X_{t_{n-1}} = S_{i_{n-1}})$. The corresponding sequence of states $\{X_t\}$ is called a Markov chain, and the above conditional probability can be interpreted as the transition probability to move from state i to state j,

$$W_{ij} = W(S_i \rightarrow S_j) = P(X_{t_n} = S_j|X_{t_{n-1}} = S_i). \quad (2.81)$$

We further require that

$$W_{ij} \geq 0, \qquad \sum_j W_{ij} = 1, \quad (2.82)$$

as usual for transition probabilities. We may then construct the total probability $P(X_{t_n} = S_j)$ that at time t_n the system is in state S_j as $P(X_{t_n} = S_j) = P(X_{t_n} = S_j|X_{t_{n-1}} = S_i)P(X_{t_{n-1}} = S_i) = W_{ij}P(X_{t_{n-1}} = S_i)$.

The master equation considers the change of this probability with time t (treating time as a continuous rather than discrete variable and writing then $P(X_{t_n} = S_j) = P(S_j, t)$)

$$\frac{dP(S_j, t)}{dt} = -\sum_i W_{ji}P(S_j, t) + \sum_i W_{ij}P(S_i, t). \quad (2.83)$$

Equation (2.83) can be considered as a 'continuity equation', expressing the fact that the total probability is conserved ($\sum_j P(S_j, t) \equiv 1$ at all times) and all probability of a state i that is 'lost' by transitions to state j is gained in the probability of that state, and vice versa. Equation (2.83) just describes the balance of gain and loss processes: since the probabilities of the events $S_j \to S_{i_1}$, $S_j \to S_{i_2}$, $S_j \to S_{i_3}$ are mutually exclusive, the total probability for a move away from the state j simply is the sum $\sum_i W_{ij} P(S_j, t)$.

Of course, by these remarks we only wish to make the master equation plausible to the reader, rather than dwelling on more formal derivations. Clearly, Eqn. (2.83) brings out the basic property of Markov processes: i.e. knowledge of the state at time t completely determines the future time evolution, there is no memory of the past (knowledge of behavior of the systems at times earlier than t is not needed). This property is obviously rather special, and only some real systems actually do have a physical dynamics compatible with Eqn. (2.83), see Section 2.3.1. But the main significance of Eqn. (2.83) is that the importance sampling Monte Carlo process (like the Metropolis algorithm which will be described in Chapter 4) can be interpreted as a Markov process, with a particular choice of transition probabilities: one must satisfy the principle of detailed balance with the equilibrium probability $P_{eq}(S_j)$,

$$W_{ji} P_{eq}(S_j) = W_{ij} P_{eq}(S_i), \tag{2.84}$$

as will be discussed later. At this point, we already note that the master equation yields

$$dP_{eq}(S_j, t)/dt \equiv 0, \tag{2.85}$$

since Eqn. (2.85) ensures that gain and loss terms in Eqn. (2.83) cancel exactly.

Finally we mention that the restriction to a discrete set of states $\{S_i\}$ is not at all important – one can generalize the discussion to a continuum of states, working with suitable probability densities in the appropriate space.

2.2.5 The 'art' of random number generation

2.2.5.1 Background

Monte Carlo methods are heavily dependent on the fast, efficient production of streams of random numbers. Since physical processes, such as white noise generation from electrical circuits, generally introduce new numbers much too slowly to be effective with today's digital computers, random number sequences are produced directly on the computer using software (Knuth, 1969). (The use of tables of random numbers is also impractical because of the huge number of random numbers now needed for most simulations and the slow access time to secondary storage media.) Since such algorithms are actually deterministic, the random number sequences which are thus produced are only 'pseudo-random' and do indeed have limitations which need to be understood. Thus, in the remainder of this book, when we refer to 'random numbers' it must be understood that we are really speaking of 'pseudo-random' numbers. These deterministic features are not always negative. For example, for testing a program it is often useful to compare the results with a previous run made using exactly the same random numbers. The explosive growth in the use of Monte Carlo simulations in diverse areas of physics has prompted extensive investigation of new methods and of the reliability of both old and new techniques. Monte Carlo simulations are subject to both statistical and systematic errors from multiple sources, some of which are well understood (Ferrenberg et al., 1991). It has long been known that poor quality random number generation can lead to systematic errors in Monte Carlo simulation (Marsaglia, 1968; Barber et al., 1985); in fact, early problems with popular generators led to the development of improved methods for producing pseudo-random numbers. For an analysis of the suitability of different random number generators see Coddington (1994). As we shall show in the following discussion both the testing as well as the generation of random numbers remain important problems that have not been fully solved. In general, the random number sequences which are needed should be uniform, uncorrelated, and of extremely long period, i.e. do not repeat over quite long intervals. Later in this chapter we shall give some guidance on the testing for these 'desirable' properties.

In the following sub-sections we shall discuss several different kinds of generators. The reason for this is that it is now clear that for optimum performance and accuracy, the random number generator needs to be matched to the algorithm and computer. Indeed, the resolution of Monte Carlo studies has now advanced to the point where *no* generator can be considered to be completely 'safe' for use with a new simulation algorithm on a new problem. The practitioner is now faced anew with the challenge of testing the random number generator for each high resolution application, and we shall review some of the 'tests' later in this section. The generators which are discussed in the next sub-sections produce a sequence of random integers. Usually floating point numbers between 0 and 1 are needed; these are obtained by carrying out a floating point divide by the largest integer N_{max} which can fit into a word.

One important topic which we shall not consider here is the question of the implementation of random number generators on massively parallel computers. In such cases one must be certain that the random number sequences on all processors are distinct and uncorrelated. As the number of processors available to single users increases, this question must surely be addressed, but we feel that at the present time this is a rather specialized topic and we shall not consider it further.

2.2.5.2 Congruential method

A simple and very popular method for generating random number sequences is the multiplicative or congruential method. Here, a fixed multiplier c is chosen along with a given seed and subsequent numbers are generated by simple multiplication:

$$X_n = (c \times X_{n-1} + a_0) \text{MOD } N_{max}, \qquad (2.86)$$

where X_n is an integer between 1 and N_{max}. It is important that the value of the multiplier be chosen to have 'good' properties and various choices have been used in the past. In addition, the best performance is obtained when the initial random number X_0 is odd. Experience has shown that a 'good' congruential generator is the 32-bit linear congruential algorithm (CONG)

$$X_n = (16807 \times X_{n-1}) \text{MOD}(2^{31} - 1). \qquad (2.87)$$

A congruential generator which was quite popular earlier turned out to have quite noticeable correlation between consecutive triplets of random numbers. Nonetheless for many uses congruential generators are acceptable and are certainly easy to implement. (Congruential generators which use a longer word length also have improved properties.)

2.2.5.3 Mixed congruential methods

Congruential generators can be mixed in several ways to attempt to improve the quality of the random numbers which are produced. One simple and relatively effective method is to use two distinct generators simultaneously: the first one generates a table of random numbers and the second generator draws randomly from this table. For best results the two generators should have different seeds and different multipliers. A variation of this approach for algorithms which need multiple random numbers for different portions of the calculations is to use independent generators for different portions of the problem.

2.2.5.4 Shift register algorithms

A fast method which was introduced to eliminate some of the problems with correlations which had been discovered with a congruential method is the shift register or Tausworthe algorithm (Kirkpatrick and Stoll, 1981). A table of random numbers is first produced and a new random number is produced by combining two different existing numbers from the table:

$$X_n = X_{n-p} \cdot \text{XOR} \cdot X_{n-q} \qquad (2.88)$$

where p and q must be properly chosen if the sequence is to have good properties. The $\cdot$XOR$\cdot$ operator is the bitwise *exclusive-OR* operator. The best choices of the pairs (p, q) are Mersine primes which satisfy the condition

$$p^2 + q^2 + 1 = \text{prime}. \qquad (2.89)$$

Examples of pairs which satisfy this condition are:

$$p = 98 \qquad q = 27$$
$$p = 250 \qquad q = 103$$
$$p = 1279 \qquad q = 216, 418$$
$$p = 9689 \qquad q = 84, 471, 1836, 2444, 4187$$

R250 for which $p = 250$, $q = 103$ has been the most commonly used generator in this class. In the literature one will find cases where X_{n-q} is used and others where X_{n-p-q} is used instead. In fact, these two choices will give the same stream of numbers but in reverse order; the quality of each sequence is thus the same. In general, higher quality of random number sequences results when large values of p and q are used although for many purposes R250 works quite well. In order for the quality of the random number sequence to be of the highest possible quality, it is important for the 'table' to be properly initialized. One simple method is to use a good congruential generator to generate the initial values; the best procedure is to use a different random number to determine each bit in succession for each entry in the initial table.

2.2.5.5 Lagged Fibonacci generators

The shift-register algorithm is a special case of a more general class of generators known as lagged Fibonacci generators. Additional generators may be produced by replacing the exclusive-or ($\cdot$XOR$\cdot$) in Eqn. (2.88) by some other operator. One generator which has been found to have good properties uses the multiplication operator:

$$X_n = X_{n-p} * X_{n-q} \tag{2.90}$$

with rather small values of the 'off-set', e.g. $p = 17$, $q = 5$. More complex generators have also been used, e.g. a 'subtract with carry generator' (Marsaglia et al., 1990) (SWC), which for 32-bit arithmetic is

$$X_n = X_{n-22} - X_{n-43} - C \tag{2.91}$$
$$\text{if } X_n \geq 0, \qquad C = 0$$
$$\text{if } X_n < 0, \qquad X_n = X_n + (2^{32} - 5), \qquad C = 1$$

and the compound generator, a combined subtract with carry-Weyl generator (Marsaglia et al., 1990) (SWCW)

$$Z_n = Z_{n-22} - Z_{n-43} - C \tag{2.92}$$
$$\text{if } Z_n \geq 0, \qquad C = 0$$
$$\text{if } Z_n < 0, \qquad Z_n = Z_n + (2^{32} - 5), \qquad C = 1$$
$$Y_n = (Y_{n-1} - 362436069) \text{ MOD } 2^{32}$$
$$X_n = (Z_n - Y_n) \text{ MOD } 2^{32}.$$

As mentioned earlier, it is known that the performance of a random number generator can be adversely affected by improper initialization of its lookup table (Kirkpatrick and Stoll, 1981) and we recommend the same initialization procedure for all generators as that described for R250. The above are only examples of a few different random number generators.

2.2.5.6　Tests for quality

Properties of random number generators have been carefully examined using a battery of mathematical tests (Marsaglia, 1968, 1985); a few simple examples of such tests are:

> *Uniformity test*: Break up the interval between zero and one into a large number of small bins and after generating a large number of random numbers check for uniformity in the number of entries in each bin.
> *Overlapping M-tuple test*: Check the statistical properties of the number of times M-tuples of digits appear in the sequence of random numbers.
> *Parking lot test*: Plot points in an m-dimensional space where the m-coordinates of each point are determined by m-successive calls to the random number generator. Then look for regular structures.

Although the 'quality' of a sequence of random numbers is notoriously difficult to assess, often all indications from standard tests are that any residual errors from random number generation should now be smaller than statistical errors in Monte Carlo studies. However, these mathematical tests are not necessarily sufficient, and an example of a 'practical' test in a Monte Carlo study of a small lattice Ising model (which can be solved exactly) will be presented later; here both 'local' and 'non-local' sampling methods were shown to yield different levels of systematic error with different 'good' generators. (The exact nature of these algorithms is not really important at this stage and will be discussed in detail in later sections.) More sophisticated, high quality generators, such as RANLUX (James, 1994; Luscher, 1994) which is based upon an algorithm by Marsaglia and Zaman (1991), are finding their way into use, but they are slow and must still be carefully tested with new algorithms as they are devised. (RANLUX includes two lags, plus a carry, plus it discards portions of the sequence of generated numbers. The complications tend to destroy short time correlations but have the negative effect of slowing down the generator.)

Problem 2.5 Suppose we have a computer with 4 bit words. Produce a sequence of random numbers using a congruential generator. What is the cycle length for this generator?

Example:

Carry out a 'parking lot' test on two different random number generators. 10 000 points are plotted using consecutive pairs of random numbers as x- and y-coor-

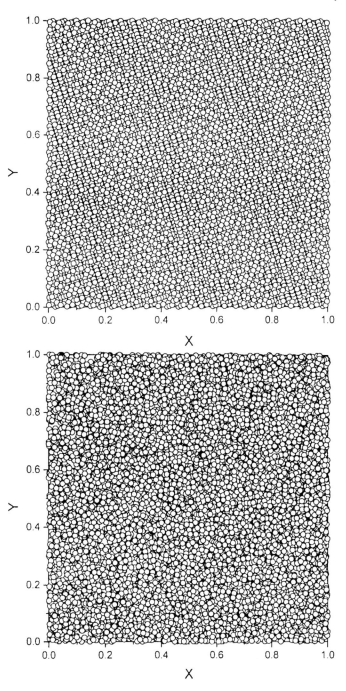

dinates. At the top is a picture of a 'bad' generator (exhibiting a striped pattern) and at the bottom are the results of a 'good' generator.

2.2.5.7 Non-uniform distributions

There are some situations in which random numbers x_i which have different distributions, e.g. Gaussian, are required. The most general way to perform this is to look at the integrated distribution function $F(x)$ of the desired distribution $f(x)$, generate a uniform distribution of random numbers y_i and then take the inverse function with the uniformly chosen random number as the variable, i.e.

$$y = F(y) = \int_0^y f(x)dx \qquad (2.93)$$

so that

$$x = F^{-1}(y). \qquad (2.94)$$

Example

Suppose we wish to generate a set of random numbers distributed according to $f(x) = x$. The cumulative distribution function is $y = F(x) = \int_0^x x'dx' = 0.5x^2$. If a random number y is chosen from a uniform distribution, then the desired random number is $x = 2.0y^{1/2}$.

An effective way to generate numbers according to a Gaussian distribution is the Box–Muller method. Here two different numbers x_1 and x_2 are drawn from a uniform distribution and then the desired random numbers are computed from

$$y_1 = (-2\ln x_1)^{1/2} \cos(2\pi x_2), \qquad (2.95a)$$
$$y_2 = (-2\ln x_1)^{1/2} \sin(2\pi x_2). \qquad (2.95b)$$

Obviously the quality of the random numbers produced depends on the quality of the uniform sequence which is generated first. Because of the extra cpu time needed for the computation of the trigonometric functions, the speed with which x_1 and x_2 are generated is not particularly important.

Problem 2.6 Given a sequence of uniformly distributed random numbers y_i, show how a sequence x_i distributed according to x^2 would be produced.

2.3 NON-EQUILIBRIUM AND DYNAMICS: SOME INTRODUCTORY COMMENTS

2.3.1 Physical applications of master equations

In classical statistical mechanics of many-body systems, dynamical properties are controlled by Newton's equations of motion for the coordinates r_i of the atoms labeled by index i, $m_i \ddot{r}_i = -\nabla_i U$, m_i being the mass of the ith particle, and U being the total potential energy (which may contain both an external potential and interatomic contributions). The probability of a point in phase space then develops according to Liouville's equation, and obviously the deterministic trajectory through phase space generated in this way has nothing to do, in general, with the probabilistic trajectories generated in stochastic processes, such as Markov processes (Section 2.2.3).

However, often one is not aiming at a fully atomistic description of a physical problem, dealing with all coordinates and momenta of the atoms. Instead one is satisfied with a coarse-grained picture for which only a subset of the degrees of freedom matters. It then is rather common that the degrees of freedom that are left out (i.e. those which typically occur on a much smaller length scale and much faster time scale) act as a heat bath, inducing stochastic transitions among the relevant (and slower) degrees of freedom. In the case of a very good separation of time scales, it is in fact possible to reduce the Liouville equation to a Markovian master equation, of the type written in Eqn. (2.83).

Rather than repeating any of the formal derivations of this result from the literature, we rather motivate this description by a typical example, namely the description of interdiffusion in solid binary alloys (AB) at low temperatures (Fig. 2.9). The solid forms a crystal lattice, and each lattice site i may be occupied by an A-atom (then the concentration variable $c_i^A = 1$, otherwise $c_i^A = 0$), by a B-atom (then $c_i^B = 1$, otherwise $c_i^B = 0$), or stay vacant. Interdiffusion then happens because A-atoms jump to a (typically nearest neighbor) vacant site, with a jump rate Γ_A, and B-atoms jump to a vacant site at jump rate Γ_B, and many such random hopping events relax any concentration gradients. The distribution of the atoms over the available sites may be completely random or correlated, and the jump rates may depend on the local neighborhood or may simply be constants, etc. Now a consideration of the potential energy in solids shows that such jump events are normally thermally activated processes, $\Gamma_{A,B} \propto \exp(-\Delta E/k_B T)$, where the energy barrier to be overcome is much higher than the thermal energy (e.g. $\Delta E \approx 1$ eV). As a result, the time a vacancy needs in order to move from one lattice site to the next one is orders of magnitude larger than the time constant of the lattice vibrations. This separation of time scales (a phonon vibration time may be of the order of 10^{-13} seconds, the time between the moves of a vacancy can be 10 orders of magnitude slower) is due to the different length scales of these motions (vibrations take only one percent of a lattice spacing at low temperatures). Thus a simulation of the

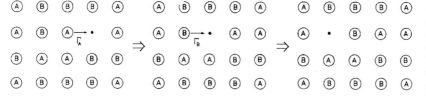

Fig. 2.9 Schematic description of interdiffusion in a model of random binary alloy (AB) with a small volume fraction of vacancies. Interdiffusion proceeds via the vacancy mechanism: A-atoms jump with rate Γ_A and B-atoms with rate Γ_B.

dynamics of these hopping processes using the molecular dynamics method which numerically integrates Newton's equations of motion, would suffer from a sampling of extremely rare events. The master equation, Eqn. (2.83), which can be straightforwardly simulated by Monte Carlo methods, allows the direct simulation of the important hopping events, completely disregarding the phonons. But it is also clear, of course, that knowledge of the basic rate constants for the slow degrees of freedom (the jump rates Γ_A, Γ_B in the case of our example) are an 'input' to the Monte Carlo simulation, rather than an 'output': the notion of 'time' for a Markov process (Section 2.2.4) does not specify anything about the *units* of this time. These units are only fixed if the connection between the slow degrees of freedom and the fast ones is explicitly considered, which usually is a separate problem and out of consideration here.

Although the conditions under which a master equation description of a physical system is appropriate may seem rather restrictive, it will become apparent later in this book that there is nevertheless a rich variety of physical systems and/or processes that can be faithfully modeled by this stochastic dynamics. (Examples include relaxation of the magnetization in spin glasses; Brownian motion of macromolecules in melts; spinodal decomposition in mixtures; growth of ordered monolayer domains at surfaces; epitaxial growth of multilayers; etc.)

2.3.2 Conservation laws and their consequences

Different situations may be examined in which different properties of the system are held constant. One interesting case is one in which the total magnetization of a system is conserved (held constant); when a system undergoes a first order transition it will divide into different regions in which one phase or the other dominates. The dynamics of first order transitions is a fascinating topic with many facets (Gunton *et al.*, 1983; Binder, 1987). It is perhaps instructive to first briefly review some of the static properties of a system below the critical point; for a simple ferromagnet a first order transition is encountered when the field is swept from positive to negative. Within the context of Landau theory the behavior can be understood by looking at the magnetization isotherm shown in Fig. 2.10. The solid portions of the curve are thermodynamically stable, while the dashed portions are metastable, and the dotted portion is unstable. The endpoints of the unstable region are termed 'spinodal points' and occur at magnetizations $\pm M_{sp}$. The spinodal

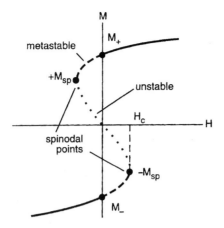

Fig. 2.10
Magnetization as a
function of magnetic
field for $T < T_c$. The
solid curves represent
stable, equilibrium
regions, the dashed
lines represent
'metastable', and the
dotted line 'unstable'
states. The values of
the magnetization at
the 'spinodal' are
$\pm M_{sp}$ and the
spinodal fields are
$\pm H_c$. M_+ and M_- are
the magnetizations at
the opposite sides of
the coexistence curve.

points occur at magnetic fields $\pm H_c$. As the magnetic field is swept, the transition occurs at $H = 0$ and the limits of the corresponding coexistence region are at $\pm M_s$. If f_{cg} is a coarse-grained free energy density, then

$$\partial^2 f_{cg}/\partial M^2 = \chi_T^{-1} \to 0 \tag{2.96}$$

at the spinodal. However, this singular behavior at the spinodal is a mean-field concept, and one must ask how this behavior is modified when statistical fluctuations are considered. A Ginzburg criterion can be developed in terms of a coarse-grained length scale L and coarse-grained volume L^d. The fluctuations in the magnetization as a function of position $M(x)$ from the mean value M must satisfy the condition

$$\langle [M(x) - M]^2 \rangle L^d / [M - M_{sp}]^2 \ll 1. \tag{2.97}$$

This leads to the condition that

$$1 \ll R^d (H_c - H)^{(6-d)/4}. \tag{2.98}$$

Thus the behavior should be mean–field-like for large interaction range R and far from the spinodal.

If a system is quenched from a disordered, high temperature state to a metastable state below the critical temperature, the system may respond in two different ways depending on where the system is immediately after the quench (see Fig. 2.11). If the quench is to a point which is close to one of the equilibrium values characteristic of the two-phase coexistence then the state evolves towards equilibrium by the nucleation and subsequent growth of 'droplets', see Fig. 2.12. (This figure is shown for pedagogical reasons and is not intended to provide an accurate view of the droplet formation in a particular physical system.) There will be a free energy barrier ΔF_l^* to the growth of clusters where l^* is the 'critical cluster size' and the nucleation rate $\mathcal{J}$ will be given by

$$\mathcal{J} \propto \exp(-\Delta F_l^*/k_B T). \tag{2.99}$$

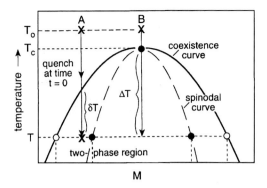

Fig. 2.11 Schematic phase coexistence diagram showing the 'spinodal' line. Paths (A) and (B) represent quenches into the nucleation regime and the spinodal decomposition regime, respectively.

Near the spinodal the argument of the exponential will be

$$\Delta F_l^* / k_B T \propto R^d (1 - T/T_c)^{(4-d)/2} [(M_{ms} - M_{sp})/(M_+ - M_-)]^{(6-d)/2}, \tag{2.100}$$

whereas near the coexistence curve

$$\Delta F_l^* / k_B T \propto R^d (1 - T/T_c)^{(4-d)/2} [(M_+ - M_{ms})/(M_+ - M_-)]^{-(d-1)}. \tag{2.101}$$

In solid mixtures the latter stages of this growth are thought to be described by the Lifshitz–Slyozov theory (Lifshitz and Slyozov, 1961). At short times a nucleation barrier must be overcome before droplets which can grow form, and at later times the process leads to a power law growth of the characteristic length scale $L(t)$, i.e.

$$L(t) \propto t^{1/3} \tag{2.102}$$

for $d \geq 2$. Scaling behavior is also predicted for both the droplet size distribution $n_l(t)$ and the structure factor $S(q, t)$:

$$n_l(t) = (\bar{l}(t))^2 \tilde{n}(l/\bar{l}(t)), \qquad (l \to \infty, t \to \infty), \tag{2.103a}$$
$$S(q, t) = (L(t))^d \tilde{S}(qL(t)), \qquad (q \to 0, t \to \infty), \tag{2.103b}$$

where $\bar{l} \propto t^{dx}$ is the mean cluster size and x is a characteristic exponent which is $1/3$ if conserved dynamics applies.

If, however, the initial quench is close to the critical point concentration, the state is unstable and the system evolves towards equilibrium by the formation of long wavelength fluctuations as shown in Fig. 2.12. The explicit shape of these structures will vary with model and with quench temperature; Fig. 2.12 is only intended to show 'typical' structures. The early stage of this process is called spinodal decomposition and the late stage behavior is termed 'coarsening'. The linearized theory (Cahn and Hilliard, 1958; Cahn, 1961) predicts

$$S(q, t) = S(q, 0)e^{2\omega(q)t} \tag{2.104}$$

Fig. 2.12 Pictorial
view of different
possible modes for
phase separation: (a)
nucleation; (b)
spinodal
decomposition. The
dark regions represent
the phase with M_-.

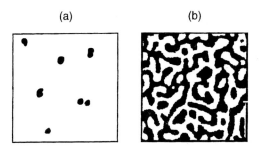

where $\omega(q)$ is zero for the critical wavevector q_c. The linearized theory is invalid for systems with short range interactions but is approximately correct for systems with large, but finite, range coupling.

2.3.3 Critical slowing down at phase transitions

As a critical point T_c is approached the large spatial correlations which develop have long temporal correlations associated with them as well (van Hove, 1954). At T_c the characteristic time scales diverge in a manner which is determined in part by the nature of the conservation laws. This 'critical slowing down' has been observed in multiple physical systems by light scattering experiments (critical opalescence) as well as by neutron scattering. The seminal work by Halperin and Hohenberg (Hohenberg and Halperin, 1977) provides the framework for the description of dynamic critical phenomena in which there are a number of different universality classes, some of which correspond to systems which only have relaxational behavior and some of which have 'true dynamics', i.e. those with equations of motion which are derived from the Hamiltonian. One consequence of this classification is that there may be different models which are in the same static universality class but which are in different dynamic classes. Simple examples include the Ising model with 'spin-flip' kinetics vs. the same model with 'spin-exchange' kinetics, and the Heisenberg model treated by Monte Carlo (stochastic) simulations vs. the same model solved by integrating coupled equations of motion. For relaxational models, such as the stochastic Ising model, the time-dependent behavior is described by a master equation

$$\partial P_n(t)/\partial t = -\sum_{n \neq m}[P_n(t)W_{n \to m} - P_m(t)W_{m \to n}], \qquad (2.105)$$

where $P_n(t)$ is the probability of the system being in state 'n' at time t, and $W_{n \to m}$ is the transition rate for $n \to m$. The solution to the master equation is a sequence of states, but the time variable is a stochastic quantity which does not represent true time. A relaxation function $\phi(t)$ can be defined which describes time correlations *within* equilibrium

$$\phi_{MM}(t) = \frac{\langle\langle M(0)M(t)\rangle - \langle M\rangle^2}{\langle M^2\rangle - \langle M\rangle^2}. \qquad (2.106)$$

When normalized in this way, the relaxation function is 1 at $t = 0$ and decays to zero as $t \to \infty$. It is important to remember that for a system in equilibrium any time in the sequence of states may be chosen as the '$t = 0$' state. The asymptotic, long time behavior of the relaxation function is exponential, i.e.

$$\phi(t) \to e^{-t/\tau} \tag{2.107}$$

where the correlation time τ diverges as T_c is approached. This dynamic (relaxational) critical behavior can be expressed in terms of a power law as well,

$$\tau \propto \xi^z \propto \varepsilon^{-\nu z} \tag{2.108}$$

where ξ is the (divergent) correlation length, $\varepsilon = |1 - T/T_c|$, and z is the dynamic critical exponent. Estimates for z have been obtained for Ising models by epsilon-expansion RG theory (Bausch $et~al.$, 1981) but the numerical estimates (Landau $et~al.$, 1988; Wansleben and Landau, 1991; Ito, 1993) are still somewhat inconsistent and cannot yet be used with complete confidence.

A second relaxation time, the integrated relaxation time, is defined by the integral of the relaxation function

$$\tau_{\text{int}} = \int_0^\infty \phi(t)dt. \tag{2.109}$$

This quantity has particular importance for the determination of errors and is expected to diverge with the same dynamic exponent as the 'exponential' relaxation time.

One can also examine the approach to equilibrium by defining a non-linear relaxation function

$$\phi_M(t) = \frac{\langle M(t) - M(\infty)\rangle}{\langle M(0)\rangle - \langle M(\infty)\rangle}. \tag{2.110}$$

The non-linear relaxation function also has an exponential decay at long times, and the characteristic relaxation time $\tau_{nl} = \int_0^\infty \phi_M(t)dt$ diverges with dynamic exponent z_{nl}. Fisher and Racz (1976) have shown, however, that there is only one independent exponent and that

$$z = z_{\text{nl}}{}^M + \beta/\nu, \tag{2.111}$$

or if the relaxation has been determined for the internal energy then

$$z = z_{\text{nl}}{}^E + (1 - \alpha)/\nu. \tag{2.112}$$

There are other systems, such as glasses and models with impurities, where the decay of the relaxation function is more complex. In these systems a 'stretched exponential' decay is observed

$$\phi \propto e^{-(t/\tau)^n}, \qquad n < 1 \tag{2.113}$$

and the behavior of τ may not be simple. In such cases, extremely long observation times may be needed to measure the relaxation time.

The properties of systems with true dynamics are governed by equations of motion and the time scale truly represents real time; since this behavior does not occur in Monte Carlo simulations it will not be discussed further at this point.

2.3.4 Transport coefficients

If some observable A is held constant and all 'flips' involve only local, e.g. nearest neighbor, changes, the Fourier components $A(q)$ can be described by a characteristic time

$$\tau_{AA}(q) = (D_{AA}q^2)^{-1} \tag{2.114}$$

where D_{AA} is a transport coefficient. In the simulation of a binary alloy, the concentrations of the constituents would be held fixed and D_{AA} would correspond to the concentration diffusivity. With different quantities held fixed, of course, different transport coefficients can be measured and we only offer the binary alloy model as an example. Equation (2.114) implies a very slow relaxation of long wavelength variations. Note that this 'hydrodynamic slowing down' is a very general consequence of the conservation of concentration and *not* due to any phase transition. If there is an unmixing critical point, see Fig. 2.11, then $D_{AA} \propto |\varepsilon|^{\gamma}$ and at T_c the relaxation time diverges as $\tau_{AA}(q) \propto q^{-(4-\eta)}$ (Hohenberg and Halperin, 1977).

2.3.5 Concluding comments: why bother about dynamics when doing Monte Carlo for statics?

Since importance sampling Monte Carlo methods correspond to a Markovian master equation by construction, the above remarks about dynamical behavior necessarily have some impact on simulations; indeed dynamical behavior can possibly affect the results for statics. For example, in the study of static critical behavior the critical slowing down will adversely affect the accuracy. In the examination of hysteresis in the study of phase diagrams, etc. the long time scales associated with metastability are an essential feature of the observed behavior. Even if one simulates a fluid in the NVT ensemble away from any phase transition, there will be slow relaxation of long wavelength density fluctuations due to the conservation of density as in Eqn. (2.114). Thus, insight into the dynamical properties of simulations always helps to judge their validity.

REFERENCES

Barber, M. N., Pearson, R. B., Toussaint, D., and Richardson, J. L. (1985), Phys. Rev. B **32**, 1720.

Bausch, R., Dohm, V., Janssen, H. K., and Zia, R. K. P. (1981), Phys. Rev. Lett. **47**, 1837.

Binder, K. (1987), Rep. Prog. Phys. **50**, 783.

Cahn, J. W. (1961), Acta Metall. **9**, 795.

Cahn, J. W. and Hilliard, J. E. (1958), J. Chem. Phys. **28**, 258.

Callen, H. (1985), *Introduction to Thermodynamics and Thermostatics* (Wiley, New York).

Chen, K., Ferrenberg, A. M., and Landau, D. P. (1993), Phys. Rev. B **48**, 239 and references therein.

Coddington, P. D. (1994), Int. J. Mod. Phys. C **5**, 547.

Domany, E., Schick, M., Walker, J. S., and Griffiths, R. B. (1980), Phys. Rev. B **18**, 2209.

Feller, W. (1968), *An Introduction to Probability Theory and its Applications*, vol. 1 (J. Wiley and Sons, New York).

Ferrenberg, A. M., Landau, D. P., and Binder, K. (1991), J. Stat. Phys. **63**, 867.

Fisher, M. E. (1974), Rev. Mod. Phys. **46**, 597.

Fisher, M. E. and Racz, Z. (1976), Phys. Rev. B **13**, 5039.

Griffiths, R. B. (1970), Phys. Rev. Lett. **24**, 715.

Gunton, J. D., San Miguel, M., and Sahni, P. S. (1983), in *Phase Transitions and Critical Phenomena*, vol. 8, eds. C. Domb and J. L. Lebowitz (Academic Press, London)7 p. 267.

Hohenberg, P. and Halperin, B. (1977), Rev. Mod. Phys. **49**, 435.

Ito, N. (1993), Physica A **196**, 591.

James, F. (1994), Comput. Phys. Commun. **79**, 111.

Kadanoff, L. P., Goetze, W., Hamblen, D., Hecht, R., Lewis, E. A. S., Palciauskas, V. V., Rayl, M., Swift, J., Aspenes, D., and Kane, J. (1967), Rev. Mod. Phys. **39**, 395.

Kalos, M. H. and Whitlock, P. A. (1986), *Monte Carlo Methods*, vol. 1 (Wiley and Sons, New York).

Kirkpatrick, S. and Stoll, E. (1981), J. Comput. Phys. **40**, 517.

Knuth, D. (1969), *The Art of Computer Programming*, vol. 2 (Addison–Wesley, Reading).

Kornblit, A. and Ahlers, G. (1973), Phys. Rev. B **8**, 5163.

Kosterlitz, J. M. and Thouless, D. J. (1973), J. Phys. C **6**, 1181.

Landau, D. P. (1983), Phys. Rev. B **27**, 5604.

Landau, D. P. (1994), J. Appl. Phys. **73**, 6091.

Landau, D. P., Tang, S., and Wansleben, S. (1988), J. de Physique **49**, C8-1525.

Landau, L. D. and Lifshitz, E. M. (1980), *Statistical Physics*, 3rd edition (Pergamon Press, Oxford).

Lawrie, I. D. and Sarbach, S. (1984), in *Phase Transitions and Critical Phenomena* , vol. **9**, eds. C. Domb and L. J. Lebowitz (Academic Press, New York), p.1.

Lifshitz, I. M. and Slyozov, V. V. (1961), J. Chem. Solids **19**, 35.

Luscher, M. (1994), Comput. Phys. Commun. **79**, 100.

Marsaglia, G. (1968), Proc. Natl. Acad. Sci. **61**, 25.

Marsaglia, G. (1985), in *Computer Science and Statistics: The Interface*, ed. L. Billard (Elsevier, Amsterdam).

Marsaglia, G. and Zaman, A. (1991), Ann. Appl. Prob. 1, 462.

Marsaglia, G., Narasimhan, B., and Zaman, A. (1990), Comput. Phys. Comm. **60**, 345.

Nelson, D. R., Kosterlitz, J. M., and Fisher, M. E. (1974), Phys. Rev. Lett. **33**, 813.

Onsager, L. (1944), Phys. Rev. **65**, 117.

Potts, R. B. (1952), Proc. Camb. Philos. Soc. **48**, 106.

Privman, V., Hohenberg, P. C., and Aharony, A. (1991), in *Phase Transitions and Critical Phenomena*, vol. **14**, eds. C. Domb and J. L. Lebowitz (Academic Press, London).

Riedel, E. K. and Wegner, F. J. (1972), Phys. Rev. Lett. **29**, 349.

Stanley, H. E. (1971), *An Introduction to Phase Transitions and Critical Phenomena* (Oxford University Press, Oxford).

Stryjewski, E. and Giordano, N. (1977), Adv. Physics **26**, 487.

van Hove, L. (1954), Phys. Rev. **93**, 1374.

Wansleben, S. and Landau, D. P. (1991), Phys. Rev. B **43**, 6006.

Wegner, F. J. and Riedel, E. K. (1973), Phys. Rev. B **7**, 248.

Wilson, K. G. (1971), Phys. Rev. B **4**, 3174, 3184.

Wu, F. Y. (1982), Rev. Mod. Phys. **54**, 235.

Yeomans, J. (1992), *Statistical Mechanics of Phase Transitions* (Oxford University Press, Oxford).

Zinn-Justin, J. and LeGuillou, J.-C. (1980), Phys. Rev. B **21**, 3976.

3   Simple sampling Monte Carlo methods

3.1 INTRODUCTION

Modern Monte Carlo methods have their roots in the 1940s when Fermi, Ulam, von Neumann, Metropolis and others began considering the use of random numbers to examine different problems in physics from a stochastic perspective (Cooper (1989), this set of biographical articles about S. Ulam provides fascinating insight into the early development of the Monte Carlo method, even before the advent of the modern computer). Very simple Monte Carlo methods were devised to provide a means to estimate answers to analytically intractable problems. Much of this work is unpublished and a view of the origins of Monte Carlo methods can best be obtained through examination of published correspondence and historical narratives. Although many of the topics which will be covered in this book deal with more complex Monte Carlo methods which are tailored explicitly for use in statistical physics, many of the early, simple techniques retain their importance because of the dramatic increase in accessible computing power which has taken place during the last two decades. In the remainder of this chapter we shall consider the application of simple Monte Carlo methods to a broad spectrum of interesting problems.

3.2 COMPARISONS OF METHODS FOR NUMERICAL INTEGRATION OF GIVEN FUNCTIONS

3.2.1 Simple methods

One of the simplest and most effective uses for Monte Carlo methods is the evaluation of definite integrals which are intractable by analytic techniques. (See the book by Hammersley and Handscomb (1964) for more mathematical details.) In the following discussion, for simplicity we shall describe the methods as applied to one-dimensional integrals, but it should be understood that these techniques are readily extended, and often most effective, when applied to multidimensional integrals. In the simplest case we wish to obtain the integral of $f(x)$ over some fixed interval:

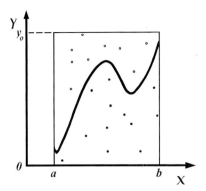

Fig. 3.1 Simple representation of 'hit-or-miss' Monte Carlo integration of a function $f(x)$, given by the solid curve, between $x = a$ and $x = b$. N points are randomly dropped into the box, N_o of them fall below the curve. The integral is estimated using Eqn. (3.2).

$$y = \int_a^b f(x)dx. \tag{3.1}$$

In Fig. 3.1 we show a pictorial representation of this problem. A straightforward Monte Carlo solution to this problem via the 'hit-or-miss' (or acceptance–rejection) method is to draw a box extending from a to b and from 0 to y_0 where $y_0 > f(x)$ throughout this interval. Using random numbers drawn from a uniform distribution, we drop N points randomly into the box and count the number, N_o, which fall below $f(x)$ for each value of x. An estimate for the integral is then given by the fraction of points which fall below the curve times the area of the box, i.e.

$$y_{\text{est}} = (N_o/N) \times [y_0(b - a)]. \tag{3.2}$$

This estimate becomes increasingly precise as $N \to \infty$ and will eventually converge to the correct answer. This technique is an example of a 'simple sampling' Monte Carlo method and is obviously dependent upon the quality of the random number sequence which is used. Independent estimates can be obtained by applying this same approach with different random number sequences and by comparing these values the precision of the procedure can be ascertained. An interesting problem which can be readily attacked using this approach is the estimation of a numerical value for π. The procedure for this computation is outlined in the example described below.

Example

How can we estimate the value of π using simple sampling Monte Carlo? Choose N points randomly in the xy-plane so that $0 < x < 1$ and $0 < y < 1$. Calculate the distance from the origin for each point and count those which are less than a distance of 1 from the origin. The fraction of the points which satisfy this condition, N_o/N, provides an estimate for the area of one-quarter of a circle so that $\pi \approx 4N_o/N$. This procedure may be repeated multiple times and the variance of the different results may be used to estimate the error. Here are some sample results for a run with 10 000 points. Note that on the right we show estimates based on up to the first 700 points; these results appear to have converged to the wrong answer but the apparent difficulty is really due simply to the use of too few points. This lesson should not be forgotten!

N	Result		N	Result
1000	3.0800		100	3.1600
2000	3.0720		200	3.0400
3000	3.1147		300	3.1067
4000	3.1240		400	3.0800
5000	3.1344		500	3.0560
6000	3.1426		600	3.0800
7000	3.1343		700	3.0743
8000	3.1242			
9000	3.1480			
10000	3.1440			

A variation of this approach is to choose the values of x in a regular, equi-distant fashion. The advantage of this algorithm is that it requires the use of fewer random numbers. For functions with very substantial variations over the range of interest, these methods are quite likely to converge slowly, and a different approach must be devised.

Another type of simple Monte Carlo method is termed the 'crude method'. In this approach we choose N values of x randomly and then evaluate $f(x)$ at each value so that an estimate for the integral is provided by

$$y_{est} = \frac{1}{N} \sum_i f(x_i) \qquad (3.3)$$

where, again, as the number of values of x which are chosen increases, the estimated answer eventually converges to the correct result. In a simple variation of this method, one can divide the interval into a set of unequal sub-intervals and perform a separate Monte Carlo integration for each sub-interval. In those regions where the function is large the sampling can be extensive and less effort can be expended on those sub-intervals over which the function is small.

3.2.2 Intelligent methods

Improved methods may be broadly classified as 'intelligent' Monte Carlo methods. In one technique, the 'control variate method', one selects a known, integrable function $f'(x)$ which has a relatively similar functional dependence on x and only integrates the difference $[f'(x) - f(x)]$ by some Monte Carlo method, i.e.

$$y_{est} = F' + \int_a^b [f'(x) - f(x)] dx \qquad (3.4)$$

where $F' = \int_a^b f'(x)dx$. The final estimate for y_{est} can be improved without additional numerical effort by an intelligent choice of $f'(x)$.

Instead of selecting all points with equal probability, one can choose them according to the anticipated importance of the value of the function at that point to the integral $p(x)$ and then weight the contribution by the inverse of the probability of choice. This is one of the simplest examples of the class of Monte Carlo methods known as 'importance sampling' which will be discussed in much greater detail in the next chapter. Using importance sampling an estimate for the integral is given by

$$y_{est} = \sum_i p^{-1}(x_i)f(x_i). \tag{3.5}$$

For functions which vary wildly over the interval of interest, this approach allows us to increase the sampling in the region in which the contribution to the integral is particularly large. Since the values of x are no longer chosen with equal probability, we begin to see the need for sequences of random numbers which are not drawn from a uniform sequence. Obviously for oddly behaved functions some expertise is needed in choosing $p(x)$, but this can be done iteratively by first carrying out a rough Monte Carlo study and improving the choice of sampling method. Intelligent importance sampling is far more effective in improving convergence than the brute force method of simply generating many more points.

Problem 3.1 Suppose $f(x) = x^{10} - 1$. Use a 'hit-or-miss' Monte Carlo simulation to determine the integral between $x = 1$ and $x = 2$.

Problem 3.2 Suppose $f(x) = x^{10} - 1$. Use an importance sampling Monte Carlo simulation to determine the integral between $x = 1$ and $x = 2$.

Problem 3.3 Estimate π using the methods described above with $N = 100\,000$ points. What is the error of your estimate? Does your estimate agree with the correct answer?

3.3 BOUNDARY VALUE PROBLEMS

There is a large class of problems which involve the solution of a differential equation subject to a specified boundary condition. As an example we consider Laplace's equation

$$\nabla^2 u = \partial^2 u/\partial x^2 + \partial^2 u/\partial y^2 = 0 \tag{3.6}$$

where the function $u(r) = f$ on the boundary. Eqn (3.6) can be re-expressed as a finite difference equation, if the increment Δ is sufficiently small,

$$\nabla^2 u = [u(x + \Delta, y) + u(x - \Delta, y) + u(x, y + \Delta)$$
$$+ u(x, y - \Delta) - 4u(x, y)]/\Delta^2 = 0 \tag{3.7}$$

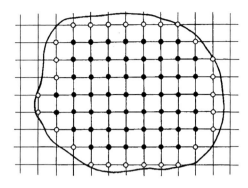

Fig. 3.2 Schematic representation of a grid superimposed upon the region of space which contains the boundary of interest. Closed circles show 'interior' positions; open circles show boundary points.

or

$$u(x, y) = [u(x + \Delta, y) + u(x - \Delta, y) + u(x, y + \Delta) + u(x, y - \Delta)]/4. \quad (3.8)$$

If we examine the behavior of the function $u(r)$ at points on a grid with lattice spacing Δ, we may give this equation a probabilistic interpretation. If we consider a grid of points in the x–y plane with a lattice spacing of Δ, then the probability of a random walk returning to the point (x, y) from any of its nearest neighbor sites is $1/4$. If we place the boundary on the grid, as shown in Fig. 3.2, a random walk will terminate at a boundary point (x', y') where the variable u has the value

$$u(x', y') = f(x', y'). \quad (3.9)$$

One can then estimate the value of $u(x, y)$ by executing many random walks which begin at the point (x, y) as the average over all N walks which have been performed:

$$u(x, y) \approx \frac{1}{N} \sum_i f(x_i', y_i'). \quad (3.10)$$

After a large number of such walks have been performed, a good estimate of $u(x, y)$ will be produced, but the estimate will depend upon both the coarseness of the grid as well as the number of random walks generated.

Example

Consider two concentric, circular conductors in a plane which are placed into the center of a square box which is 20 cm on a side. The inner conductor has a radius of 4 cm and carries a potential of 2 V; the outer conductor has a radius of 16 cm and has a potential of 4 V. What is the potential halfway between the two conductors? Consider a square box with an $L \times L$ grid. Execute N random walks and follow the estimates for the potential as a function of N for different grid sizes L. Note that the variation of the estimates with grid size is not simple.

N	$L = 10$	$L = 20$	$L = 40$	$L = 80$	$L = 160$	$L = 320$
500	3.6560	3.3000	3.2880	3.3240	3.3400	3.3760V
5 000	3.5916	3.2664	3.3272	3.3372	3.3120	3.3172
10 000	3.6318	3.2886	3.3210	3.3200	3.3128	3.3222
50 000	3.6177	3.2824	3.3149	3.3273	3.3211	3.3237
100 000	3.6127	3.2845	3.3211	3.3240	3.3243	3.3218

exact value $= 3.3219$ V.

Of course, with these comments and the preceding example we only wish to provide the flavor of the idea – more detailed information can be found in a recent book (Sabelfeld, 1991).

3.4 SIMULATION OF RADIOACTIVE DECAY

One of the simplest examples of a physical process for which the Monte Carlo method can be applied is the study of radioactive decay. Here one begins with a sample of N nuclei which decay at rate λ sec^{-1}. We know that the physics of the situation specifies that the rate of decay is given by

$$dN/dt = -\lambda N, \qquad (3.11)$$

where the nuclei which decay during the time interval dt can be chosen randomly. The resultant time dependence of the number of undecayed nuclei is

$$N = N_o e^{-\lambda t}, \qquad (3.12)$$

where N_o is the initial number of nuclei and λ is related to the 'half-life' of the system. In the most primitive approach, the position of the nuclei plays no role and only the number of 'undecayed' nuclei is monitored. Time is divided into discrete intervals, and each undecayed nucleus is 'tested' for decay during the first time interval. The number of undecayed nuclei is determined, time is then incremented by one step, and the process is repeated so that the number of undecayed nuclei can be determined as a function of time. The time discretization must be done intelligently so that a reasonable number of decays occur in each time step or the simulation will require too much cpu time to be effective. On the other hand, if the time step is chosen to be too large, then so many decays occur during a given interval that there is very little time resolution. This entire process may be repeated many times to obtain a series of independent 'experiments' and the mean value of N, as well as an error bar, may be determined for each value of time. Note that since each 'sample' is independent of the others, measurements for each value of time are uncorrelated even though there may be correlations between different times for a single sample. The extension to systems with multiple decay paths is straightforward.

Problem 3.4 Given a sample with 10 000 radioactive nuclei each of which decays at rate p per sec, what is the half-life of the sample if $p = 0.2$? (Hint: The most accurate way to determine the half-life is not to simply determine the time which it takes for each sample to decay to half its original size. What does physics tell you about the expected nature of the decay for all times?)

3.5 SIMULATION OF TRANSPORT PROPERTIES

3.5.1 Neutron transport

Historically the examination of reactor criticality was among the first problems to which Monte Carlo methods were applied. The fundamental question at hand is the behavior of large numbers of neutrons inside the reactor. In fact, when neutrons travelling in the moderator are scattered, or when a neutron is absorbed in a uranium atom with a resultant fission event, particles fly off in random directions according to the appropriate differential cross-sections (as the conditional probabilities for such scattering events are called). In principle, these problems can be described by an analytic theory, namely integro-differential equations in a six-dimensional space (Davison, 1957); but this approach is rather cumbersome due to the complicated, inhomogeneous geometry of a reactor that is composed of a set of fuel elements surrounded by moderator, shielding elements, etc. In comparison, the direct simulation of the physical processes is both straightforward and convenient. (Note that such types of simulations, where one follows the trajectories of individual particles, belong to a class of methods that is called 'event-driven Monte Carlo'.)

To begin with we consider a neutron with energy E that is at position $\mathbf{r}$ at time t and moving with constant velocity in the direction of the unit vector $\mathbf{u}$. The neutron continues to travel in the same direction with the same energy until at some point on its straight path it collides with some atom of the medium. The probability that the particle strikes an atom on an infinitesimal element of its path is $\sigma_c \delta s$, where σ_c is the cross-section for the scattering or absorption event. The value of σ_c depends on E and the type of medium in which the neutron is travelling. If we consider a path of length s which is completely inside a single medium (e.g. in the interior of a uranium rod, or inside the water moderator, etc.), the cumulative distribution of the distances s that the particle travels before it hits an atom of the medium is $P_c(s) = 1 - \exp(-\sigma_c s)$.

In the Monte Carlo simulation we now simply keep track of the particles from collision-to-collision. Starting from a state $(E, \mathbf{u}, \mathbf{r})$, we generate a distance s with the probability $P_c(s)$ (if the straight line from $\mathbf{r}$ to $\mathbf{r} + s\mathbf{u}$ does not intersect any boundary between different media). Now the particle has a collision at the point $\mathbf{r}' = \mathbf{r} + s\mathbf{u}$. If there is a boundary, one only allows the particle to proceed up to the boundary. If this is the outer boundary, this means that the neutron has escaped to the outside world and it is not considered further. If it is an interior boundary between regions, one repeats the above procedure, replacing $\mathbf{r}$ by the boundary position, and adjusts σ_c to be the appropriate value for the new region that the neutron has entered. This procedure is valid because of the Markovian character of the distribution $P_c(s)$. Note that E determines the velocity $\mathbf{v}$ of each neutron, and thus the time t' of the next event is uniquely determined. The collision process itself is determined by an appropriate differential cross-section, e.g. for an inelastic scattering event it is $d^2\sigma/d\Omega\,d\omega$, where Ω is the solid angle of the scattering (with the z-axis in the direction of $\mathbf{u}$) and $\hbar\omega = E' - E$ the energy change. These cross-sections are considered to be known quantities because they can be determined by suitable experiments. One then has to sample E' and the angles $\Omega = (\theta, \varphi)$ from the appropriate conditional probability.

Now, one problem in reactor criticality is that the density function $\rho(E, \mathbf{u}, r)$ will develop in time with a factor $\exp[\mu(t' - t)]$: if $\mu > 0$, the system is supercritical, whereas if $\mu < 0$, it is subcritical. In order to keep the number of tracks from either decreasing or increasing too much, reweighting techniques must be used. Thus, if μ is rather large, one randomly picks out a neutron and discards it with probability p. Otherwise, the neutron is allowed to continue, but its weight in the sample is increased by a factor $(1 - p)^{-1}$. The value p can be adjusted such that the size of the sample (i.e. the number of neutron tracks that are followed) stays asymptotically constant.

3.5.2 Fluid flow

The direct simulation Monte Carlo method (Bird, 1987; Watanabe *et al.*, 1994) has proven to be useful for the simulation of fluid flow from an atomistic perspective. The system is divided into a number of cells, and trajectories of particles are followed for short time intervals by decoupling interparticle collisions. Collision subcells are used in which interparticle collisions are treated on a probabilistic basis. The size of the collision subcells must be monitored so that it is smaller than the mean free path of the particles; otherwise atomistic information is lost. (Thus, the method is well suited to the study of gases but should not be expected to work well for very dense fluids.) This method has succeeded in delivering information about a number of different systems. For example, this technique produces vortices in a flow field. The direct simulation Monte Carlo method has also been used to study the transition from conduction to convection in a Rayleigh–Bénard system, complete with the formation of convection rolls, as the bottom plate is heated. The results for this problem compared quite favorably with those from solution of the Navier–Stokes equation. Typically a system of 40×20 sampling cells each of which contained 5×5 collision cells was used. Each collision cell contained between 16 and 400 particles. One result of this study was the discovery that semi-slip boundary conditions at the top and bottom are inadequate; instead strict diffuse boundary conditions must be used.

3.6 THE PERCOLATION PROBLEM

A geometric problem which has long played a significant role in statistical mechanics is that of 'percolation'. Percolation processes are those in which, by the random addition of a number of objects, a contiguous path which spans the entire system is created. In general, particles may be distributed continuously in space and the overlap between particles determines the connected paths; however, for our purposes in the first part of this discussion we shall confine ourselves to lattice systems in which the random creation of bonds eventually leads to a connected 'cluster' which spans the lattice. We shall briefly discuss some aspects of percolation here. Percolation has a long history of study by various numerical methods, and for the reader who is interested in obtaining a more thorough knowledge of various aspects of percolation theory, we emphasize that other literature will provide further information (Stauffer, 1985).

3.6.1 Site percolation

A lattice is composed of a periodic array of potential occupation sites. Initially the lattice is empty, i.e. none of the sites are actually occupied. Sites are then randomly occupied with probability p and clusters are formed of spins which occupy neighboring sites, i.e. bonds are drawn between all occupied nearest neighbor sites. The smallest cluster can then be a single site if none of the nearest neighbor sites are occupied. Two different properties of the system can be determined directly. First of all, for each value of p the probability P_{span} of having a spanning, or 'infinite' cluster may be determined by generating many realizations of the lattice and counting the fraction of those cases in which a spanning cluster is produced. As the lattice size becomes infinite, the probability that a spanning cluster is produced becomes zero for $p < p_{\mathrm{c}}$ and unity for $p > p_{\mathrm{c}}$. Another important quantity is the order parameter M which corresponds to the fraction of occupied sites in the lattice which belong to the infinite cluster. The simplest way to determine M through a simulation is to generate many different configurations for which a fraction p of the sites is occupied and to count the fraction of states for which an infinite cluster appears. For relatively sparsely occupied lattices M will be zero, but as p increases eventually we reach a critical value $p = p_{\mathrm{c}}$ called the 'percolation threshold' for which $M > 0$. As p is increased still further, M continues to grow. The behavior of the percolation order parameter near the percolation threshold is given by an expression which is reminiscent of that for the critical behavior of the order parameter for a temperature induced transition given in Section 2.1.2.

$$M = B(p - p_{\mathrm{c}})^{\beta} \qquad (3.13)$$

where $(p - p_{\mathrm{c}})$ plays the same role as $(T_{\mathrm{c}} - T)$ for a thermal transition. Of course, for a finite L^{d} lattice in d-dimensions the situation is more complicated since it is possible to create a spanning cluster using just dL bonds as

Fig. 3.3 Site
percolation clusters on
an $L \times L$ lattice: (a)
simplest 'infinite
cluster'; (b) random
infinite cluster.

(a)

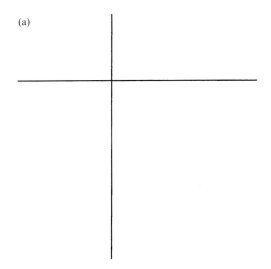

(b)

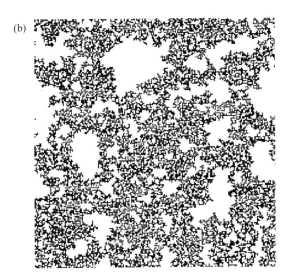

shown in Fig. 3.3. Thus, as soon as $p = d/L^{d-1}$ the percolation probability
becomes non-zero even though very few of the clusters percolate. For random placement of sites on the lattice, clusters of all different sizes are formed
and percolation clusters, if they exist, are quite complex in shape. (An example is shown in Fig. 3.3b.) The characteristic behavior of M vs. p is shown for
a range of lattice sizes in Fig. 3.4. As the lattice size increases, the finite size
effects become continuously smaller. We see that M (defined as P_∞ in the
figure) rises smoothly for values of p that are distinctly smaller than p_c rather
than showing the singular behavior given by Eqn. (3.13). As L increases,
however, the curves become steeper and steeper and eventually Eqn. (3.13)

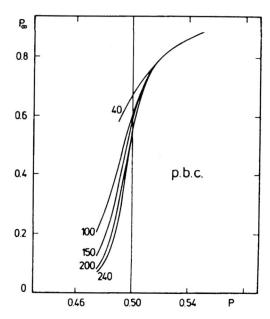

Fig. 3.4 Variation of the percolation order parameter M with p for bond percolation on $L \times L$ lattices with periodic boundary conditions. The solid curves show finite lattice results and the vertical line shows the percolation threshold. From Heermann and Stauffer (1980).

emerges for macroscopically large lattices. Since one is primarily interested in the behavior of macroscopic systems, which clearly cannot be simulated directly due to limitations on cpu time and storage, a method must be found to extrapolate the results from lattice sizes L which are accessible to $L \to \infty$. We will take up this issue again in detail in Chapter 4. The moments of the cluster size distribution also show critical behavior. Thus, the equivalent of the magnetic susceptibility may be defined as

$$\chi = \sum_c s^2 n(s) \tag{3.14}$$

where $n(s)$ is the number of clusters of size s and the sum is over all clusters. At the percolation threshold the cluster size distribution $n(s)$ also has characteristic behavior

$$n(s) \propto s^{-\tau}, \qquad s \to \infty, \tag{3.15}$$

which implies that the sum in Eqn. (3.14) diverges for $L \to \infty$.

The implementation of the Monte Carlo method to this problem is, in principle, quite straightforward. For small values of p it is simplest to begin with an empty lattice, and randomly fill the points on the lattice, using pairs (in two dimensions) of random integers between 1 and L, until the desired occupation has been reached. Clusters can then be found by searching for connected pairs of nearest neighbor occupied sites. For very large numbers of occupied sites it is easiest to start with a completely filled lattice and randomly empty the appropriate number of sites. In each case it is necessary to check that a point is not chosen twice, so in the 'interesting' region where the system is neither almost empty nor almost full, this method becomes inefficient and a different strategy must be found. Instead one can go through an initially empty lattice, site by site, filling each site with probability p. At the end of this sweep the actual concentration of filled sites is liable to be different from p, so a few sites will need to be randomly filled or emptied until the desired value of p is reached. After the desired value of p is reached the properties of the system are determined. The entire process can be repeated many times so that we can obtain mean values of all quantities of interest as well as determine the error bars of the estimates.

Problem 3.5 Consider an $L \times L$ square lattice with $L = 16$, 32, and 64. Determine the percolation probability for site percolation as a function of p. Estimate the percolation threshold.

3.6.2 Cluster counting: the Hoshen–Kopelman algorithm

In order to identify the clusters in a system and to determine the largest cluster and see if it is a percolating cluster, a rapid routine must be devised. A very fast 'single-pass' routine by Hoshen and Kopelman (1976) is simple to implement and quite efficient. It is rather easy to identify clusters by going through each row of the lattice in turn and labeling each site which is con-nected to a nearest neighbor with a number. Thus the cluster label $L_{i,j} = n$ for each occupied site, where n is the cluster number which is assigned when looking to see if previously inspected sites are nearest neighbors or not. This process is shown for the first row of a square lattice in Fig. 3.5. The difficulty which arises from such a direct approach becomes obvious when we consider the third row of the lattice at which point we realize that those sites which were initially assigned to cluster 1 and those assigned to cluster 2 actually belong to the same cluster. A second pass through the lattice may be used to eliminate such errors in the cluster assignment, but this is a time consuming process. The Hoshen–Kopelman method corrects such mislabeling 'on the fly' by introducing another set of variables known as the 'labels of the labels', N_n. The 'label of the label' keeps track of situations in which we discover that two clusters actually belong to the same cluster, i.e. that an occupied site has two nearest neighbors which have already been assigned different cluster numbers. When this happens the 'label of the label' which is larger is set to the negative of the value of the smaller one (called the 'proper' label) so that both 'clusters' are identified as actually belonging to the same cluster and the proper label is set equal to the total size of the cluster. Thus in Fig. 3.5 we

Fig. 3.5 Labeling of clusters for site percolation on a square lattice. The question mark shows the 'conflict' which arises in a simple labeling scheme.

see that after examination of the third row has been completed, $N_1 = 7$, and $N_2 = -1$. The Hoshen–Kopelman method finds a wide range of application beyond the simple percolation problem mentioned here.

Of course, there are many other properties of the clusters which are interesting. As an example we mention the 'backbone', which is that portion of the cluster which forms a connected path with no dangling ends between the two most distant points. This information is lost during implementation of the Hoshen–Kopelman algorithm, but other types of 'depth first' and 'breadth first' searches may be used, see e.g. Babalievski (1998), which retain more information. These generally sacrifice the very efficient use of storage in order to keep more detail.

Problem 3.6 Use the Hoshen–Kopelman method to determine the cluster size dependence for the site dilution problem with $L = 64$ and $p = 0.59$. Can it be described in terms of a power law?

3.6.3 Other percolation models

The simplest variation of the percolation model discussed above is the case where the bonds are thrown on the lattice randomly and clusters are formed directly from connected bonds. All of the formalism applied to the site problem above is also valid, and 'bond percolation' problems have been studied quite extensively in the past. The major difference is that clusters, defined in terms of connected lattice size, may have a minimum size of 2. A physical motivation for the study of such models comes from the question of the nature of the conductivity of disordered materials ('random resistor networks'). Another class of models results if we remove the restriction of a lattice and allow particles to occupy positions which vary continuously in space. 'Continuum percolation', as it is called, suffers from the added complication that tricks which can sometimes be used on lattice models cannot be applied. A quite different process is known as 'invasion percolation'; its invention was prompted by attempts to understand flow in porous media by Wilkinson and Willemsen (1983). Random number are assigned to each site of a lattice. Choose a site, or sites, on one side of the lattice and draw a bond to the neighbor which has the lowest random number assigned to it. (The growing cluster represents the invading fluid with the remainder of the sites representing the initial, or defending, fluid.) This process continues until the cluster reaches the other side (i.e. the exit).

3.7 FINDING THE GROUNDSTATE OF A HAMILTONIAN

For systems with Hamiltonians the groundstate is usually a relatively unique, minimum energy state. If the groundstate of a system is not known, a simple Monte Carlo simulation can be used to find states of low energy, and hopefully that of lowest energy. For purposes of discussion we will consider a system of Ising spins. Some initial, randomly chosen state of the system is selected and then one proceeds through the lattice determining the change in energy of the system if the spin is overturned; if the energy is lowered the spin is overturned, otherwise it is left unchanged and one proceeds to the next site. The system is swept through repeatedly, and eventually no spin-flips occur; the system is then either in the groundstate or in some metastable state. This process can be repeated using different initial configurations, and one tests to see if the same state is reached as before or if a lower energy state is found. For systems with very complicated energy landscapes (i.e. the variation of the energy as some parameter x is changed) there may be many energy minima of approximately the same depth and a more sophisti-cated strategy will have to be chosen. This situation will be discussed in the next chapter. In some cases relatively non-local metastable structures, e.g. anti-phase domains, are formed and cannot be removed by single spin flips. (Anti-phase domains are large regions of well ordered structures which are 'shifted' relative to each other and which meet at a boundary with many unsatisfied spins.) It may then be helpful to introduce multiple spin flips or other algorithmic changes as a way of eliminating these troublesome defects. In all cases it is essential to begin with diverse initial states and check that the same 'groundstate' is reached.

Example

Consider an $L \times L$ Ising square lattice in which all spins to the left of a diagonal are initially up and all those to the right are down. All portions of the system are in their lowest energy state except for those spins which are in the domain wall between the up-spin and down-spin regions. Since the spins in the domain wall have equal numbers of up and down neighbors they cannot lower their energy by overturning, but if we allow those spins to flip with 50% probability, we provide the method with a way of eventually eliminating the domain structure.

3.8 GENERATION OF 'RANDOM' WALKS

3.8.1 Introduction

In this sub-section we shall briefly discuss random walks on a lattice which is a special case of the full class of random walks. A random walk consists of a connected path formed by randomly adding new bonds to the end of the existing walk, subject to any restrictions which distinguish one kind of ran-dom walk from another. The mean-square end-to-end distance $\langle R^2 \rangle$ of a walk with N steps may diverge as N goes to infinity as (de Gennes, 1979)

$$\langle R^2(N) \rangle = aN^{2\nu}(1 + bN^{-\Delta} + \cdots)$$ (3.16)

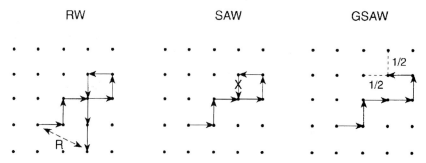

Fig. 3.6 Examples of different kinds of random walks on a square lattice. For the RW every possible new step has the same probability. For the SAW the walk dies if it touches itself. The GSAW walker recognizes the danger and takes either of the two steps shown with equal probability.

where ν is a 'critical exponent' that determines the universality class. Here a and b are some 'non-universal' constants which depend on the model and lattice structure chosen and $\varDelta$ is a 'correction to scaling' exponent. In such cases there is a strong analogy to critical behavior in percolation or in temperature driven transitions in systems of interacting particles. The equivalent of the partition function for a system undergoing a temperature driven transition is given by the quantity Z_N which simply counts the number of distinct random walks on the lattice and which behaves as

$$Z_N \propto N^{\gamma-1} q_{\text{eff}}^N \tag{3.17}$$

as $N \to \infty$. γ is another critical exponent and q_{eff} is an effective coordination number which is related to the exchange constant in a simple magnetic model. The formalism for describing this geometric phenomenon is thus the same as for temperature driven transitions, even including corrections to scaling in the expression for the mean-square end-to-end distance as represented by the term in $N^{-\varDelta}$ in Eqn. (3.16). The determination of ν and γ for different kinds of walks is essential to the classification of these models into different universality classes. We now know that the lattice dimensionality as well as the rules for the generation of walks affect the critical exponents and thus the universality class (Kremer and Binder, 1988). Examples of several kinds of walks are shown in Fig. 3.6.

3.8.2 Random walks

For simple, random walks (RW) the walker may cross the walk an infinite number of times with no cost. In d dimensions the end-to-end distance diverges with the number of steps N according to

$$\sqrt{\langle R^2(N) \rangle} \propto N^{\frac{1}{2}}. \tag{3.18}$$

A simulation of the simple random walk can be carried out by picking a starting point and generating a random number to determine the direction of each subsequent, additional step. After each step the end-to-end distance can be calculated. Errors may be estimated by carrying out a series of independent random walks and performing a statistical analysis of the resultant distribution. Thus, the simple RW has the trivial result $\nu = 1/2$ but is not really very useful in understanding physical properties of polymers in dilute solution; but random walks have great significance for the description of diffusion phenomena – the number of steps N is then related to time.

At this point we briefly mention a simple variant of the RW for which the choice of the $(n + 1)$ step from the nth step of a return to the point reached at the $(n - 1)$ step, i.e. an 'immediate reversal', is forbidden. Although for this so-called 'non-reversal random walk' (NRRW) the exponents remain unchanged, i.e. $\nu = 1/2$, $\gamma = 1$, as for the ordinary RW, prefactors change. This means that in Eqn. (3.17) $q_{\mathrm{eff}} = (q - 1)$ for the NRRW whereas $q_{\mathrm{eff}} = q$ for the ordinary RW, etc. This NRRW model represents, in fact, a rather useful approach for the modeling of polymer configurations in dense melts, and since one merely has to keep track of the previous step and then choose one of the remaining $q - 1$ possibilities, it is straightforward to implement. Furthermore, this NRRW model is also a good starting point for the simulation of 'self-avoiding walks', a topic to which we shall turn in the next section.

Problem 3.7 Perform a number of random walk simulations to estimate the value of ν for a simple random walk on a square lattice. Give error bars and compare your result with the exact answer in Eqn. (3.18).

3.8.3 Self-avoiding walks

In contrast to the simple random walk, for a self-avoiding walk (SAW), the walker dies when attempting to intersect a portion of the already completed walk. (Immediate reversals are inherently disallowed.) There has been enormous interest in this model of SAWs since this is the generic model used to probe the large scale statistical properties of the configurations of flexible macromolecules in good solvents. Although it is possible to carry out an exact enumeration of the distribution of walks for small N, it is in general not possible to extract the correct asymptotic behavior for the range of N which is accessible by this method. Monte Carlo methods have also been used to study much larger values of N for different kinds of walks, but even here very slow crossover as a function of N has complicated the analysis. After each step has been added, a random number is used to decide between the different possible choices for the next step. If the new site is one which already contains a portion of the walk, the process is terminated at the Nth step. Attrition becomes a problem and it becomes difficult to generate large numbers of walks with large N. The most simple minded approach to the analysis of the data is to simply make a plot of $\log \langle R^2(N) \rangle$ vs. $\log N$ and to calculate ν from the slope. If corrections to scaling are present, the behavior of the data may become quite subtle and a more sophisticated approach is needed. The results can instead be analyzed using traditional 'ratio methods' which have been successful in analyzing series expansions. In this manner we can calculate an 'effective exponent' by forming the ratio

$$v(N) = \frac{1}{2} \frac{\ln\left[\langle R^2(N+i)\rangle / \langle R^2(N-i)\rangle\right]}{\ln[(N+i)/(N-i)]} \tag{3.19}$$

for different values of $i \ll N$. The values of i must chosen to be large enough to help eliminate 'short time' noise in the comparison of nearby values but small enough that the effects of correction terms do not infect the effective exponent estimate. The effective exponent is then related to the true value, i.e. for $N = \infty$, by

$$v(N) = v - 1/2 \; bN^{-\Delta} + \cdots . \tag{3.20}$$

Thus, by extrapolation to $N \to \infty$ we can extract a rather accurate estimate for the (asymptotic) exponent. This method, which is introduced here for convenience, is not restricted to SAWs and can be applied to many problems involving enumeration. For SAWs the current estimates for v are (Kremer and Binder, 1988)

$$v = 3/4, \qquad d = 2, \tag{3.21a}$$
$$v \approx 0.588, \qquad d = 3, \tag{3.21b}$$
$$v = 1/2, \qquad d \geq 4. \tag{3.21c}$$

The exponent γ is also of great interest and numerical estimates can be made by comparing the values of the 'partition function' which are obtained for two successive values of N, i.e. using Eqn. (3.16):

$$\ln\left(\frac{Z(N)}{Z(N-1)}\right) = \ln q_{\text{eff}} + (\gamma - 1)/N. \tag{3.22}$$

Here, too, a more sensitive analysis can be made by using 'symmetric' values in step number by looking not only at $Z(N)$ but also $Z(N+i)$ and $Z(N-i)$ so that

$$\ln \frac{Z(N)}{Z(N-i)} - \ln \frac{Z(N+i)}{Z(N)} = (\gamma - 1) \ln \frac{N^2}{(N-i)(N+i)} \tag{3.23}$$
$$\to (1 - \gamma)i^2/N^2.$$

Once again, i must be chosen to be sufficiently large that the effects of 'short time' fluctuations are minimized but small enough that curvature effects do not enter.

Although these techniques are very straightforward, many research problems of current interest remain that one can solve with them. For example, consider the case of a star polymer adsorbed with its core on a wall as shown in Fig. 3.7a. While in two dimensions we expect that the size of a polymer scales with the number of monomers as $R \sim N^v = N^{3/4}$, for a star polymer we have the additional question of how the number of arms f affects the scaling in the macromolecular object. This question was studied using a simple sampling Monte Carlo method by Ohno and Binder (1991). To

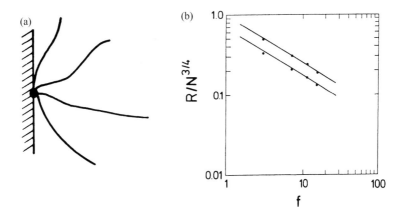

Fig. 3.7 (a) A two-dimensional star polymer consisting of $f = 4$ flexible arms covalently linked together in a core (dot) adsorbed at a one-dimensional repulsive wall (shaded). (b) Log–log plot of $R/N^{3/4}$ for center-adsorbed stars plotted vs. the number of arms, where $N = fl$ is the total number of monomers, $l = 50$ is the number of monomers per arm, and R is the center-end distance of the arms (upper set of points) or the mean distance of a monomer from the center (lower set of points). Straight lines illustrate agreement with the theoretical prediction $R/N^{3/4} \propto f^{-1/2}$. From Ohno and Binder (1991).

remedy the attrition problem mentioned earlier, they used a variation of simple sampling known as the enrichment method. Treating each arm as a self-avoiding walk on a lattice with q-fold coordination, and avoiding immediate reversals, they added each new bond to randomly connect to one of the $(q - 1)$ neighbor sites. Thus, for example, on a square lattice the probability that the self-avoiding walk does not 'die' in this step is $(3/q_{\text{eff}})^{-f} \approx 0.880^f$. For large f the probability of growing a star polymer with long arms would be vanishingly small. Thus, the recipe is to attempt to add a bond to each arm not just once but many times, i.e. on average $m = (3/q_{\text{eff}})^f$ times, and keeping track of the survivors. In this way a 'population' of star polymers that is grown in parallel from $\mathcal{N}$ centers neither dies out nor explodes in number as bonds are added consecutively to create arms of length l. Of course there is a price that must be paid: different star polymers that 'survive' in the final 'generation' are not, in general, statistically independent of each other. Nevertheless, this method is useful in a practical sense.

3.8.4 Growing walks and other models

Because of the attrition, the generation of long SAWs is quite difficult. An alternative strategy is to attempt to develop 'smart walks' which find a way to avoid death. An example is the growing self-avoiding walk (GSAW) (Lyklema and Kremer, 1986). The process proceeds at first as for SAWs, but a walker senses a 'trap' and chooses instead between the remaining 'safe' directions so that it can continue to grow. Other, still 'smarter' walks have been studied numerically (Lyklema, 1985) and a number of sophisticated methods have been devised for the simulation of polymeric models (Baumgärtner, 1992).

To a great extent modeling has concentrated on the 'good solvent' case in which polymers are treated as pure SAWs (self-avoiding walks); however, in θ-solutions the solvent–monomer repulsion leads to a net attraction between the chain monomers. Thus, the SAW can be generalized by introducing an energy that is won if two nearest neighbor sites are visited by the walk. Of course, the weighting of configurations then requires appropriate Boltzmann factors (Kremer and Binder, 1988). Exactly at the θ–point the SAW condition and the attraction cancel and the exponents become those of the simple random walk. The θ-point may then be viewed as a kind of tricritical point, and for $d = 3$ the exponents should thus be mean-field-like.

3.9 FINAL REMARKS

In closing this chapter, we wish to emphasize that there are related applications of Monte Carlo 'simple sampling' techniques outside of statistical physics which exist in broad areas of applied mathematics, also including the so-called 'quasi-Monte Carlo methods' (Niederreiter, 1992). These applications deal with mathematical problems (Monte Carlo algorithms for calculating eigenvalues, or for solving integro-differential equations, etc.) and various applications ranging from economy to technology (option pricing, radiosity and illumination problems, computer graphics, road visibility in fog, etc.). Such problems are completely outside of the scope of our presentation; however, we direct the interested reader to Niederreiter *et al.* (1998) for a series of recent review articles.

REFERENCES

Babalievski, F. (1998), J. Mod. Phys. C **9**, 43.

Baumgärtner, A. (1992), in *The Monte Carlo Method in Condensed Matter Physics*, ed. K. Binder (Springer Verlag, Heidelberg).

Bird, G. A. (1987), Phys. Fluids **30**, 364.

Cooper, N. G. (ed.) (1989), *From Cardinals to Chaos* (Cambridge University Press, Cambridge).

Davison, B. (1957), *Neutron Transport Theory* (Oxford University Press, Oxford).

de Gennes, P. G. (1979), *Scaling Concepts in Polymer Physics* (Cornell University Press, Ithaca, New York).

Hammersley, J. H. and Handscomb, D. C. (1964), *Monte Carlo Methods* (Wiley, New York).

Heermann, D. W. and Stauffer, D. (1980), Z. Phys. B **40**, 133.

Hoshen, J. and Kopelman, R. (1976), Phys. Rev. B **14**, 3428.

Kremer, K. and Binder, K. (1988), Comput. Phys. Rep. 7, 261.

Lyklema, J. W. (1985), J. Phys. A **18**, L617.

Lyklema, J. W. and Kremer, K. (1986), J. Phys. A **19**, 279.

Niederreiter, H. (1992), *Random Number Generation and Quasi-Monte Carlo Methods* (SIAM, Philadelphia).

Niederreiter, H., Hellekalek, P., Larcher, G., and Zinterhof, P. (1998), *Monte Carlo and Quasi-Monte Carlo Methods* (Springer, New York, Berlin).

Ohno, K. and Binder, K. (1991), J. Stat. Phys. **64**, 781.

Sabelfeld, K. K. (1991), *Monte Carlo Methods in Boundary Value Problems* (Springer, Berlin).

Stauffer, D. (1985), *Introduction to Percolation Theory* (Taylor and Francis, London).

Watanabe, T., Kaburaki, H., and Yokokawa, M. (1994), Phys. Rev. E **49**, 4060.

Wilkinson, D. and Willemsen, J. F. (1983), J. Phys. A **16**, 3365.

4 Importance sampling Monte Carlo methods

4.1 INTRODUCTION

In this chapter we want to introduce simple importance sampling Monte Carlo techniques as applied in statistical physics and which can be used for the study of phase transitions at finite temperature. We shall discuss details, algorithms, and potential sources of difficulty using the Ising model as a paradigm. It should be understood, however, that virtually all of the discussion of the application to the Ising model is relevant to other models as well, and a few such examples will also be discussed. Other models as well as sophisticated approaches to the Ising model will be discussed in later chapters. The Ising model is one of the simplest lattice models which one can imagine, and its behavior has been studied for about three-quarters of a century. The simple Ising model consists of spins which are confined to the sites of a lattice and which may have only the values $+1$ or -1. These spins interact with their nearest neighbors on the lattice with interaction constant $\mathcal{J}$; the Hamiltonian for this model was given in Eqn. (2.19) but we repeat it again here for the benefit of the reader:

$$\mathcal{H} = -\mathcal{J}\sum_{i,j}\sigma_i\sigma_j - H\sum_i\sigma_i \qquad (4.1)$$

where $\sigma_i = \pm 1$. The Ising model has been solved exactly in one dimension and as a result it is known that there is no phase transition. In two dimensions Onsager obtained exact results (Onsager, 1944) for the thermal properties of $L \times M$ lattices with periodic boundary conditions in zero field which showed that there is a second order phase transition with divergences in the specific heat, susceptibility, and correlation length. In Fig. 4.1 we show configurations for finite $L \times L$ Ising lattices in zero field; these states show the model almost in the groundstate, near the phase transition, and at high temperatures where there are virtually no correlations between spins. Note that in zero field the model has up–down symmetry so that overturning all the spins produces a degenerate state. At high temperature all the clusters of like spins are small, near the transition there is a broad distribution of clusters, and at low temperatures there is a single large cluster of ordered spins and a number of small clusters of oppositely directed spins.

  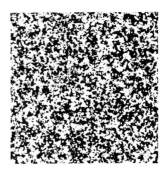

Fig. 4.1 Typical spin configurations for the two-dimensional Ising square lattice: (left) $T \ll T_c$; (center) $T \sim T_c$; (right) $T \gg T_c$.

In principle, the Ising model can be simulated using the simple sampling techniques discussed in the previous chapter: spin configurations could be generated completely randomly and their contribution weighted by a Boltzmann factor. Unfortunately most of the configurations which are produced in this fashion will contribute relatively little to the equilibrium averages, and more sophisticated methods are required if we are to obtain results of sufficient accuracy to be useful.

Problem 4.1 Suppose we carry out a simple sampling of the Ising model configurations on an $L \times L$ lattice at $k_B T/J = 1.5$. What is the distribution of the magnetization M of the states that are generated? How large is the probability that a state has a magnetization $M > M_0$, where M_0 is some given value of order unity, e.g. the spontaneous magnetization for $T < T_c$. Use your result to explain why simple sampling is not useful for studying the Ising model.

4.2 THE SIMPLEST CASE: SINGLE SPIN-FLIP SAMPLING FOR THE SIMPLE ISING MODEL

The nearest neighbor Ising model on the square lattice plays a special role in statistical mechanics – its energy, spontaneous magnetization, and correlations in zero magnetic field can be calculated exactly, and this fact implies that the static critical exponents are also known. Critical exponents are known exactly for only a small number of models. The most notable of the exactly soluble models is the two-dimensional Ising square lattice (Onsager, 1944) for which the exact solution shows that the critical exponents which were discussed in Chapter 2 are

$$\alpha = 0, \qquad \beta = 1/8, \quad \text{and} \quad \gamma = 7/4. \tag{4.2}$$

We shall first discuss techniques which are suitable for simulating this model so that there are exact results with which the data from the Monte Carlo simulations may be compared.

4.2.1 Algorithm

In the classic, Metropolis method (Metropolis *et al.*, 1953) configurations are generated from a previous state using a transition probability which depends on the energy difference between the initial and final states. The sequence of states produced follows a time ordered path, but the time in this case is referred to as 'Monte Carlo time' and is non-deterministic. (This can be seen from an evaluation of the commutator of the Hamiltonian and an arbitrary spin; the value, which gives the time dependence of the spin, is zero.) For relaxational models, such as the (stochastic) Ising model (Kawasaki, 1972), the time-dependent behavior is described by a master equation (cf. Section 2.2.4)

$$\frac{\partial P_n(t)}{\partial t} = -\sum_{n \neq m} [P_n(t) W_{n \to m} - P_m(t) W_{m \to n}], \tag{4.3}$$

where $P_n(t)$ is the probability of the system being in state n at time t, and $W_{n \to m}$ is the transition rate for $n \to m$. In equilibrium $\partial P_n(t)/\partial t = 0$ and the two terms on the right-hand side of Eqn. (4.3) must be equal! The resultant expression is known as 'detailed balance', as mentioned previously in Eqn. (2.83)

$$P_n(t) W_{n \to m} = P_m(t) W_{m \to n}. \tag{4.4}$$

The probability of the nth state occurring in a classical system is given by

$$P_n(t) = e^{-E_n/k_B T}/Z \tag{4.5}$$

where Z is the partition function. This probability is usually not exactly known because of the denominator; however, one can avoid this difficulty by generating a Markov chain of states, i.e. generate each new state directly from the preceding state. If we produce the nth state from the mth state, the relative probability is the ratio of the individual probabilites and the denominator cancels. As a result, only the energy difference between the two states is needed, e.g.

$$\Delta E = E_n - E_m. \tag{4.6}$$

Any transition rate which satisfies detailed balance is acceptable. The first choice of rate which was used in statistical physics is the Metropolis form (Metropolis *et al.*, 1953)

$$W_{n \to m} = \tau_0^{-1} \exp(-\Delta E/k_b T), \qquad \Delta E > 0 \tag{4.7a}$$
$$= \tau_0^{-1}, \qquad \Delta E < 0 \tag{4.7b}$$

where τ_0 is the time required to attempt a spin-flip. (Often this 'time unit' is set equal to unity and hence suppressed in the equations.) The way the Metropolis algorithm is implemented can be described by a simple recipe:

Metropolis importance sampling Monte Carlo scheme

(1) Choose an initial state
(2) Choose a site i
(3) Calculate the energy change ΔE which results if the spin at site i is overturned
(4) Generate a random number r such that $0 < r < 1$
(5) If $r < \exp(-\Delta E / k_B T)$, flip the spin
(6) Go the next site and go to (3)

After a set number of spins have been considered, the properties of the system are determined and added to the statistical average which is being kept. Note that the random number r must be chosen *uniformly* in the interval [0,1], and successive random numbers should be uncorrelated. We shall have a great deal more to say about random numbers shortly. The 'standard' measure of Monte Carlo time is the Monte Carlo step/site (MCS/site) which corresponds to the consideration of every spin in the system once. With this algorithm states are generated with a probability proportional to Eqn. (4.5) once the number of states is sufficiently large that the initial transients (see Fig. 4.2) are negligible. Then, the desired averages $\langle A \rangle = \sum_n P_n A_n$ of a variable A simply become arithmetic averages over the entire sample of states which is kept. Note that if an attempted spin-flip is rejected, the old state is counted again for the averaging.

A typical time development of properties of the system is shown in Fig. 4.2. For early times the system is relaxing towards equilibrium and both the internal energy and order parameter are changing, but with different characteristic time scales. There is a second range of times in which the system is in equilibrium and the properties merely show thermodynamic fluctuations, and at still longer times one can observe global spin inversion; in a finite system this will occur in equilibrium between states of equal energy and spontaneous magnetization which differs only in sign. Of course, the precise results will depend upon many factors including temperature, lattice size, boundary conditions, etc., and all of these considerations will be discussed in forthcoming sections. Fig. 4.2 simply provides a starting point for these presentations. In a more complex problem one might not know what the groundstate looks like or what the relevant time scales are. It is thus always wise to take precautions before interpreting the data. Prudent steps to take include repeating a given run with different initial states to see if the same equilibrium distribution is reached and to repeat runs with different random numbers. By working initially with small systems one can generally keep the characteristic times short so that it is easy to make 'long' runs.

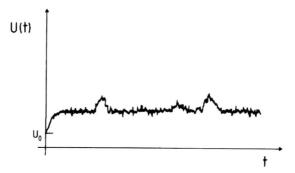

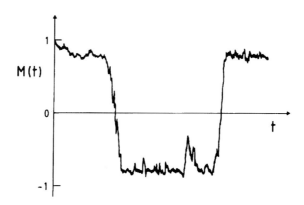

Fig. 4.2 Schematic variation of internal energy and spontaneous magnetization with time for a Monte Carlo simulation of an Ising square lattice in zero field.

A minor variation on the simple Metropolis algorithm described above involves the random selection of sites in the lattice to be considered. If this procedure is used for a system with N sites, 1 MCS/site corresponds to the consideration of N randomly chosen sites. Note that it is likely that some spins will be chosen more than once and some not at all during 1 MCS/site. The time development of the system will look just like that shown in Fig. 4.2, but the explicit variation and time scales will differ from those for the Metropolis method. This random selection of sites must be used if one is not just interested in static equilibrium properties but wishes to record dynamic correlation functions of the corresponding stochastic model.

As shown in the 'principle of detailed balance', Eqn. (4.4), the Metropolis flipping probability is not a unique solution. An alternative method, commonly referred to as 'Glauber dynamics' (Glauber, 1963), uses the single spin–flip transition rate

$$W_{n \to m} = \tau_0^{-1}[1 + \sigma_i \tanh(E_i/k_B T)], \qquad (4.8)$$

where $\sigma_i E_i$ is the energy of the ith spin in state n. Unlike the Metropolis method, the Glauber rate is antisymmetric about 0.5 for $E_i \to -E_i$. Müller–Krumbhaar and Binder (1973) showed that both Glauber and Metropolis algorithms are just special cases of a more general transition rate form. In most situations the choice between Glauber and Metropolis dynamics is somewhat arbitrary; but in at least one instance there is a quite important difference. At very high temperatures the Metropolis algorithm will flip a spin on every attempt because the transition probability approaches 1 for $\Delta E > 0$. Thus, in one sweep through the lattice every spin overturns, and in the next sweep every spin overturns again. The process has thus become non-ergodic (see Section 2.1.3) and the system just oscillates between the two states. With the Glauber algorithm, however, the transition probability approaches $1/2$ in this instance and the process remains ergodic.

Simplifications are possible for the Ising model which greatly reduce the amount of computer resources needed. For each spin there are only a small number of different environments which are possible, e.g. for a square lattice with nearest neighbor interactions, each spin may have 4, 3, 2 , 1 or 0 nearest neighbors which are parallel to it. Thus, there are only 5 different energy changes associated with a successful spin-flip and the probability can be computed for each possibility and stored in a table. Since the exponential then need not be computed for each spin-flip trial, a tremendous saving in cpu time results. Although the rapid increase in available computer memory has largely alleviated the problem with storage, large Ising systems may be compressed into a relatively small number of words by packing different spins into a single word. Each bit then describes the state of a spin so that e.g. only a single 32 bit word is needed to describe a 32-spin system. For models with more degrees of freedom available at each site, these simplifications are not possible and the simulations are consequently more resource consumptive.

The Ising model as originally formulated and discussed above may be viewed as a spin-S model with $S = 1/2$, but the definition can be extended to the case of higher spin without difficulty. For spin $S = 1/2$ there are only two states possible at each site, whereas for $S = 1$ there are three possible states, 1, 0, and -1. This means that a nearest neighbor pair can have *three* possible states with different energies and the total space of possible lattice configurations is similarly enlarged. (For higher values of S there will, of course, be still more states.) The spin-S Ising model can be simulated using the method just described with the modification that the 'new' state to which a given spin attempts to flip must be chosen from among multiple choices using another random number. After this is done, one proceeds as before.

One feature of a Monte Carlo algorithm which is important if the method is to be vectorized (these techniques will be discussed in the next chapter) is that the lattice needs to be subdivided into non-interacting, interpenetrating sublattices, i.e. so that the spins on a single sublattice do not interact with each other. This method, known as the 'checkerboard decomposition', can be used without difficulty on scalar computers as well as on vector platforms. If one wishes to proceed through the lattice in order using the checkerboard decomposition, one simply examines each site in turn in a single sublattice before proceeding to the second sublattice. (We mention this approach here simply because the checkerboard decomposition is referred to quite often in the literature.)

4.2.2 Boundary conditions

4.2.2.1 Periodic boundary conditions

Since simulations are performed on finite systems, one important question which arises is how to treat the 'edges' or boundaries of the lattice. These boundaries can be effectively eliminated by wrapping the d-dimensional lattice on a $(d+1)$-dimensional torus. This boundary condition is termed a 'periodic boundary condition' (pbc) so that the first spin in a row 'sees' the last spin in the row as a nearest neighbor and vice versa. The same is true for spins at the top and bottom of a column. Figure 4.3 shows this procedure for a square lattice. This procedure effectively eliminates boundary effects, but the system is still characterized by the finite lattice size L since the maximum value of the correlation length is limited to $L/2$, and the resultant properties of the system differ from those of the corresponding infinite lattice. (These effects will be discussed at length in the next section.) The periodic boundary condition must be used with care, since if the ordered state of the system has spins which alternate in sign from site to site, a 'misfit seam' can be introduced if the edge length is not chosen correctly. Of course, for off-lattice problems periodic boundary conditions are also easily introduced and equally useful for the elimination of edge effects.

4.2.2.2 Screw periodic boundary conditions

The actual implementation of a 'wraparound' boundary condition is easiest by representing the spins on the lattice as entries in a one-dimensional vector which is wrapped around the system. Hence the last spin in a row sees the first spin in the next row as a nearest neighbor (see Fig. 4.3). In addition to limiting the maximum possible correlation length, a result of this form of periodic boundary is that a 'seam' is introduced. This means that the properties of the system will not be completely homogeneous. In the limit of infinite lattice size this effect becomes negligible, but for finite systems there will be a systematic difference with respect to fully periodic boundary conditions which may not be negligible!

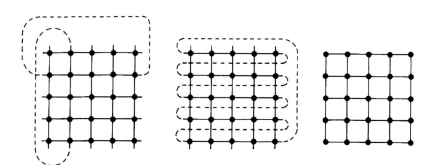

Fig. 4.3 Application of typical boundary conditions for the two-dimensional Ising model: (left) periodic boundary; (center) screw periodic; (right) free edges.

4.2.2.3 Antiperiodic boundary conditions

If periodic boundary conditions are imposed with the modification that the sign of the coupling is reversed at the boundary, an interface is introduced into the system. This procedure, known as antiperiodic boundary conditions, is not useful for making the system seem more infinite, but has the salutory effect of allowing us to work with a single interface in the system. (With periodic boundary conditions interfaces could only exist in pairs.) In this situation the interface is not fixed at one particular location and may wander back and forth across the boundary. By choosing a coordinate frame centered in the local interface center one can nevertheless study the interfacial profile undisturbed by any free edge effects (Schmid and Binder, 1992). Of course, one chooses this antiperiodic boundary condition in only one (lattice) direction, normal to the interface that one wishes to study, and retains periodic boundary conditions in the other direction(s).

In the above example the interface was parallel to one of the surfaces, whereas in a more general situation the interface may be inclined with respect to the surface. This presents no problem for simulations since a tilted interface can be produced by simply taking one of the periodic boundaries and replacing it by a skew boundary. Thus, spins on one side of the lattice see nearest neighbors on the other side which are one or more rows below, depending on the tilt angle of the interface. We then have the interesting situation that the boundary conditions are different in each Cartesian direction and are themselves responsible for the change in the nature of the problem being studied by a simple Monte Carlo algorithm. This is but one example of the clever use of boundary conditions to simplify a particular problem; the reader should consider the choice of the boundary conditions before beginning a new study.

4.2.2.4 Antisymmetric boundary conditions

This type of periodic boundary condition was introduced explicitly for $L \times L$ systems with vortices. (Vortices are topological excitations that occur most notably in the two-dimensional XY-model, see e.g. Section 4.2.2.5. A vortex looks very much like a whirlpool in two-dimensional space.) By connecting the last spin in row n antiferromagnetically with the first spin in row $(L - n)$, one produces a geometry in which a single vortex can exist; in contrast with pbc only vortex–antivortex pairs can exist (Kawamura and Kikuchi, 1993) on a lattice. This is a quite specialized boundary condition which is only useful for a limited number of cases, but it is an example of how specialized boundaries can be used for the study of unusual excitations.

4.2.2.5 Free edge boundary conditions

Another type of boundary does not involve any kind of connection between the end of a row and any other row on the lattice. Instead the spins at the end of a row see no neighbor in that direction (see Fig. 4.3). This free edge boundary not only introduces finite size smearing but also surface and corner effects due to the 'dangling bonds' at the edges. (Very strong changes may occur near the surfaces and the behavior of the system is not homogeneous.) In some cases, however, the surface and corner behavior themselves become the subjects of study. In some situations free edge boundaries may be more realistic, e.g. in modeling the behavior of superparamagnetic particles or grains, but the properties of systems with free edge boundaries usually differ from those of the corresponding infinite system by a much greater amount than if some sort of periodic boundary is used. In order to model thin films, one uses pbc in the directions parallel to the film and free edge boundary conditions in the direction normal to the film. In such cases, where the free edge boundary condition is thought to model a physical free surface of a system, it may be appropriate to also include surface fields, modified surface layer interactions, etc. (Landau and Binder, 1990). In this way, one can study phenomena such as wetting, interface localization–delocalization transitions, surface induced ordering and disordering, etc. This free edge boundary condition is also very common for off-lattice problems (Binder, 1983; Landau, 1996).

4.2.2.6 Mean-field boundary conditions

Another way to reduce finite size effects is to introduce an effective field which acts only on the boundary spins and which is adjusted to keep the magnetization at the boundary equal to the mean magnetization in the bulk. The resultant critical behavior is quite sharp, although sufficiently close to T_c the properties are mean-field-like. Such boundary conditions have been applied only sparingly, e.g. for Heisenberg magnets in the bulk (Binder and Müller-Krumbhaar, 1973) and with one free surface (Binder and Hohenberg, 1974).

4.2.2.7 Hyperspherical boundary conditions

In the case of long range interactions, periodic boundary conditions may become cumbersome to apply because each degree of freedom interacts with all its periodic images. In order to sum up the interactions with all periodic images, one has to resort to the Ewald summation method (see Chapter 6). An elegant alternative for off-lattice problems is to put the degrees of freedom on the d-dimensional surface of a $(d + 1)$-dimensional sphere (Caillol, 1993).

Problem 4.2 Perform a Metropolis Monte Carlo simulation for a 10 × 10 Ising model with periodic boundary conditions. Plot the specific heat (calculated from the fluctuations of the internal energy, see Chapter 2) and the order parameter (estimated as the absolute value of the magnetization) as a function of temperature.

Problem 4.3 Perform a Metropolis Monte Carlo simulation for a 10×10 Ising model with free edge boundary conditions. Plot the specific heat and the order parameter as a function of temperature.

4.2.3 Finite size effects

4.2.3.1 Order of the transition

In the above discussion we have briefly alluded to the fact that the effects of the finiteness of the system could be dramatic. (The reader who has actually worked out Problems 4.1 and 4.2 will have noted that in a 10×10 lattice the transition is completely smeared out!) Since our primary interest is often in determining the properties of the corresponding infinite system, it is important that we have some sound, theoretically based methods for extracting such behavior for the results obtained on the finite system. One fundamental difficulty which arises in interpreting simulational data, is that the equilibrium, thermodynamic behavior of a finite system is smooth as it passes through a phase transition for *both* first order and second order transitions. The question then becomes, 'How do we distinguish the order of the transition?' In the following sections we shall show how this is possible using finite size scaling.

4.2.3.2 Finite size scaling and critical exponents

At a second order phase change the critical behavior of a system in the thermodynamic limit can be extracted from the size dependence of the singular part of the free energy which, according to finite size scaling theory (Fisher, 1971; Privman, 1990; Binder, 1992), is described by a scaling ansatz similar to the scaling of the free energy with thermodynamic variables T, H (see Chapter 2). Assuming homogeneity and using L and T as variables, we find that

$$F(L, T) = L^{-(2-\alpha)/\nu} \mathcal{F}(\varepsilon L^{1/\nu}), \qquad (4.9)$$

where $\varepsilon = (T - T_c)/T_c$. It is important to note that the critical exponents α and ν assume their infinite lattice values. The choice of the scaling variable $x = \varepsilon L^{1/\nu}$ is motivated by the observation that the correlation length, which diverges as $\varepsilon^{-\nu}$ as the transition is approached, is limited by the lattice size L. (L 'scales' with ξ; but rather than $L/\xi \propto \varepsilon^{\nu} L$, one may also choose $\varepsilon L^{1/\nu}$ as the argument of the function $\mathcal{F}$. This choice has the advantage that $\mathcal{F}$ is analytic since F is analytic in T for finite L.) Appropriate differentiation of the free energy yields the various thermodynamic properties which have corresponding scaling forms, e.g.

$$M = L^{-\beta/\nu} \mathcal{M}^{\circ}(\varepsilon L^{1/\nu}), \qquad (4.10a)$$

$$\chi = L^{\gamma/\nu} \chi^{\circ}(\varepsilon L^{1/\nu}), \qquad (4.10b)$$

$$C = L^{\alpha/\nu} C^{\circ}(\varepsilon L^{1/\nu}), \qquad (4.10c)$$

where $\mathcal{M}^{o}(x)$, $\chi^{o}(x)$, and $C^{o}(x)$ are scaling functions. In deriving these relations, Eqns. (4.10a–c), one actually uses a second argument $HL^{(\gamma+\beta)/\nu}$ in the scaling function $\mathcal{F}$ in Eqn. (4.9), where H is the field conjugate to the order parameter. After the appropriate differentiation has been completed H is then set to zero. Scaling relations such as $2-\alpha = \gamma + \beta$ are also used. Note that the finite size scaling ansatz is valid only for sufficiently large L and temperatures close to T_c. Corrections to scaling and finite size scaling must be taken into account for smaller systems and temperatures away from T_c. Because of the complexity of the origins of these corrections they are not discussed in detail here; readers are directed elsewhere (Liu and Fisher, 1990; Ferrenberg and Landau, 1991) for a detailed discussion of these corrections and techniques for including them in the analysis of MC data. As an example of finite size behavior, in Fig. 4.4 we show data for the spontaneous magnetization of $L \times L$ Ising square lattices with pbc. The raw data are shown in the left-hand portion of the figure, and a finite size scaling plot, made with the exact values of the critical temperature and critical exponents is shown in the right-hand portion of the figure. Note that the large scatter of data points in this plot is characteristic of early Monte Carlo work – the computational effort entailed in producing these data from Landau (1976) is easily within the capability of everyone's PC today, and with any moderately fast workstation *today* one can do far better. Exactly at the transition the thermodynamic properties then all exhibit power law behavior, since the scaling functions $\mathcal{M}^{o}(0)$, $\chi^{o}(0)$, $C^{o}(0)$ just reduce to proportionality constants, i.e.

$$M \propto L^{-\beta/\nu}, \tag{4.11a}$$

$$\chi \propto L^{\gamma/\nu}, \tag{4.11b}$$

$$C \propto L^{\alpha/\nu}, \tag{4.11c}$$

which can be used to extract estimates for the ratio of certain critical exponents. The power law behavior for the order parameter is verified in Fig. 4.4 (right) directly noting that for small x all data approach a constant, which is then an estimate of $\mathcal{M}^{o}(0)$. Note that the scaling functions that appear in Eqn. (4.10) are universal, apart from scale factors for their arguments. The prefactors in Eqn. (4.11) are thus also of interest for the estimation of universal amplitude ratios (Privman *et al.*, 1991).

In addition to these quantities, which are basically just first or second order moments of the probability distribution of order parameter or energy, we may obtain important, additional information by examining higher order moments of the finite size lattice probability distribution. This can be done quite effectively by considering the reduced fourth order cumulant of the order parameter (Binder, 1981). For an Ising model in zero field, for which all odd moments disappear by symmetry, the fourth order cumulant simplifies to

$$U_4 = 1 - \frac{\langle m^4 \rangle}{3 \langle m^2 \rangle^2}. \tag{4.12}$$

Fig. 4.4 (top)
Spontaneous
magnetization for
$L \times L$ Ising square
lattices with periodic
boundary conditions;
(bottom) finite size
scaling plot for the
data shown at the top.
From Landau (1976).

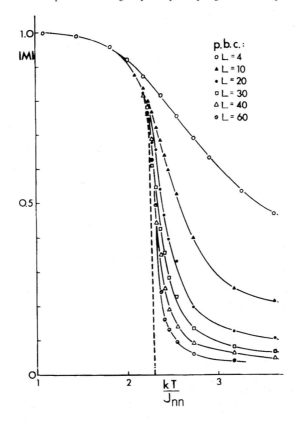

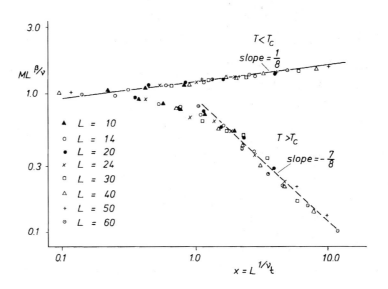

As the system size $L \to \infty$, $U_4 \to 0$ for $T > T_c$ and $U_4 \to 2/3$ for $T < T_c$. For large enough lattice size, curves for U_4 cross as a function of temperature at a 'fixed point' value U^* (our terminology here is used in a renormalization group sense, where the rescaling transformation $L' = bL$ with a scale factor $b > 1$ is iterated) and the location of the crossing 'fixed point' is the critical point. Hence, by making such plots for different size lattices one can make a preliminary identification of the universality class from the value of U_4^* and obtain an estimate for T_c from the location of the crossing point. Of course, if the sizes used are too small, there will be correction terms present which prevent all the curves from having a common intersection. Nonetheless there should then be a systematic variation with increasing lattice size towards a common intersection. (The same kind of behavior will also be seen for other models, although the locations of the crossings and values of U_4 will obviously be different.) An example of the behavior which can be expected is shown in Fig. 4.5 for the Ising square lattice in zero field.

Another technique which can be used to determine the transition temperature very accurately, relies on the location of peaks in thermodynamic derivatives, for example the specific heat. For many purposes it is easier to deal with inverse temperature so we define the quantity $K = J/k_B T$ and use K for much of the remainder of this discussion. The location of the peak defines a finite-lattice (or effective) transition temperature $T_c(L)$, or equivalently $K_c(L)$, which, taking into account a correction term of the form L^{-w}, varies with system size like

$$T_c(L) = T_c + \lambda L^{-1/\nu}(1 + bL^{-w}), \tag{4.13a}$$

$$K_c(L) = K_c + \lambda' L^{-1/\nu}(1 + b'L^{-w}), \tag{4.13b}$$

where λ, b, or λ', b' are some (model dependent) constants, and where the exponents will be the same in the two formulations but the prefactors will differ. Because each thermodynamic quantity has its own scaling function, the peaks in different thermodynamic derivatives occur at different tempera-

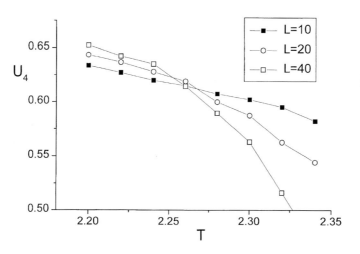

Fig. 4.5 Temperature dependence of the fourth order cumulant for $L \times L$ Ising square lattices with periodic boundary conditions.

tures for finite systems, some with positive λ, some with negative λ. To use Eqn. (4.13) to determine the location of the infinite lattice transition it is necessary to have both an accurate estimate for ν and accurate values for finite lattice 'transitions' $K_c(L)$. In a case where neither K_c, ν, nor w are known beforehand, a fit using Eqn. (4.13) involves 5 adjustable parameters. Hence, a reliable answer is only obtained if data with very good statistical accuracy are used and several quantities are analyzed simulataneously since they must all have the same K_c, ν, and w (see e.g. Ferrenberg and Landau (1991) for an example).

It has been notoriously difficult to determine ν from MC simulation data because of the lack of quantities which provide a direct measurement. We now understand that it is useful to examine several thermodynamic derivatives including that of the fourth order magnetization cumulant U_4 (Binder, 1981). In the finite size scaling region, the derivative varies with L like

$$\frac{\partial U_4}{\partial K} = aL^{1/\nu}(1 + bL^{-w}). \qquad (4.14)$$

Additional estimates for ν can be obtained by considering less traditional quantities which should nonetheless possess the same critical properties. For example, the logarithmic derivative of the nth power of the magnetization is

$$\frac{\partial \ln\langle m^n\rangle}{\partial K} = (\langle m^n E\rangle/\langle m^n\rangle) - \langle E\rangle \qquad (4.15)$$

and has the same scaling properties as the cumulant slope (Ferrenberg and Landau, 1991). The location of the maxima in these quantities also provides us with estimates for $K_c(L)$ which can be used in Eqn. (4.13) to extrapolate to K_c. For the three-dimensional Ising model consideration of the logarithmic derivatives of $|m|$ and m^2, and the derivative of the cumulant to determine ν proved to be particularly effective.

Estimates for other critical exponents, as well as additional values for $K_c(L)$, can be determined by considering other thermodynamic quantities such as the specific heat C and the finite-lattice susceptibility

$$\chi' = KL^d(\langle m^2\rangle - \langle |m|\rangle^2). \qquad (4.16)$$

Note that the 'true' susceptibility calculated from the variance of m, $\chi = KL^d(\langle m^2\rangle - \langle m\rangle^2)$, cannot be used to determine $K_c(L)$ because it has no peak. For sufficiently long runs at any temperature $\langle m\rangle = 0$ for $H = 0$ so that any peak in χ is merely due to the finite statistics of the simulation. For runs of modest length, $\langle m\rangle$ may thus have quite different values, depending on whether or not the system overturned completely many times during the course of the run. Thus, repetition of the run with different random number sequences may yield a true susceptibility χ which varies wildly from run to run below T_c. While for $T > T_c$ the 'true' susceptibility must be used if one wishes to estimate not only the critical exponent of χ but also the prefactor, for $T < T_c$ it is χ' and not χ that converges smoothly to the susceptibility of a state that has a spontaneous magnetization in the thermodynamic limit. For $T > T_c$ it is then better to use the result $\langle m \rangle = 0$ for $H = 0$ and estimate χ from $\chi = KL^d \langle m^2 \rangle$.

4.2.3.3 Finite size scaling and first order transitions

If the phase transition is first order, so that the correlation length does not diverge, a different approach to finite size scaling must be used. We first consider what happens if we fix the temperature $T < T_c$ of the Ising square lattice ferromagnet and cross the phase boundary by sweeping the magnetic field H. The subsequent magnetization curves are shown schematically in Fig. 4.6. The simplest, intuitive description of the behavior of the probability distribution of states in the system is plotted in the lower part of the figure. In the infinite system, in equilibrium, the magnetization changes discontinuously at $H = 0$ from a value $+M_{sp}$ to a value $-M_{sp}$. If, however, L is finite, the system may jump back and forth between two states (see Fig. 4.2) whose most probable values are $\pm M_L$, and the resultant equilibrium behavior is given by the continuous, solid curve. We start the analysis of the finite size

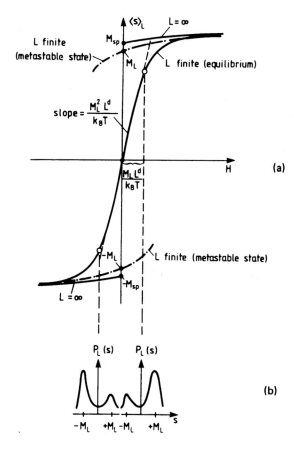

Fig. 4.6 Variation of the magnetization in a finite ferromagnet with magnetic field H. The curves include the infinite lattice behavior, the equilibrium behavior for a finite lattice, and the behavior when the system is only given enough time to relax to a metastable state. From Binder and Landau (1984).

behavior by approximating this distribution by two Gaussian curves, one centered on $+M_L$ and one at $-M_L$. In this (symmetric) case, the probability distribution $P_L(s)$ for the magnetization s then becomes

$$P_L(s) = \tfrac{1}{2}L^{d/2}(2\pi k_B T \chi_{(L)})^{-1/2} \times \left\{ \exp\left[-(s - M_L)^2 L^d / (2k_B T \chi_{(L)})\right] \right.$$

$$\left. + \exp\left[-(s + M_L)^2 L^d / (2k_B T \chi_{(L)})\right] \right\}.$$

$$(4.17)$$

If a magnetic field H is now applied then

$$P_L(s) = A \left(\exp\left\{ -\left[(s - M_{sp})^2 - 2\chi s H\right] L^d / 2k_B T \chi \right\} \right.$$

$$\left. + \exp\left\{ -\left[(s + M_{sp})^2 - 2\chi s H\right] L^d / 2k_B T \chi \right\} \right),$$

$$(4.18)$$

where χ is the susceptibility if the system stays in a single phase. The transition is located at the field for which the weights of the two Gaussians are equal; in the Ising square lattice this is, of course, at $H = 0$. It is now straightforward to calculate the moments of the distribution and thus obtain estimates for various quantities of interest. Thus,

$$\langle s \rangle_L \approx \chi H + M_{sp} \tanh\left[\frac{H M_{sp} L^d}{k_B T}\right]$$

$$(4.19)$$

and the susceptibility is ($\chi_L = K L^d (\langle s^2 \rangle_L - \langle s \rangle_L^2)$) is defined in analogy with the 'true' susceptibility)

$$\chi_L = \chi + M_{sp}^2 L^d \left/ \left[k_B T \cosh^2\left(\frac{H M_{sp} L^d}{k_B T}\right) \right]\right..$$

$$(4.20)$$

This expression shows that length enters only via the lattice volume, and hence it is the dimensionality d which now plays the essential role rather than a (variable) critical exponent as is the case with a second order transition. In Fig. 4.7 we show finite size scaling plots for the susceptibility below T_c and at T_c for comparison. The scaling is quite good for sufficiently large lattices and demonstrates that this 'thermodynamic' approach to finite size scaling for a first order transition works quite well. Note that the approach of χ_L to the thermodynamic limit is quite subtle, because the result depends on the order in which limits are taken: $\lim\limits_{H \to 0} \lim\limits_{L \to \infty} \chi_L = \chi$ (as required for the 'true' susceptibility) but $\lim\limits_{L \to \infty} \lim\limits_{H \to 0} \chi_L / L^d = M_{sp}^2 / k_B T$.

In other cases the first order transition may involve states which are not related by any particular symmetry (Binder, 1987). An example is the two-dimensional q-state Potts model (see Eqn. (2.38)) for $q > 4$ in which there is a temperature driven first order transition. At the transition the disordered state has the same free energy as the q-fold degenerate ordered state. Again one can describe the distribution of states by the sum of two Gaussians,

(a)

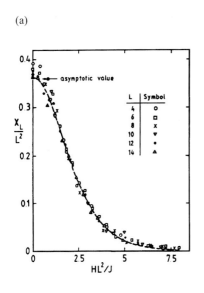

(b)

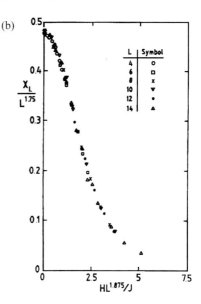

Fig. 4.7 Finite size scaled susceptibility vs. scaled field for the two-dimensional Ising model along paths of constant temperature: (a) $k_B T/\mathcal{J} = 2.1$; (b) $k_B T/\mathcal{J} = 2.269 = T_c$. From Binder and Landau (1984).

but these two functions will now typically have rather different parameters. (Challa *et al.*, 1986). The probability distribution function for the internal energy E per lattice site is (E_+, C_+, and E_-, C_- are energy and specific heat in the high temperature phase or low temperature phase right at the transition temperature T_c, respectively, and $\Delta = T - T_c$)

$$
P_L(E) = A\left[\frac{a_+}{\sqrt{C_+}}\exp\left[\frac{-[E-(E_+ + C_+\Delta T)]^2 L^d}{2k_B T^2 C_+}\right]\right.
$$
$$
\left. + \frac{a_-}{\sqrt{C_-}}\exp\left[\frac{-[E-(E_- + C_-\Delta T)]^2 L^d}{2k_B T^2 C_-}\right]\right].
\tag{4.21}
$$

Here A is a normalizing constant and the weights a_+, a_- are given by

$$
a_+ = e^x, \qquad a_- = qe^{-x}
\tag{4.22}
$$

where $x = (T - T_c(\infty))(E_+ - E_-)L^d/(2k_B TT_c)$. Originally, Challa *et al.* (1986) had assumed that at the transition temperature $T_c(\infty)$ of the infinite system each peak of the q ordered domains and the disordered phase has equal *height*, but now we know that they have equal *weight* (Borgs and Kotecký, 1990). From Eqn. (4.22) we find that the specific heat maximum occurs at

$$
\frac{T_c(L) - T_c}{T_c} = \frac{k_B T_c \ln[q]}{(E_+ - E_-)L^d}
\tag{4.23}
$$

and the value of the peak is given by

$$
C_L\bigg|_{max} \approx \frac{C_+ + C_-}{2} + \frac{(E_+ - E_-)^2 L^d}{4k_B T_c^2}.
\tag{4.24}
$$

Challa *et al.* (1986) also proposed that a reduced fourth order cumulant of the energy, i.e.

$$V_L = 1 - \frac{\langle E^4 \rangle_L}{3 \langle E^2 \rangle_L^2} \tag{4.25}$$

has a minimum at an effective transition temperature which also approaches the infinite lattice value as the inverse volume of the system. The behavior of V_L for the $q = 10$ Potts model in two dimensions is shown in Fig. 4.8. Thus, even in the asymmetric case it is the volume L^d which is important for finite size scaling. Effective transition temperatures defined by extrema of certain quantities in general differ from the true transition temperature by corrections of order $1/L^d$, and the specific heat maximum scales proportional to L^d (the prefactor being related to the latent heat $E_- - E_-$ at the transition. see Eqn. (4.24)).

This discussion was included to demonstrate that we understand, in principle, how to analyze finite size effects at first order transitions. In practice, however, this kind of finite size analysis is not always useful, and the use of free energy integrations may be more effective in locating the transition with modest effort. Other methods for studying first order transitions will be presented in later sections.

Fig. 4.8 Variation of the 'reduced' fourth order energy cumulant V_L with temperature for the $q = 10$ Potts model in two dimensions. The vertical arrow shows the transition temperature for $L = \infty$. After Challa *et al.* (1986).

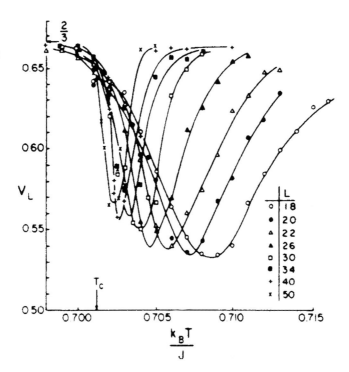

4.2.3.4 Finite size subsystem scaling

A theoretical approach to the understanding of the behavior of different systems in statistical physics has been to divide the system into sub-blocks and coarse-grain the free energy to derive scaling laws. We shall see this approach carried out explicitly in Chapter 9 where we discuss Monte Carlo renormalization group methods. The behavior of a sub-block of length scale L'/b in a system of size L' will be different from that of a system of size L'/b because the correlation length can be substantially bigger than the size of the system sub-block. In this case it has been shown (Binder, 1981) that the susceptiblity at the transition actually has an energy-like singularity

$$\langle s^2 \rangle_L \propto L^{2\beta/\nu}\Big[f_2(\infty) - g_2(\xi/L)^{-(1-\alpha)/\nu}\Big]. \tag{4.26}$$

The block distribution function for the two-dimensional Ising model, shown in Fig. 4.9, has a quite different behavior below and above the critical point. For $T < T_c$ the distribution function can be well described in terms of M_{sp}, χ, and the interface tension F_s, while for $T > T_c$ the distribution becomes Gaussian with a width determined by χ. In addition, the advantage of studying subsystems is that in a single run one can obtain information on size effects on many length scales (smaller than the total size of the simulated system, of course).

4.2.3.5 Field mixing

Up to this point our examples for finite size scaling at critical points have involved the Ising model which is a particularly 'symmetric' model. When viewed as a lattice gas this model has particle–hole symmetry. In more realistic models of fluids, however, this symmetry is lost. Such models consider particles which may move freely in space and which interact via the well known Lennard-Jones form

$$\phi(r) = 4w[(\sigma/r)^{12} - (\sigma/r)^6], \tag{4.27}$$

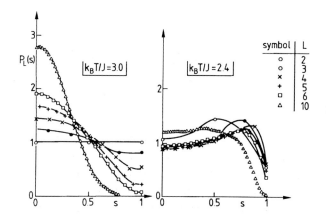

Fig. 4.9 Block distribution function for the two-dimensional Ising model for $L \times L$ sub-blocks: (left) $T > T_c$; (right) $T < T_c$. From Binder (1981).

where r is the distance between particles, σ gives the characteristic range of the interaction and w gives the potential well depth. The critical point of the Lennard-Jones system is described by *two* non-trivial parameter values, the critical chemical potential μ_c and the critical well depth w_c. In general, then, the scaling fields which are appropriate for describing the critical behavior of the system contain linear combinations of the deviations from these critical values:

$$\tau = w_c - w + s(\mu - \mu_c), \qquad (4.28a)$$

$$h = \mu - \mu_c + r(w_c - w), \qquad (4.28b)$$

where r and s depend upon the system (Wilding and Bruce, 1992). (For the Ising model $r = s = 0$.) We can now define two relevant densities which are conjugate to these scaling fields

$$\langle \mathcal{E} \rangle = L^{-d} \partial \ln Z_L / \partial \tau = [u - r\rho]/(1 - sr), \qquad (4.29a)$$

$$\langle \mathcal{M} \rangle = L^{-d} \partial \ln Z_L / \partial h = [\rho - su]/(1 - sr), \qquad (4.29b)$$

which are linear combinations of the usual energy density and particle density. Thus, a generalized finite size scaling hypothesis may be formulated in terms of these generalized quantities, i.e.

$$p_L(\mathcal{M}, \mathcal{E}) \approx \Lambda_{\mathcal{M}}^+ \Lambda_{\mathcal{E}}^+ \tilde{p}_{\mathcal{M},\mathcal{E}}(\Lambda_{\mathcal{M}}^+ \delta\mathcal{M}, \Lambda_{\mathcal{E}}^+ \Delta\mathcal{E}, \Lambda_{\mathcal{M}} h, \Lambda_{\mathcal{E}} \tau) \qquad (4.30)$$

where

$$\Lambda_{\mathcal{E}} = a_E L^{1/\nu}, \qquad \Lambda_{\mathcal{M}} = a_M L^{d-\beta/\nu}, \qquad \Lambda_{\mathcal{M}}^+ \Lambda_{\mathcal{M}} = \Lambda_{\mathcal{E}}^+ \Lambda_{\mathcal{E}} = L^d \quad (4.31)$$

and

$$\delta\mathcal{M} = \mathcal{M} - \langle \mathcal{M} \rangle_c, \qquad \delta\mathcal{E} = \mathcal{E} - \langle \mathcal{E} \rangle_c. \qquad (4.32)$$

Note that precisely at criticality Eqn. (4.30) simplifies to

$$p_L(\mathcal{M}, \mathcal{E}) \approx \Lambda_{\mathcal{M}}^+ \Lambda_{\mathcal{E}}^+ \tilde{p}_{\mathcal{M},\mathcal{E}}^*(\Lambda_{\mathcal{M}}^+ \delta\mathcal{M}, \Lambda_{\mathcal{E}}^+ \delta\mathcal{E}), \qquad (4.33)$$

so that by taking appropriate derivatives one may recapture power law behavior for the size dependence of various quantities. Surprises occur, however, and because of the field mixing contributions one finds that for critical fluids the specific heat

$$C_V = L^d (\langle u^2 \rangle - \langle u \rangle^2)/k_B T^2 \sim L^{\gamma/\nu} \qquad (4.34)$$

which is quite *different* from that found in the symmetric case. In Fig. 4.10 we show the parameter distribution at criticality as a function of the scaling variables.

4.2.3.6 Finite size effects in simulations of interfaces

As has been discussed in Section 4.2.2, one can deliberately stabilize interfaces in the system by suitable choice of boundary conditions. Such simulations are done with the intention to characterize the interfacial profile between coexisting phases, for instance. Figure 4.11 summarizes some of

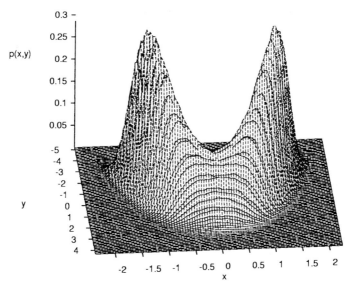

Fig. 4.10 Joint order parameter–energy distribution for an asymmetric lattice gas model as a function of scaling variables $x = a_M^{-1} L^{\beta/\nu} (\mathcal{M} - \mathcal{M}_c)$, $y = a_\varepsilon^{-1} L^{(1-\alpha)/\nu} (\mathcal{E} - \mathcal{E}_c)$. From Wilding (1995).

the standard simulation geometries that have been used for such a purpose, taking the Ising model again as simple example. Since directions parallel and perpendicular to an interface clearly are not equivalent, it also is no longer natural to choose the same value for the linear dimensions of the simulation box in the parallel and perpendicular directions. Thus, Fig. 4.11 assumes a linear dimension D in the direction across the interface, and another linear dimension L parallel to it. In case (a), the system has periodic boundary conditions in the parallel direction, but free boundaries in the perpendicular directions, with surface magnetic fields (negative ones on the left boundary, positive ones on the right boundary) to stabilize the respective domains, with an interface between them that on average is localized in the center of the film and runs parallel to the boundaries where the surface fields act. We disregard here the possibility that the interface may become 'bound' to one of the walls, and assume high enough temperature so the interface is a 'rough', fluctuating object, not locally localized at a lattice plane (in $d = 3$ dimensions where the interface is two-dimensional). In case (b), an analogous situation with a simple interface is stabilized by an antiperiodic boundary condition, while in case (c), where fully periodic boundary conditions are used, only an even number of interfaces can exist in the system. (In order to avoid the problem that one kind of domain, say the + domain, completely disappears because the interfaces meet and annihilate each other, we require a simulation at constant magnetization, see Section 4.4.1 below.)

The pictures in Fig. 4.11 are rather schematic, on a coarse-grained level, where both magnetization fluctuations in the bulk of the domains and small-scale roughness of the interface are ignored. But we emphasize the long wavelength fluctuations in the local position of the interface, because these fluctuations give rise to important finite size effects. It turns out that interfaces in a sense are soft objects, with a correlation length of fluctuations

Fig. 4.11 Schematic sketch of three possible geometries to study interfaces in Ising systems: (a) the 'surface field' boundary condition; (b) the antiperiodic boundary condition; and (c) the fully periodic boundary condition. Interfaces between coexisting phases of positive (+) and negative (−) spontaneous magnetization are shown schematically as dash-dotted lines.

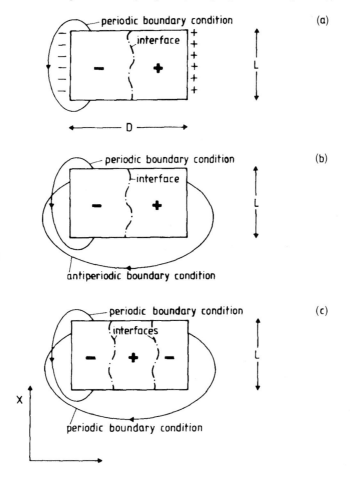

parallel to the interface ($\xi_\parallel$) that diverges if L and D tend to infinity: thus the interface is like a system at a critical point!

These fluctuations can be qualitatively accounted for by the concept of 'capillary wave' excitations, i.e. harmonic distortions of the local interface position z away from the average. For $D \to \infty$, one finds that the mean square width of the interface scales with the parallel linear dimension L (Jasnow, 1984)

$$w^2 \equiv \langle z^2 \rangle - \langle z \rangle^2 \propto \begin{cases} L & d = 2 & \text{dimensions} \\ \ln L & d = 3 & \text{dimensions,} \end{cases} \quad (4.35)$$

while in the opposite limit where L is infinite and D is varied one finds that ξ_n is finite (Parry and Evans, 1992)

$$\xi_\parallel \propto \begin{cases} D^2, & d = 2 \\ \exp(\kappa D), & d = 3, \end{cases} \quad \kappa = \text{constant.} \quad (4.36)$$

Then w^2 also becomes independent of L for large L, but rather depends on the perpendicular linear dimension D,

$$w^2 \propto \begin{cases} D^2, & d = 2, \\ D, & d = 3, \end{cases} \quad D \to \infty. \tag{4.37}$$

Of course, all these relations for the interfacial width make sense only for rather large linear dimensions L and D, such that w in Eqns. (4.35) and (4.37) is much larger than the 'intrinsic width' of the interface. If D is not very large, it is possible that the intrinsic width itself is squeezed down, and one then encounters a regime where $w \propto D$ in $d = 3$ dimensions.

Thus, simulations of interfaces are plagued by various finite size effects. More details and an example (interfaces in binary polymer mixtures) can be found in Werner *et al.* (1997).

4.2.3.7 Final thoughts

In many cases it is possible to perform a direct enumeration of states for a sufficiently small system. Generally this is possible only for systems which are so small that corrections to finite size scaling are important. The results should nonetheless lie on a smooth curve delineating the finite size behavior and can be useful in attempting to extract correction terms. Small lattices play another important role. Since exact results may be obtained for small systems, very useful checks of the correctness of the program may be made. Experience has shown that it is quite easy to make small errors in implementing the different boundary conditions discussed above, particularly at the corners. For large lattices such errors produce quite small imperfections in the data, but for small lattices the boundary spins make up a substantial fraction of the total system and errors in the data become larger. Thus programming mistakes and other subtle errors are often most visible for small systems.

4.2.4 Finite sampling time effects

When one plans a computer simulation study of a given model using a fixed 'budget' of computer resources, one must make a choice between performing long simulations of small systems or shorter simulations of larger systems. In order to use the available computer time as efficiently as possible, it is important to know the sources of both systematic and statistical errors. One source of systematic errors, finite size effects, was treated in the previous section; here we consider how such errors depend on the number of updates performed, i.e. the length of the run.

4.2.4.1 Statistical error

Suppose N successive observations A_μ, with $\mu = 1, \dots, N$ of a quantity A have been stored in a simulation, with $N \gg 1$. We consider the expectation value of the square of the statistical error

$$\langle(\delta A)^2\rangle = \left\langle\left[\frac{1}{N}\sum_{\mu=1}^{N}(A_\mu - \langle A\rangle)\right]^2\right\rangle$$

$$= \frac{1}{N^2}\sum_{\mu=1}^{N}\langle(A_\mu - \langle A\rangle)^2\rangle + \frac{2}{N^2}\sum_{\mu_1=1}^{N}\sum_{\mu_2=\mu_1+1}^{N}(\langle A_{\mu_1}A_{\mu_2}\rangle - \langle A\rangle^2).$$

(4.38)

In order to further explain what this means we now invoke the 'dynamic interpretation' of Monte Carlo sampling in terms of the master equation (Müller-Krumbhaar and Binder, 1973). The index μ which labels each successive configuration then plays the role of a 'time' variable (which may or may not be related to physical time, as discussed elsewhere in this book (see e.g. Sections 2.2.3, 2.3, 4.4, 5.2, etc.). If the states $\{X_\mu\}$ of the system from which the observations $\{A_\mu\}$ are taken are distributed according to a Boltzmann equilibrium distribution, the origin of this 'time' is indistinguishable from any other instant of this 'time', i.e. there is translational invariance with respect to this 'time' variable so that $\langle A_{\mu_1}A_{\mu_2}\rangle = \langle A_0 A_{\mu_2-\mu_1}\rangle$. Of course, this invariance would not hold in the first part of the Monte Carlo run (see Fig. 4.2), where the system starts from some arbitrary initial state which is not generally characteristic for the desired equilibrium – this early part of the run (describing the 'relaxation towards equilibrium') is hence not considered here and is omitted from the estimation of the average $\langle A\rangle$. The state $\mu = 1$ in Eqn. (4.38) refers to the first state that is actually included in the computation of the average, and not the first state that is generated in the Monte Carlo run.

Using this invariance with respect to the origin of 'time', we can change the summation index μ_2 to $\mu_1 + \mu$ where $\mu \equiv \mu_2 - \mu_1$, and hence

$$\langle(\delta A)^2\rangle = \frac{1}{N}\left[\langle A^2\rangle - \langle A\rangle^2 + 2\sum_{\mu=1}^{N}\left(1 - \frac{\mu}{N}\right)(\langle A_0 A_\mu\rangle - \langle A\rangle^2)\right]. \quad (4.39)$$

Now we explicitly introduce the 'time' $t = \mu\delta t$ associated with the Monte Carlo process where δt is the time interval between two successive observations A_μ, $A_{\mu+1}$. It is possible to take $\delta t = \tau_s/N$, where N is the number of degrees of freedom, and τ_s is a time constant used to convert the transition probability of the Metropolis method to a transition probability per unit time: this would mean that every Monte Carlo 'microstate' is included in the averaging. Since subsequent microstates are often highly correlated with each other (e.g. for a single spin-flip Ising simulation they differ at most by the orientation of one spin in the lattice), it typically is much more efficient to take δt much larger than τ_s/N, i.e. $\delta t = \tau_s$. (This time unit then is called '1

Monte Carlo step/spin (MCS)', which is useful since it has a sensible thermodynamic limit.) But often, in particular near critical points where 'critical slowing down' (Hohenberg and Halperin, 1977) becomes pronounced, even subsequent states $\{X_\mu\}$ separated by $\delta t = 1$ MCS are highly correlated, and it may then be preferable to take $\delta t = 10$ MCS or $\delta t = 100$ MCS, for instance, to save unnecessary computation. (When we discuss reweighting techniques in Chapter 7 we shall see that this is not always the case.)

Assuming, however, that the 'correlation time' between subsequent states is much larger than δt, we may transform the summation over the discrete 'times' $t = \delta t \, \mu$ to an integration, $t = \delta t \, N$,

$$
\begin{aligned}
\langle (\delta A)^2 \rangle &= \frac{1}{N}\left[\langle A^2 \rangle - \langle A \rangle^2 + \frac{2}{\delta t}\int_0^t \left(1 - \frac{t'}{t}\right)\left[\langle A(0)A(t')\rangle - \langle A \rangle^2\right] dt' \right] \\
&= \frac{1}{N}\left(\langle A^2 \rangle - \langle A \rangle^2\right)\left[1 + \frac{2}{\delta t}\int_0^t \left(1 - \frac{t'}{t}\right)\phi_A(t')\, dt'\right],
\end{aligned}
$$

$$(4.40)$$

where we define the normalized time autocorrelation function (also called 'linear relaxation function') $\phi_A(t)$ as

$$
\phi_A(t) = \frac{\left[\langle A(0)A(t)\rangle - \langle A \rangle^2\right]}{\left[\langle A^2 \rangle - \langle A \rangle^2\right]}.
$$

$$(4.41)$$

For the magnetization M of an Ising model, this function has already been discussed in Eqns. 2.72 and 2.70. Note that $\phi_A(t=0)=1$, $\phi_A(t\to\infty)=0$, and $\phi_A(t)$ decays monotonically with increasing time t. We assume that the time integral of $\phi_A(t)$ exists, i.e.

$$
\tau_A \equiv \int_0^\infty \phi_A(t)\, dt,
$$

$$(4.42)$$

and τ_A then can be interpreted as the 'relaxation time' of the quantity A (cf. Eqn. (2.73)).

Let us now assume that the simulation can be carried out to times $t \gg \tau_A$. Since $\phi_A(t)$ is essentially non-zero only for $t' \le \tau_A$, the term t'/t in Eqn. (4.37) then can be replaced by infinity. This yields (Müller-Krumbhaar and Binder, 1973)

$$
\langle (\delta A)^2 \rangle = \frac{1}{N}\left[\langle A^2 \rangle - \langle A \rangle^2\right]\left(1 + 2\frac{\tau_A}{\delta t}\right).
$$

$$(4.43)$$

We see that $\langle (\delta A)^2 \rangle$ is in general *not* given by the simple sampling result $[\langle A^2 \rangle - \langle A \rangle^2]/N$, but is rather enhanced by the factor $(1 + 2\tau_A/\delta t)$. This factor is called the 'statistical inefficiency' of the Monte Carlo method and may become quite large, particularly near a phase transition. Obviously, by calculating $\langle A \rangle$ and $\langle A^2 \rangle$, as well as $\langle (\delta A)^2 \rangle$ we can estimate the relaxation time τ_A. Kikuchi and Ito (1993) demonstrated that for kinetic Ising model simulations such an approach is competitive in accuracy to the standard method where one records $\phi_A(t)$ (Eqn. (4.41)) and obtains τ_A by numerical integration (see Eqn. (4.42)). Of course, if $\tau_A \gg \delta t$, then Eqn. (4.43) may be further simplified by neglecting the unity in the bracket and, using $N\delta t = t$,

$$\langle (\delta A)^2 \rangle = [\langle A^2 \rangle - \langle A \rangle^2](2\tau_A/t). \tag{4.44}$$

This means that the statistical error is independent of the choice of the time interval δt, it only depends on the ratio of relaxation time (τ_A) to observation time (t). Conversely, if δt is chosen to be so large that subsequent states are uncorrelated, we may put $\langle A_0 A_\mu \rangle \approx \langle A \rangle^2$ in Eqn. (4.38) to get $\langle (\delta A)^2 \rangle = [\langle A^2 \rangle - \langle A \rangle^2]/\mathcal{N}$. For many Monte Carlo algorithms τ_A diverges at second order phase transitions ('critical slowing down', see Sections 2.3.3 and 5.2.3), and then it becomes very hard to obtain sufficiently high accuracy, as is obvious from Eqn. (4.44). Therefore the construction of algorithms that reduce (or completely eliminate) critical slowing down by proper choice of global moves (rather than single spin-flips) is of great significance. Such algorithms, which are not effective in all cases, will be discussed in Section 5.1.

Problem 4.4 From a Monte Carlo simulation of an $L = 10$ Ising square lattice, determine the order parameter correlation time at $T = 3.0\,J/k_B$ and at $T = 2.27\,J/k_B$.

Problem 4.5 Perform a Metropolis Monte Carlo simulation for a 10×10 Ising model with periodic boundary conditions. Include the magnetic field H in the simulation and plot both $\langle M \rangle$ and $\langle |M| \rangle$ as a function of field for $k_B T/J = 2.1$. Choose the range from $H = 0$ to $H = 0.05\,J$. Do you observe the behavior which is sketched in Fig. 2.10? Interpret your results!

4.2.4.2 Biased sampling error: Ising criticality as an example

The finite sampling time is not only the source of the statistical error, as described above, but can also lead to systematic errors (Ferrenberg et $al.$, 1991). For example, in the Monte Carlo sampling of response functions the latter are systematically underestimated. This effect comes simply from the basic result of elementary probability theory (see Section 2.2) that in estimating the variance s^2 of a probability distribution using n independent samples, the expectation value $E(s^2)$ of the variance thus obtained is systematically lower than the true variance σ^2 of the distribution, by a factor $(1 - 1/n)$:

$$E(s^2) = \sigma^2(1 - 1/n). \tag{4.45}$$

Since we may conclude from Eqn. (4.43) that for $t \gg \tau_A$ we have $n = \mathcal{N}/(1 + 2\tau_A/\delta t)$ independent 'measurements', we may relate the calculated susceptibility $\chi_{\mathcal{N}}$ of a spin system to that which we would obtain from a run of infinite length χ_∞ by

$$\chi_{\mathcal{N}} = \chi_\infty \left(1 - \frac{1 + 2\tau_M/\delta t}{\mathcal{N}} \right), \tag{4.46}$$

τ_M being the relaxation time of the magnetization, i.e. $A = M$ in Eqns. (4.34–4.43).

This effect becomes particularly important at T_c, where one uses the values of χ from different system sizes ($N = L^d$ in d dimensions, where L is the linear dimension and the lattice spacing is taken to be unity) to estimate the critical exponent ratio γ/ν (see Section 4.2.3). The systematic error resulting from Eqn. (4.46) will generally vary with L, since the relaxation time τ_M may depend on the system size quite dramatically ($\tau_M \propto L^z$, z being the 'dynamic exponent', see Section 2.3.3).

While finite size scaling analyses are now a standard tool, the estimation of errors resulting from Eqn. (4.46) is generally given inadequate attention. To emphasize that neglect of this biased sampling error is not always warranted, we briefly review here some results of Ferrenberg *et al.* (1991) who performed calculations for the nearest neighbor ferromagnetic Ising model. The Monte Carlo simulations were carried out right at the 'best estimate' critical temperature T_c of the infinite lattice model ($T_c^{-1} = 0.221\,654\,k_B/\mathcal{J}$) for system sizes ranging from $16 \le L \le 96$. Well over 10^6 MCS were performed, taking data at intervals $\delta t = 10$ MCS, and dividing the total number of observations $\mathcal{N}_{tot}$ into g bins of 'bin length' $\mathcal{N}$, $\mathcal{N}_{tot} = g\mathcal{N}$, and calculating $\chi_{\mathcal{N}}$ from the fluctuation relation. Of course, in order to obtain reasonable statistics they had to average the result over all $g \gg 1$ bins. Figure 4.12 shows the expected strong dependence of $\chi_{\mathcal{N}}$ on both $\mathcal{N}$ and L: while for $L = 16$ the data have settled down to an $\mathcal{N}$-dependent plateau value for $\mathcal{N} \ge 10^3$, for $L = 48$ even the point for $\mathcal{N} = 10^4$ still falls slightly below the plateau, and for $L = 96$ the asymptotic behavior is only reached for $\mathcal{N} \ge 10^5$. (Note that in this calculation a very fast vectorizing multispin coding (Section 5.3.1) single spin flip algorithm was used.) Thus with a constant number $\mathcal{N}$ as large as $\mathcal{N} = 10^4$ for a finite size scaling analysis, one would *systematically* underestimate the true finite system susceptibility for large L, and an incorrect value of γ/ν in the relation $\ln \chi_{\mathcal{N}}(L) = (\gamma/\nu)\ln L$ would result. However, if we measure τ_M for the different values of L and use Eqn. (4.11), we can correct for this

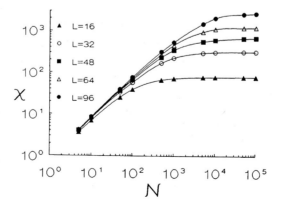

Fig. 4.12 Variation of χ_N for the susceptibility of $L \times L \times L$ ferromagnetic nearest neighbor Ising lattices at the critical temperature as a function of the 'bin length' $\mathcal{N}$. Different symbols indicate various values of L, as indicated in the figure. From Ferrenberg *et al.* (1991).

Fig. 4.13 Scaled
susceptibility vs.
scaled bin length n.
The solid line is the
function
$\chi_\infty L^{-\gamma/\nu}(1-1/n)$,
using the accepted
value $\gamma/\nu = 1.97$. In
the insert the reduced
systematic error, $\Delta\chi_n$
$=(\chi_\infty - \chi_N)/\chi_N$ is
plotted vs. n^{-1} to
highlight the large bin
length behavior (the
solid line, with slope
unity, is predicted by
Eqn. (4.46)). From
Ferrenberg *et al.*
(1991).

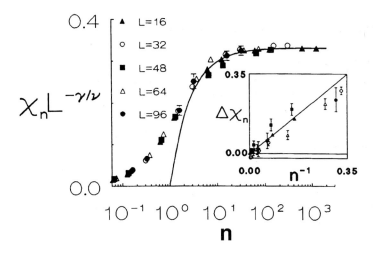

effect. In the present example, the appropriate correlation time τ_M is
$\tau_M = 395,\ 1640,\ 3745,\ 6935$ and $15\,480$, for $L = 16,\ 32,\ 48,\ 64$ and 96,
respectively. Using these values we can rewrite Eqn. (4.45) as
$\chi_N = \chi_\infty(1 - 1/n)$, computing n as $n = \mathcal{N}/(1 + 2\tau_M/\delta t)$. Figure 4.13
shows that when $\chi_N L^{-\gamma/\nu}$ is plotted vs. n, all the data collapse on a
universal function, which for $n \geq 5$ is compatible with the simple result
$(1 - 1/n)$.

Problem 4.6 Carry out Monte Carlo simulations for an $L = 10$ Ising square
lattice with different run lengths for $T = 2.8\,J/k_B$. Calculate the susceptibility and plot it vs. run length. Extract an estimate for the infinite lattice susceptibility.

4.2.4.3 Relaxation effects

When one starts a simulation run, typically equilibrium states for the system
are not yet known. The Metropolis algorithm requires some initial state of the
system, however, this choice will probably not be characteristic for the equilibrium that one wishes to study. For example, one may intend to study the
critical region of an Ising ferromagnet but one starts the system for example
in a state where all spins are perfectly aligned, or in a random spin configuration. Then it is necessary to omit the first $\mathcal{N}_0$ configurations from the
averages, since they are not yet characteristic for equilibrium states of the
system (see Fig. 4.2): Therefore any Monte Carlo estimate $\overline{A}$ of an average
$\langle A \rangle$ actually reads

$$\overline{A} = \frac{1}{\mathcal{N} - \mathcal{N}_0} \sum_{\mu=\mathcal{N}_0+1}^{\mathcal{N}} A(\mathbf{X}_\mu) = \frac{1}{t - t_0} \int_{t_0}^{t} A(t')\,dt', \qquad (4.47)$$

where $t_0 = \mathcal{N}_0 \, \delta t$. Time-displaced correlation functions $\langle A(t)B(0)\rangle$ as they appear in Eqn. (4.40) are actually estimated as

$$\overline{A(t')B(0)} = \frac{1}{t - t' - t_0} \int_{t_0}^{t-t'} A(t' + t'')B(t'') \, dt'', \qquad t - t' > t_0. \quad (4.48)$$

As emphasized above, times t_0 must be chosen which are large enough that thermal equilibrium has been achieved, and therefore time averages along the Monte Carlo 'trajectory' in phase space, as defined in Eqns. (4.47) and (4.48), make sense.

However, it is also interesting to study the non-equilibrium relaxation process by which equilibrium is approached, starting from a non-equilibrium initial state In this process, $A(t') - \overline{A}$ depends on the observation time t' systematically, and an ensemble average $\langle A(t')\rangle - \langle A(\infty)\rangle$ ($\lim_{t \to \infty} \overline{A} = \langle A\rangle = \langle A(\infty)\rangle$ if the system is ergodic) is non-zero. Hence we define

$$\langle A(t)\rangle = \sum_{\{X\}} P(\mathbf{X}, t)A(\mathbf{X}) = \sum_{\{X\}} P(\mathbf{X}, 0)A\{\mathbf{X}(t)\}. \quad (4.49)$$

In the second step of this equation we have used the fact that the ensemble average involved is actually an average weighted by the probability distribution $P(\mathbf{X}, 0)$ of an ensemble of initial states $\{\mathbf{X}(t = 0)\}$ which then evolve as described by the master equation of the associate Monte Carlo process. In practice, Eqn. (4.49) means an average over $m \gg 1$ independent runs

$$\overline{A(t)} = \frac{1}{m} \sum_{l=1}^{m} A^{(l)}(t), \quad (4.50)$$

with $A^{(l)}(t)$ being the observable A observed at time t in the lth run of this non-equilibrium Monte Carlo averaging (these runs also differ in practice by use of different random numbers for each realization (l) of the time evolution).

Using Eqn. (4.49) we can define a non-linear relaxation function which was already considered in Eqn. (2.110)

$$\phi_A^{nl}(t) = [\langle A(t)\rangle - \langle A(\infty)\rangle]/[\langle A(0)\rangle - \langle A(\infty)\rangle] \quad (4.51)$$

and its associated relaxation time

$$\tau_A^{(nl)} = \int_0^{\infty} \phi_A^{(nl)}(t) \, dt. \quad (4.52)$$

The condition that the system is well equilibrated then simply reads

$$t_0 \gg \tau_A^{(nl)}. \quad (4.53)$$

This inequality must hold for all physical observables A, and hence it is important to focus on the slowest relaxing quantity (for which $\tau_A^{(nl)}$ is largest) in order to estimate a suitable choice of t_0. Near second order phase transitions, the slowest relaxing quantity is usually the order parameter M of the transition, and not the internal energy. Hence the 'rule-of-thumb' published in some Monte Carlo investigations that the equilibration of the system is established by monitoring the time evolution of the internal energy is clearly not a reliable procedure. This effect can readily be realized by examining the finite size behavior of the times $\tau_M^{(nl)}$, $\tau_E^{(nl)}$, at criticality, cf. Eqns. (2.111) and (2.112)

$$\tau_M^{(nl)} \propto L^{z-\beta/\nu}, \qquad \tau_E^{(nl)} \propto L^{z-(1-\alpha)/\nu}, \tag{4.54}$$

where the exponent of the order parameter is β, of the critical part of the energy is $1 - \alpha$, and of the correlation length is ν. Typically β/ν is much less than $(1 - \alpha)/\nu$ and the correlation time associated with the magnetization diverges much faster than that of the internal energy.

We also wish to emphasize that starting the system in an arbitrary state, switching on the full interaction parameters instantly, and then waiting for the system to relax to equilibrium is not always a very useful procedure. Often this approach would actually mean an unnecessary waste of computing time. For example, in systems where one wishes to study ordered phases at low temperature, it may be hard to use fully disordered states as initial configurations since one may freeze in long-lived multidomain configurations before the system relaxes to the final monodomain sample. In glass-like systems (spin glass models, etc.) it is advisable to produce low temperature states by procedures resembling slow cooling rather than fast quenching. Sometimes it may be preferable to relax some constraints (e.g. self-avoiding walk condition for polymers) first and then to switch them on gradually. There are many 'tricks of the trade' for overcoming barriers in phase space by suitably relaxing the system by gradual biased changes in its state, gradually switching on certain terms in the Hamiltonian, etc., which will be mentioned from time to time later.

4.2.4.4 Back to finite size effects again: self-averaging

Suppose we observe a quantity A in n statistically independent observations made while the system is in equilibrium, and calculate its error

$$\Delta_A(n, L) = \sqrt{(\langle A^2 \rangle - \langle A \rangle^2)/n}, \qquad n \gg 1. \tag{4.55}$$

We now ask, does this error go to zero if $L \to \infty$? If it does, A is called 'self-averaging', while if it yields an L-independent non-zero limit, we say A exhibits 'lack of self-averaging'. In pure phases away from phase boundaries, extensive quantities (energy per site E, magnetization per site M, etc.) have a Gaussian distribution whose variance scales inversely with the volume L^d,

$$P_L(A) = L^{d/2}(2\pi C_A)^{-1/2} \exp[-(A - \langle A \rangle)^2 L^d/2C_A]. \tag{4.56}$$

If, for example, $A = M$ then $C_A = k_B T \chi$, and if $A = E$, then $C_E = k_B T^2 C$ with C being the specific heat, etc. For these quantities, we hence see that errors scale as $\Delta_A(n, L) \propto (nL^d)^{-1/2}$. This property is called 'strong self-averaging' (Milchev *et al.*, 1986), in contrast to the behavior at critical points where the exponent governing the power law for the size dependence is smaller, $\Delta_A(n, L) \propto (nL^{x_1^A})^{-1/2}$ ($x_1^M = 2\beta/\nu$, $x_1^E = 2(1-\alpha)/\nu$; this situation is termed 'weak self-averaging').

The situation differs drastically if we consider quantities that are sampled from fluctuation relations (such as C, χ, . . .), rather than quantities that are spatial averages of a simple density (such as E, M, . . .). We still can formally use Eqn. (4.55), but we have to replace A by $C_A = (\delta A)^2 L^d$ in this case,

$$\Delta_{C_A}(n, L) = L^d n^{-1/2} \sqrt{\langle (\delta A)^4 \rangle - \langle (\delta A)^2 \rangle^2}, \qquad \delta A = A - \langle A \rangle. \qquad (4.57)$$

Since for the Gaussian distribution, Eqn. (4.55), $\langle (\delta A)^4 \rangle = 3\langle (\delta A)^2 \rangle^2$, Eqn. (4.56) reduces to ($C_A = L^d \langle (\delta A)^2 \rangle$)

$$\Delta_{C_A}(n, L) = L^d n^{-1/2} \langle (\delta A)^2 \rangle \sqrt{2} = C_A \sqrt{2/n}. \qquad (4.58)$$

Consequently, the size L^d cancels out precisely, and the relative error Δ_{C_A} $(n, L)/C_A = \sqrt{2/n}$ is completely universal. It only depends on the number n of statistically independent observations. Thus, increasing L at fixed n will strongly improve the accuracy of quantities such as E and M, but nothing is gained with respect to the accuracy of χ, C, etc. Thus, it is more economical to choose the smallest size which is still consistent with the condition $L \gg \xi$ and to increase n rather than L to improve the accuracy. For those researchers who feel that the best approach is to study the largest system size possible, we believe that an analysis of fluctuation relations in subsystems (Section 4.2.3.5) is mandatory!

4.2.5 Critical relaxation

The study of critical slowing down in spin models has formed an extremely active area of research, and Monte Carlo simulations have played an important role in developing an understanding of critical relaxation. The basic features of the underlying theory were presented in Section 2.3.3 and we now wish to examine the implementation of these ideas within the context of computer simulation. As we shall see below, the interest in this problem extends well beyond the determination of the dynamic critical exponent for a particular sampling algorithm since in any simulation there are multiple time scales for different quantities which must be understood even if the topic of interest is the static behavior.

Critical relaxation has been studied for many years for a number of different spin models but with uncertain results. Thus, in spite of the fact that the static behavior of the two-dimensional Ising model is known exactly, the determination of the critical relaxation has remained a rather elusive goal. As shown in Fig. 4.14, there have been estimates made for the dynamic critical exponent z over a period of more than 30 years using a number of different theoretical and numerical methods, and we may only just be coming to an

Fig. 4.14 Variation of the estimate of the dynamic exponent z for the two-dimensional Ising model as a function of time. The horizontal dashed line shows the value $\gamma/\nu = 1.75$ which is a lower bound.

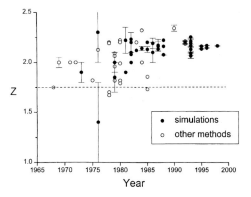

accurate knowledge of the exponent for a few models (Landau *et al.*, 1988; Wansleben and Landau, 1991; Ito, 1993). In the following sub-sections we shall briefly examine the different features associated with critical relaxation and the different ways that Monte Carlo data can be used to extract an estimate for z.

4.2.5.1 Non-linear relaxation

As we have already seen, the approach of a thermodynamic property A to its equilibrium behavior occurs in a characteristic fashion and is described by a simple, non-linear relaxation function, $\phi_A(t)$, given by Eqn. (2.110). The accurate determination of this relaxation function is non-trivial since knowledge of the equilibrium value of the quantity being studied is needed. This necessitates performing simulations which are long compared to the non-linear relaxation time to insure that an equilibrium value can be measured; however, to guarantee that the statistical errors are small for the non-linear relaxation function it is also necessary to make many equivalent runs with different random number sequences and average the data together. As a result some balance between the number and length of the runs must be achieved. Finally, the long time behavior of the non-linear relaxation function can be fitted by an exponential function to determine the asymptotic relaxation time $\tau \propto \xi^z$ (Eqn. (2.108)) while the integral of ϕ_A can be used to estimate the non-linear relaxation time τ_{nl}. The variation of τ_{nl} with temperature as the critical point is approached may then be used to estimate the dynamic exponent, although finite size effects will become important quite close to T_c. From Eqns.(2.111) and (2.112) we recall that $\tau_{nl} \propto \xi^{z_{nl}^A}$ with an exponent that is always smaller than z but is related to z by a scaling law, $z_{nl}^A = z - \beta^A/\nu$, β^A being the exponent of the 'critical part' of the quantity A. ($\beta^A = \beta$ if A is the order parameter and $\beta^A = 1 - \alpha$ if A is the energy, etc.) How is it possible that $z_{nl}^A < z$ although the asymptotic decay of $\phi_A(t)$ occurs with the 'linear' relaxation time τ which is governed by the exponent z? The solution to this puzzle is that the asymptotic decay sets in only when $\phi_A(t)$ has decayed down to values of the order of the static critical part of A, i.e. is of the order of $\xi^{-\beta_A/\nu} \sim \varepsilon^{\beta_A}$. Near T_c these values are small and accuracy is hard to obtain. Alternatively, if the critical temperature is well known, the critical exponent can be determined from the finite size behavior at T_c.

For an infinite system at T_c the magnetization will decay to zero (since this is the equilibrium value) as a power law

$$m(t) \propto t^{\beta/z\nu}, \tag{4.59}$$

where β and ν are the static critical exponents which are known exactly for the two-dimensional Ising model. Eventually, for a finite lattice the decay will become exponential, but for sufficiently large lattices and sufficiently short times, a good estimate for z can be determined straightforwardly using Eqn. (4.59). (The study of multiple lattice sizes to insure that finite size effects are not becoming a problem is essential!) Several different studies have been successfully carried out using this technique. For example, Ito (1993) used multilattice sampling and carefully analyzed his Monte Carlo data, using Eqn. (4.59), for systems as large as $L = 1500$ to insure that finite size effects were not beginning to appear. (A skew periodic boundary was used in one direction and this could also complicate the finite size effects.) From this study he estimated that $z = 2.165(10)$. Stauffer (1997) examined substantially larger lattices, $L = 496\,640$, for times up to $t = 140$ MCS/site using this same method and concluded that $z = 2.18$. Although these most recent values appear to be well converged, earlier estimates varied considerably. For a recent review of the problems of non-linear relaxation in the Ising model see Wang and Gan (1998). We also note that there exists yet another exponent which appears in non-equilibrium relaxation at criticality when we start the system at T_c in a random configuration. The magnetization then has a value of $\sim N^{-1/2}$, and increases initially like $M(t) \propto t^\theta$ with a new exponent θ (Janssen et al., 1989; Li et al., 1994).

4.2.5.2 Linear relaxation

Once a system is in equilibrium the decay of the time-displaced correlation function is described by a linear relaxation function (cf. Eqn. (2.106)). The generation of the data for studying the linear relaxation can be carried out quite differently than for the non-linear relaxation since it is possible to make a single long run, first discarding the initial approach to equilibrium, and then treating many different points in the time sequence as the starting point for the calculation of the time-displaced correlation function. Therefore, for a Monte Carlo run with N successive configurations the linear correlation function at time t can be computed from

$$\phi_{AA}(t) = \Gamma \left(\frac{1}{N-t} \sum_{t'}^{N-t} A(t')A(t'+t) - \frac{1}{(N-t)^2} \sum_{t'}^{N-t} A(t') \sum_{t''}^{N-t} A(t'') \right),$$

$$\tag{4.60}$$

where $\Gamma = (\langle A^2 \rangle - \langle A \rangle^2)^{-1}$. From this expression we see that there will indeed be many different estimates for short time displacements, but the number of values decreases with increasing time displacement until there is only a single value for the longest time displacement. The characteristic behavior of the time-displaced correlation function shown in Fig. 4.15 indicates that there are three basic regions of different behavior. In the early stages of the decay (Region I) the behavior is the sum of a series of exponential decays. Actually it is possible to show that the initial slope of $\phi_{AA}(t)$, $(d\phi_{AA}(t))/dt)_{t=0} = \tau_I^{-1}$, defines a time τ_I which scales as the static fluctuation, $\tau_I \propto (\langle A^2 \rangle - \langle A \rangle^2)$. Since $\phi_{AA}(t)$ is non-negative, this result implies that $\tau > \tau_I$, and hence the inequality $z > \gamma/\nu$ results when we choose $A = M$, i.e. the order parameter. If instead $A = E$, i.e. the energy, the initial decay is rather rapid since $\tau_I \propto C \propto \varepsilon^{-\alpha}$, where α is the specific heat exponent. Nevertheless, the asymptotic decay of $\phi_{EE}(t)$ is governed by an exponential relaxation $e^{-t/\tau}$ where τ diverges with the same exponent z as the order parameter relaxation time. For a more detailed discussion see Stoll *et al.* (1973). In Region II the time dependence of the relaxation function can be fitted by a single exponential described by a correlation time τ which diverges as the critical point is approached. Finally, in Region III the statistical errors become so large that it becomes impossible to perform a meaningful fit. The difficulty, of course, is that it is never completely obvious when the data have entered the regime where they are described by a single exponential, so any analysis must be performed carefully. Generally speaking, the early time regime is much more pronounced for the internal energy as compared to the order parameter. This is clear from the above remark that the initial relaxation time for the energy scales like the specific heat. In order to compare the decay of different quantities, in Fig. 4.16 we show a semi-logarithmic plot for the three-dimensional Ising model. From this figure we can see that the magnetization decay is quite slow and is almost perfectly linear over the entire range. In contrast, the internal energy shows quite pronounced contributions

Fig. 4.15 Schematic behavior of the time-displaced correlation function as defined by Eqn. (4.60).

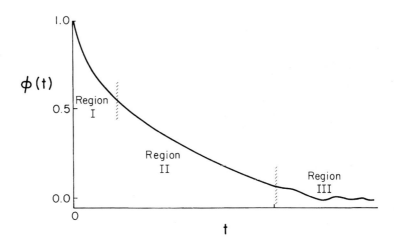

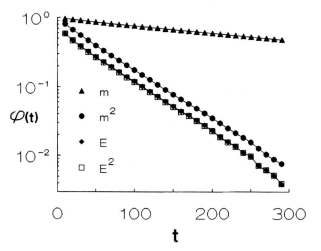

Fig. 4.16 Linear relaxation function for different quantities for the three-dimensional Ising model at the critical temperature with $L = 16$ and periodic boundary conditions. From Ferrenberg *et al.* (1991).

from multiple decay modes at short times and has a much shorter relaxation time in the asymptotic regime. Note that both m^2 and E^2 have time reversal symmetry (but m does not) and have the same asymptotic relaxation time as does E. For time displacements greater than about 250 MCS/site the statistical fluctuations begin to grow quite quickly and it becomes difficult to analyze the data in the asymptotic regime.

Although the general approach is straightforward, there are nonetheless considerable subtleties in this kind of analysis. The use of skew periodic boundary conditions simplifies the computer code but introduces a 'seam' into the model which provides a correction for small lattices. The relaxation function is a biased estimator so the length of the individual runs must also be quite long to eliminate another source of small corrections. (In fact, for finite length runs the relaxation function will oscillate about a small *negative* value at very long times.) Lastly, it is often necessary to perform least squares fits over different ranges of time to ascertain where noise is becoming a problem at long times.

A completely different approach to the analysis of the correlations in equilibrium which does not require the computation of the relaxation function is through the determination of the 'statistical inefficiency' described for example by Eqn. (4.43). A 'statistical dependence time' τ_{dep} is calculated by binning the measurements in time and calculating the variance of the mean of the binned values; as the size of the bins diverges, the estimate τ_{dep} approaches the correlation time. Kikuchi and Ito (1993) used this approach to study the three-dimensional Ising model and found that $z = 2.03\,(4)$.

4.2.5.3 Integrated vs. asymptotic relaxation time

As we saw earlier in this chapter, an integrated correlation time may be extracted by integrating the relaxation function; and it is this correlation time, given in Eqn. (4.42), which enters into the calculation of the true statistical error. The resulting integrated correlation time also diverges as the critical point is approached, but the numerical value may be different in magnitude from the asymptotic correlation time if there is more than one exponential that contributes significantly to the relaxation function. This is relatively easy to see if we look at the behavior of the internal energy E with time shown in Fig. 4.16: from this figure we can see that both E and m^2 have the same asymptotic relaxation time, but m^2 will have a much larger integrated relaxation time. When one examines all of the response functions it becomes clear that there are a number of different correlation times in the system, and the practice of only measuring quantities at well separated intervals to avoid wasting time on correlated data may actually be harmful to the statistical quality of the results for some quantities.

4.2.5.4 Dynamic finite size scaling

The presence of finite size effects on the dynamic (relaxational) behavior can be used to estimate the dynamic critical exponent. Dynamic finite size scaling for the correlation time τ can be written

$$\tau = L^z \mathcal{F}(\varepsilon L^{1/\nu}), \tag{4.61}$$

so at the critical point the correlation time diverges with increasing lattice size as

$$\tau \propto L^z. \tag{4.62}$$

As in the case of statics, this finite size scaling relation is valid only as long as the lattice size L is sufficiently large that corrections to finite size scaling do not become important. The behavior of the correlation time for the order parameter and the internal energy may be quite different. For example, in Fig. 4.17 we show the finite size behavior of both correlation times for the three-dimensional Ising model. As a result we see that the asymptotic dynamic exponents for both quantities are consistent, but the amplitudes of the divergencies are almost an order of magnitude different.

Of course, it is also possible to extract an estimate for z using finite size data and Eqn. 4.61. In this approach a finite size scaling plot is made in the same manner as for static quantities with the same requirement that data for different sizes and temperatures fall upon a single curve. Here too, when the data are too far from T_c, scaling breaks down and the data no longer fall upon the same curve. In addition, when one tries to apply dynamic finite size scaling, it is important to be aware of the fact that $\phi_{MM}(t)$ does not decay with a single relaxation time but rather with an entire spectrum, i.e.

$$\phi_{MM}(t \to \infty) = c_1 e^{-t/\tau_1} + c_3 e^{-t/\tau_3} + \cdots, \qquad \tau_1 > \tau_3 > \cdots \tag{4.63}$$

where $c_1, c_3, \ldots$ are amplitudes, and all times $\tau_1 \propto \tau_3 \propto \ldots L^z$. Only the amplitudes $\hat{\tau}_n (\tau_n = \hat{\tau}_n L^z)$ decrease with increasing n. Note that we have used odd indices here because M is an 'odd operator', i.e. it changes sign under spin reversal. The second largest relaxation time τ_2 actually appears for

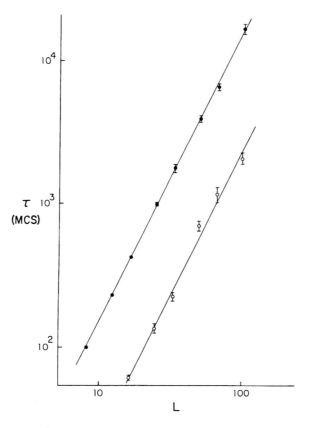

Fig. 4.17 Dynamic finite size scaling analysis for the three-dimensional Ising model at T_c. Closed circles are for the order parameter, open circles are for the internal energy. From Wansleben and Landau (1991).

the leading asymptotic decay of the 'even operators' such as E or M^2 which are invariant under spin reversal.

All of these relaxation times, τ_n, have a scaling behavior as written in Eqn. (4.61); however, it is important to note that τ_1 is distinct from all other relaxation times because it increases monotonically as the temperature is lowered through T_c, while all other τ_n have their maximum somewhere in the critical region (Koch and Dohm, 1998; Koch et al., 1996). The reason for this uninterrupted increase of τ_1, is that below T_c it develops into the ergodic time τ_e which describes how long it takes for the system to tunnel between regions of phase space with positive and negative magnetizations. This process must occur through a high energy barrier ΔF between the two regions and $\tau_e \propto L^z \exp(\Delta F/k_B T)$. Actually, ΔF can be estimated for an Ising system (for a simulation geometry of an L^d system with periodic boundary conditions) as $2\sigma L^{d-1}$, where σ is the interfacial free energy of the system. This corresponds to the creation of a domain with two walls running through the entire simulation box to reverse the sign of the spontaneous magnetization. Thus, we obtain the estimate

$$\tau_1 = \tau_e \propto L^z \exp(2L^{d-1}\sigma/k_B T). \tag{4.64}$$

This monotonic increase of τ_1 with decreasing T corresponds to the increase in the fluctuation $\langle M^2 \rangle - \langle M \rangle^2 \ (= \langle M^2 \rangle$ for $H = 0)$. Remember, however, that below T_c we have to use $\langle M^2 \rangle - \langle |M| \rangle^2$ to take into account the symmetry breaking; and in the same vein, below T_c it is the next relaxtion time τ_3 which characterizes the decay of magnetization fluctuations in a state with non-zero spontaneous magnetization.

4.2.5.5 Final remarks

In spite of the extensive simulational work done on critical relaxation, the quality of the estimates of the dynamic exponent z is not nearly as high as that of the estimates for static exponents. The diverse techniques described above are simple in concept but complicated in their implementation. Nonetheless a reasonably good consensus is beginning to emerge for the two–dimensional Ising model between the 'best' estimates from Monte Carlo simulation, series expansion, and a recent, novel analysis based on variational approximations of the eigenstates of the Markov matrix describing heat-bath single spin-flip dynamics (Nightingale and Blöte, 1998).

4.3 OTHER DISCRETE VARIABLE MODELS

4.3.1 Ising models with competing interactions

The Ising model with nearest neighbor interactions has already been discussed several times in this book; it has long served as a testing ground for both new theoretical methods as well as new simulational techniques. When additional couplings are added the Ising model exhibits a rich variety of behavior which depends on the nature of the added interactions as well as the specific lattice structure. Perhaps the simplest complexity can be introduced by the addition of next-nearest neighbor interactions, J_{nnn}, which are of variable strength and sign so that the Hamiltonian becomes

$$\mathcal{H} = -J_{nn} \sum_{i,j} \sigma_i \sigma_j - J_{nnn} \sum_{i,k} \sigma_i \sigma_k - H \sum_i \sigma_i, \qquad (4.65)$$

where the first sum is over nearest neighbor pairs and the second sum over next-nearest pairs. It is straightforward to extend the single spin-flip Metropolis method to include J_{nnn}: the table of flipping probabilities becomes a two-dimensional array and one must sum separately over nearest and next-nearest neighbor sites in determining the flipping energy. In specialized cases where the magnitudes of the couplings are the same, one can continue to use a one-dimensional flipping probability array and simply include the contribution of the next-nearest neighbor site to the 'sum' of neighbors using the appropriate sign. If the checkerboard algorithm is being used, the next-nearest neighbor interaction will generally connect the sublattices; in this situation the system need merely be decomposed into a greater number of sublattices so that the spins on these new sublattices do not interact. An example is given below for the Ising square lattice.

Example

For the Ising square lattice with nearest neighbor coupling the simplest checkerboard decomposition is shown on the left. If next-nearest neighbor coupling is added the simplest possible checkerboard decomposition is shown on the right.

1	2	1	2		1	2	1	2
2	1	2	1		3	4	3	4
1	2	1	2		1	2	1	2
2	1	2	1		3	4	3	4

If both nearest and next-nearest neighbor interactions are ferromagnetic, the system will only undergo a transition to a ferromagnetic state and there are seldom complexities. (One simple case which may lead to difficulties is when there are only nearest neighbor interactions which are quite different in magnitude in different directions. This may then lead to a situation in which well ordered chains form at some relatively high temperature, and long range order sets in only at a much lower temperature. In this case it becomes very difficult for chains to overturn to reach the ground state because each individual spin in the chain is effectively 'held in place' by its neighbors. (Graim and Landau, 1981)) If, however, the couplings are both antiferromagnetic, or of opposite sign, there may be multiple configurations of quite similar free energy which are separated from each other by a significant free energy barrier. The resultant sequence of states may then also have a complicated time dependence. For the simple case of nearest neighbor, antiferromagnetic interactions only, below the transition temperature the system may alternate between two different states, one in which sublattice 1 is up and sublattice 2 is down, and one in which all spins are reversed. If a strong, antiferromagnetic next-nearest neighbor interaction is added it will be necessary to decompose the system into four interpenetrating, next-nearest neighbor sublattices, and there will be *four* different ordered states as shown below:

state	s.1.1	s.1.2	s.1.3	s.1.4
1	+	+	−	−
2	−	−	+	+
3	+	−	+	−
4	−	+	−	+

One important consequence of this behavior is that the relevant order parameter changes! For some range of couplings it is not immediately clear which kind of order will actually result and multiple order parameters (and their finite size behavior) must then be determined. Even if the simple antiferromagnetic states are lowest in free energy, the states shown above may be close in free energy and may appear due to fluctuations. The net result is that one must pay close attention to the symmetry of the states which are produced and to the resultant time dependence.

With the inclusion of third nearest neighbor interactions the number of different states which appear becomes larger still. Other, metastable domain states, also become prevalent. In Fig. 4.18 we show a number of different possible spin configurations for the Ising square lattice with competing inter-

Fig. 4.18 Ising square lattice with nn-, nnn-, and third nnn-couplings: (top) possible spin configurations; (bottom) phase diagrams for different ratios of $R = \mathcal{J}_{nnn}/\mathcal{J}_{nn}$ and $R' = \mathcal{J}_{3nn}/\mathcal{J}_{nn}$. From Landau and Binder (1985).

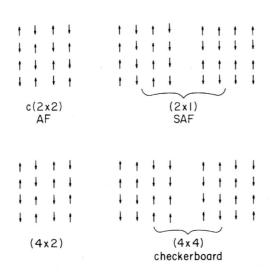

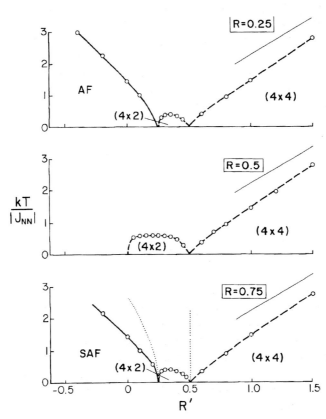

actions. In the bottom part of this figure we then show phase diagrams, deduced from Monte Carlo studies, for three different values of nnn-neighbor coupling as the 3nn-interaction is varied. For different regions of couplings, different states become lowest in free energy, and unit cells as large as 4×4 are needed to index them. When the interactions become complex, it may well be possible that entropic effects play a substantial role in determining which states actually appear. It may then be helpful to calculate multiple order parameters in order to determine which states are actually realized.

Interesting new physics may arise from competing interactions. In one of the simplest such examples, the addition of antiferromagnetic nnn-coupling to an nn–Ising square lattice antiferromagnet produces the degenerate 'super-antiferromagnetic' state described earlier with non-universal critical exponents (those of the XY-model with fourth order anisotropy). The order parameter must then be redefined to take into account the degeneracy of the ordered state, but the finite size analyses which were described in Section 4.2.3 of this chapter can still be applied. For example, the crossing of the fourth order cumulant still occurs but at a different value than for the simple Ising model. Monte Carlo data were used to determine the variation of the critical temperature as well as the change in critical exponents with coupling. In Fig. 4.19 we show the comparison between the Monte Carlo estimates for

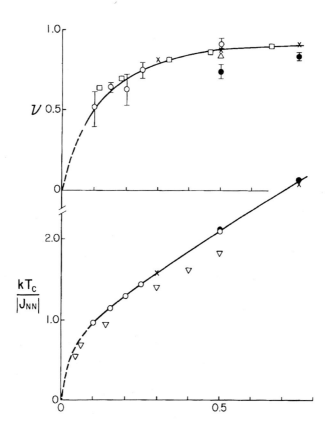

Fig. 4.19 Critical behavior for the superantiferromagnetic state in the Ising square lattice. ($\bigcirc$) Results of the Monte Carlo block distribution analysis; ($\triangle$) Monte Carlo results using finite size scaling; ($\times$) MCRG results; ($\bullet$) series expansion estimates; ($\square$) finite strip width RG; (∇) real space RG results. From Landau and Binder (1985).

T_c, as well as for ν, obtained from an analysis of the fourth order cumulant. For comparison, results obtained from a number of other methods are shown. Finite size scaling of the fourth order cumulant (block spin scaling) data showed quite clearly that the critical behavior was non-universal. This study is now rather old and higher resolution could be easily obtained with modern computing equipment; but even these data suffice to show the variation with coupling and to test other theoretical predictions. For a detailed study of the critical behavior of this model, see Landau and Binder (1985).

A very interesting case occurs when a competing antiferromagnetic interaction is added in only one lattice direction to an Ising ferromagnet to produce the so-called ANNNI model (Selke, 1992). For sufficiently strong antiferromagnetic interaction, the model exhibits a phase transition from the disordered phase to a 'modulated' phase in which the wavelength of the ordering is incommensurate with the lattice spacing! In $d = 2$ dimensions this phase is a 'floating phase' with zero order parameter and a power law decay of the correlation function; in $d = 3$ the ordered region contains a multitude of transitions to high-order commensurate phases, i.e. phases with order which has periods which are much larger than the lattice spacing. The detailed behavior of this model to date is still incompletely understood.

4.3.2 q-state Potts models

Another very important lattice model in statistical mechanics in which there are a discrete number of states at each site is the q-state Potts model (Potts, 1952) with Hamiltonian

$$\mathcal{H} = -\mathcal{J} \sum_{i,j} \delta_{\sigma_i \sigma_j} \tag{4.66}$$

where $\sigma_i = 1, 2, \ldots, q$. Thus a bond is formed between nearest neighbors o nly if they are in the same state. From the simulations perspective this model is also quite easy to simulate; the only complication is that now there are multiple choices for the new orientation to which the spin may 'flip'. The easiest way to proceed with a Monte Carlo simulation is to randomly choose one of the $q - 1$ other states using a random number generator and then to continue just as one did for the Ising model. Once again one can build a table of flipping probabilities, so the algorithm can be made quite efficient. Simple q-state Potts models on periodic lattices are known to have first order transitions for $q > 4$ in two dimensions and for $q > 2$ in three dimensions. For q close to the 'critical' values, however, the transitions become very weakly first order and it becomes quite difficult to distinguish the order of the transition without prior knowledge of the correct result. These difficulties are typical of those which arise at other weakly first order transitions; hence, Potts models serve as very useful testing grounds for new techniques.

Problem 4.7 Perform a Monte Carlo simulation of a $q = 3$ Potts model on a square lattice. Plot the internal energy as a function of temperature. Estimate the transition temperature.

Problem 4.8 Perform a Monte Carlo simulation of a $q = 10$ Potts model on a square lattice. Plot the internal energy as a function of temperature. Estimate the transition temperature. How do these results compare with those in Problem 4.7?

4.3.3 Baxter and Baxter–Wu models

Another class of simple lattice models with discrete states at each site involves multispin couplings between neighbors. One of the simplest examples is the Baxter model (1972) which involves Ising spins on two interpenetrating (next-nearest neighbor) sublattices on a square lattice; the two sublattices are coupled by a (nearest neighbor) four spin interaction so that the total Hamiltonian reads:

$$\mathcal{H} = -\mathcal{J}_{nnn} \sum_{i,k} \sigma_i \sigma_k - \mathcal{J}_{nnn} \sum_{j,l} \sigma_j \sigma_l - \mathcal{J}_{nn} \sum_{i,j,k,l} \sigma_i \sigma_j \sigma_k \sigma_l, \qquad (4.67)$$

where the first two sums are over nnn-pairs and the last sum is over nn-plaquettes. Once again, there are only a discrete number of possible states involving each site, i.e. the number of 'satisfied' next-nearest neighbor pairs and the number of four spin plaquettes, so that tables of flipping probabilities can be constructed. There are obviously multiple degenerate states because of the different possible orientations of each of the sublattices, so the order parameter must be carefully constructed. The critical behavior of the Baxter model is non-universal, i.e. it depends explicitly on the values of the coupling constants.

Another simple, discrete state lattice model with somewhat subtle microscopic behavior considers Ising spins on a triangular lattice with nearest neighbor three-spin coupling; the model, first proposed by Baxter and Wu (1973), has the Hamiltonian

$$\mathcal{H} = -\mathcal{J}_{nn} \sum_{i,j,k} \sigma_i \sigma_j \sigma_k. \qquad (4.68)$$

Even though the model is extremely simple, in a Monte Carlo simulation it has surprisingly complex behavior because different fluctuations occur at different time scales. The groundstate for this system is four-fold degenerate as shown in Fig. 4.20. This also means that the order parameter is complicated and that regions of the system may be in states which look quite different. If clusters of different ordered states 'touch' each other, a domain wall-like structure may be created with the result that the energy of the system is increased by an amount which depends upon the size of the overlap. The energy fluctuations then contain multiple kinds of excitations with different time scales, and care must be taken to insure that all characteristic

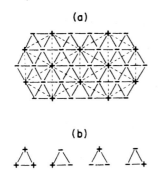

Fig. 4.20. Degenerate groundstates for the Baxter–Wu model: (a) ordered ferrimagnetic groundstate (solid lines connect nearest neighbors, dashed lines are between next-nearest neighbors); (b) elementary (nearest neighbor) plaquettes showing the four different degenerate groundstates.

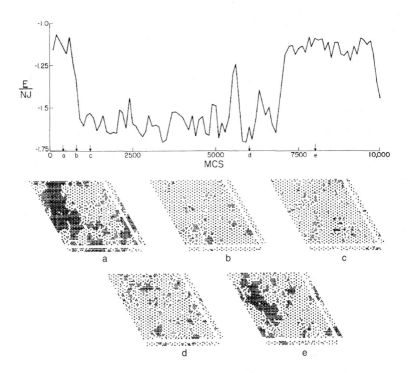

Fig. 4.21 Time dependence of the internal energy of the Baxter–Wu model and the development of 'domain-like states'. Periodic boundaries are copied from one side to another as shown in the lower portion of the figure. From Novotny and Landau (1981).

fluctuations are sampled. The correlation between the time dependence of the energy and the microscopic behavior is shown in Fig. 4.21 which clearly underscores the utility of even simple scientific visualization techniques to guide our understanding of numerical results. (These data are also rather old and using modern computers it is easy to make much longer runs; they nonetheless represent an example of complexity which may also occur in other systems.) This behavior also demonstrates the advantages of making occasional very long runs to test for unexpected behavior.

4.3.4. Clock models

Models with spins which may assume a continuous range of directions will be discussed in the next chapter, but a set of models which may be thought of as limiting cases of such continuous spin models with anisotropy in two dimensions are the so-called 'clock' models. In the q-state clock model the spins can only point in one of the q possible directions on a clock with q hours on it. The Hamiltonian then looks very much like that of a continuous spin model, but we must remember that the spins may only point in a discrete number of positions:

$$\mathcal{H} = -\mathcal{J} \sum_{i,j} S_i \cdot S_j. \tag{4.69}$$

As $q \to \infty$ the model becomes a continuous spin model. Just as in the case of a high spin Ising model, the number of possible nearest neighbor states can become quite large and the flip probability table can become big. Nonetheless the Monte Carlo algorithm proceeds as before, first using a random number to select a possible new state and then calculating the energy change which a 'flip' would produce. It can also be shown that for $q = 4$, the clock model becomes exactly identical to an Ising model with interaction $\mathcal{J}/2$, so the program can be tested by comparing with the known behavior for finite Ising models. For $q > 4$, the clock model becomes a limiting case for the XY-model with q-fold anisotropy. This model has two Kosterlitz–Thouless transitions and the interpretation of the data, and location of the transitions, becomes a quite subtle matter (Challa and Landau, 1986). It is possible to use a very large value of q to approximate a continuous spin XY-model and thus take advantage of the tricks that one can employ when dealing with a model with discrete states. One must not forget, however, that asymptotically near to the transition the difference between the two models becomes evident.

4.3.5 Ising spin glass models

The field of spin glasses has a voluminous literature and the reader is directed elsewhere for an in-depth coverage (see e.g. Binder and Young, 1986). Spin glasses are magnetic systems with competing interactions which result in frozen-in disorder reminiscent of that which occurs in ordinary glass. Thus, although there is no long range order, there will be short range order with a resultant cusp in the magnetic susceptibility. Below the spin glass temperature T_f there is hysteresis and a pronounced frequency dependence when a small field is applied. These effects arise because the geometry and/or interactions give rise to 'frustration', i.e. the inability of the system to find an ordered state which satisfies all interacting neighbors. One of the simplest spin glass models (with short range interactions) employs Ising spins σ_i with Hamiltonian

$$\mathcal{H} = -\sum_{i,j} \mathcal{J}_{ij} \sigma_i \sigma_j - H \sum_i \sigma_i, \tag{4.70}$$

where the distribution $P(\mathcal{J}_{ij})$ of 'exchange constants' $\mathcal{J}_{ij}$ is of the Edwards–Anderson form

$$P(\mathcal{J}_{ij}) = [2\pi((\Delta\mathcal{J}_{ij})^2)]^{-1/2} \exp\left[-(\mathcal{J}_{ij} - \bar{\mathcal{J}}_{ij})^2/2(\Delta\mathcal{J}_{ij})^2\right] \qquad (4.71)$$

or the $\pm\mathcal{J}$ form

$$P(\mathcal{J}_{ij}) = p_1\delta(\mathcal{J}_{ij} - \mathcal{J}) + p_2\delta(\mathcal{J}_{ij} + \mathcal{J}). \qquad (4.72)$$

Explicit distributions of bonds are placed on the system and Monte Carlo simulations can be performed using techniques outlined earlier; however, near the spin glass freezing temperature T_f and below the time scales become very long since there is a very complicated energy landscape and the process of moving between different 'local' minima becomes difficult. Of course, the final properties of the system must be computed as an average over multiple distributions on bonds. One complication which arises from spin glass behavior is that the spontaneous magnetization of the system is no longer a good order parameter. One alternative choice is the Edwards–Anderson parameter

$$q = \overline{\langle\sigma_i\rangle^2} \qquad (4.73)$$

where $\langle\cdots\rangle$ denotes the expectation value for a single distribution of bonds and the $\overline{\cdots}$ indicates an average over all bond distributions. Another choice is the local parameter

$$q = \frac{1}{N}\sum_i \sigma_i\varphi_i^l \qquad (4.74)$$

where ϕ_i^l represents the spin state of site i in the lth groundstate. The Monte Carlo simulations reveal extremely long relaxation times, and the data are often difficult to interpret. (For a recent view of the 'state of the art' see Young and Kawashima (1996).) In the next chapter we shall discuss improved methods for the study of spin glasses.

4.3.6 Complex fluid models

In this section we discuss briefly the application of Monte Carlo techniques to the study of microemulsions, which are examples of complex fluids. Microemulsions consist of mixtures of water, oil, and amphiphilic molecules and for varying concentrations of the constituents can form a large number of structures. These structures result because the amphiphilic molecules tend to spontaneous formation of water–oil interfaces (the hydrophilic part of the molecule being on the water-rich side and the hydrophobic part on the oil-rich side of the interface). These interfaces may then be arranged regularly (lamellar phases) or randomly (sponge phases), and other structures (e.g. vesicles) may form as well. Although real complex fluids are best treated using sophisticated off-lattice models, simplified, discrete state lattice models have been used quite successfully to study oil–water–amphiphilic systems (see, e.g. Gompper and Goos (1995)). Models studied include the Ising model with nn- and nnn-interaction and multispin interactions and the

Blume–Emery–Griffiths (BEG) model with three spin coupling. These models can be easily studied using the methods described earlier in this chapter, although because of the complicated structures which form, relaxation may be slow and the system may remain in metastable states. These systems have also been studied using a Ginzburg–Landau functional and spatial discretization. Thus the free energy functional

$$F\{\Phi\} = \int d^3 r \left(c(\nabla^2\Phi)^2 + g(\Phi)(\nabla\Phi)^2 + f(\Phi) - \mu\Phi \right) \qquad (4.75)$$

for a scalar order parameter Φ becomes

$$F(\Phi(r_{ij})) = c \sum_i \left(\sum_{k=1}^{3} \frac{\phi(\bar{X}_i + \hat{e}_k) - 2\phi(\bar{X}_i) + \phi(\bar{X}_i - \hat{e}_k)}{a_o^2} \right)^2$$

$$+ \sum_{ij} g\{\tfrac{1}{2}[\phi(X_i) + \phi(X_j)]\} \left[\frac{\phi(X_i) - \phi(X_j)}{a_o} \right]^2 + f(\phi) - \mu\phi$$

$$(4.76)$$

where a_o is the lattice constant and the $\hat{e}_k$s are the lattice vectors. Monte Carlo moves are made by considering changes in the local order parameter, i.e.

$$\Phi \rightarrow \Phi + \Delta\Phi \qquad (4.77)$$

with the usual Metropolis criterion applied to determine if the move is accepted or not. Monte Carlo simulations have been used to determine phase diagrams for this model as well as to calculate scattering intensities for neutron scattering experiments.

4.4 SPIN-EXCHANGE SAMPLING

4.4.1 Constant magnetization simulations

For the single spin-flipping simulations described above, there were no conserved quantities since both energy and order parameter could change at each flip. A modification of this approach in which the magnetization of the system remains constant may be easily implemented in the following fashion. Instead of considering a single spin which may change its orientation, one chooses a pair of spins and allows them to attempt to exchange positions. This 'spin-exchange' or Kawasaki method (Kawasaki, 1972) is almost as easy to implement as is spin-flipping. In its simplest form, spin-exchange involves nearest neighbor pairs, but this constraint is not compulsory. (If one is not interested in simulating the time dependence of a model for a physical system, it may even be advantageous to allow more distant neighbor interchanges.) One examines the interacting near neighbors of both spins in the pair and determines the change in energy if the spins are interchanged. This energy difference is then used in the acceptance procedure described above. Obviously, a pair of spins has a greater number of near neighbors than does a single spin, and even with nearest-neighbor coupling only a checkerboard decomposition requires more than two sublattices. Nonetheless, spin-exchange is straightforward to implement using table building and other tricks which can be used for spin-flip Monte Carlo. The behavior which results when this method is used is quite different from that which results using spin-flipping and will be discussed in the next several sections.

Problem 4.9 Simulate an $L = 10$ Ising square lattice using Kawasaki dynamics. Choose an initially random state and quench the system to $T = 2.0 J/k_B$. Plot the internal energy as a function of time. Make a 'snapshot' of the initial configuration and of the last configuration generated.

4.4.2 Phase separation

At a first order transition the system separates into two distinct regions, each of which is typical of one of the two co-existing phases. (The basic ideas have been introduced in Section 2.3.) If, for example, a disordered system is quenched from some high temperature to below the critical temperature, the disordered state becomes unstable. If this is done in an AB binary alloy in which the number of each kind of atom is fixed, phase separation will occur (Gunton et al., 1983). Because of the Ising-lattice gas-binary alloy equivalence, a Monte Carlo simulation can be carried out on an Ising model at fixed magnetization using spin-exchange dynamics. The structure factor $S(k, t)$ can be extracted from the Fourier transform of the resultant spin configurations and used to extract information about the nature of the phase separation. As a specific example we consider the physical situation described by Fig. 2.9 in which a binary alloy containing vacancies may evolve in time by the diffusion of atoms and vacancies. A vacancy site is chosen at random and it attempts to exchange position with one of its nearest neighbors. The probability of a jump which involves an energy change $\delta\mathcal{H}$ in which the vacancy exchanges site with an A-atom (B-atom) is denoted $W_A(W_B)$ and is given by

$$W_A = \begin{cases} \Gamma_A & \text{if } \delta\mathcal{H} < 0 \\ \Gamma_A \exp(-\delta\mathcal{H}/k_B T) & \text{if } \delta\mathcal{H} > 0 \end{cases} \tag{4.78}$$

$$W_B = \begin{cases} \Gamma_B & \text{if } \delta\mathcal{H} < 0 \\ \Gamma_B \exp(-\delta\mathcal{H}/k_B T) & \text{if } \delta\mathcal{H} > 0. \end{cases} \tag{4.79}$$

As an example of the results which are obtained from this Monte Carlo procedure we show characteristic results which are obtained for the structure factor for four different jump rates in Fig. 4.22. Data are shown for five different times following the quench and show the evolution of the system. For wavevectors that are small enough ($k < k_c$) the equal-time structure factor grows with time: this is the hallmark of spinodal decomposition (see Section 2.3.2). Another important property of the developing system which

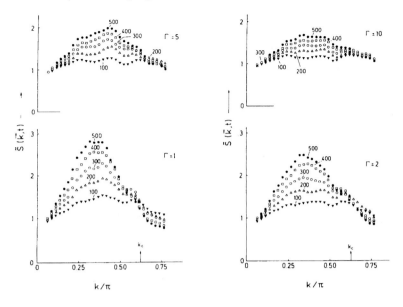

Fig. 4.22 Smoothed structure factor of an AB binary alloy with vacancies: $c_A = c_B = 0.48$, $c_V = 0.04$. From Yaldram and Binder (1991).

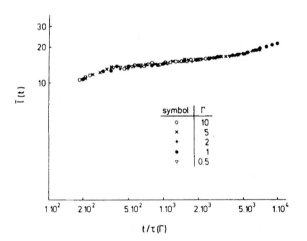

Fig. 4.23 Log–log plot of the mean cluster size vs. scaled time for phase separation in the AB binary alloy. From Yaldram and Binder (1991).

needs to be understood is the development of the mean cluster size $\bar{l}$ as a function of time where

$$\bar{l}(t) = \sum_{l \geq 10} l\, n_l(t) \bigg/ \sum_{l \geq 10} n_l(t) \qquad (4.80)$$

and n_l is the number of clusters of size l. In Fig. 4.23 we show the mean cluster size against the scaled time for five different values of the jump rate. The scaling time $\tau(\Gamma)$ not only describes the behavior of the mean cluster size but is also appropriate to describe the scaling of the internal energy.

4.4.3 Diffusion

In this section we consider lattice gas models which contain two species A
and B, as well as vacancies which we denote by the symbol V. The sum of the
concentrations of each species c_A, c_B, c_V is held fixed and the total of all the
components is unity, i.e. $c_A + c_B + c_V = 1$. In the simulations particles are
allowed to change positions under various conditions and several different
types of behavior result. (See Fig. 2.9 for a schematic representation of
interdiffusion in this model.)

First we consider non-interacting systems. In the simplest case there is
only one kind of particle in addition to vacancies, and the particles undergo
random exchanges with the vacancies. Some particles are tagged, i.e. they are
followed explicitly, and the resultant diffusion constant is given by

$$D_t = f_c V D_{sp}, \tag{4.81}$$

where D_{sp} is the single particle diffusion constant in an empty lattice, V is the
probability that a site adjacent to an occupied site is vacant, and f_c is the
(backwards) correlation factor which describes the tendency of a particle
which has exchanged with a vacancy to exchange again and return to its
original position. This correlation can, of course, be measured directly by
simulation. The process of interdiffusion of two species is a very common
process and has been studied in both alloys and polymer mixtures. By
expressing the free energy density f of the system in terms of three non-
trivial chemical potentials μ_A, μ_B, μ_V, i.e.

$$f = \mu_A c_A + \mu_B c_B + \mu_V c_V, \tag{4.82}$$

we can write a Gibbs–Duhem relation, valid for an isothermal process:

$$c_A d\mu_A + c_B d\mu_B + c_V d\mu_V = 0. \tag{4.83}$$

The conservation of species leads to continuity equations

$$\partial c_A/\partial t + \bar\nabla \cdot \bar j_A = 0; \quad \partial c_B/\partial t + \bar\nabla \cdot \bar j_B = 0; \quad \frac{\partial c_V}{\partial t} + \bar\nabla \cdot \bar j_V = 0. \tag{4.84}$$

The constitutive linear equations relating the current densities to the gradients of the chemical potentials are ($\beta = 1/k_B T$)

$$j_A = -\beta\lambda_{AA}\nabla\mu_A - \beta\lambda_{AB}\nabla\mu_B - \beta\lambda_{AV}\nabla\mu_V,$$
$$j_B = -\beta\lambda_{BA}\nabla\mu_A - \beta\lambda_{BB}\nabla\mu_B - \beta\lambda_{BV}\nabla\mu_V, \tag{4.85}$$
$$j_V = -\beta\lambda_{VA}\nabla\mu_A - \beta\lambda_{VB}\nabla\mu_B - \beta\lambda_{VV}\nabla\mu_V,$$

where the λ_{ij} are known as Onsager coefficients. The Onsager symmetry
relations reduce the number of independent parameters since
$\lambda_{AB} = \lambda_{BA}, \ldots$ and the conservation of the total number of 'particles' allows
us to eliminate the Onsager coefficients connected to the vacancies. The
remaining Onsager coefficients can be estimated from Monte Carlo simulations of their mobilities when forces acted on one of the species. In Fig. 4.24
we show a schematic view of how to set up a model. A combination of a
chemical potential gradient and judicious choice of boundary conditions

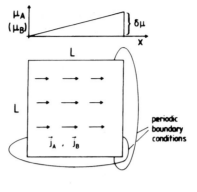

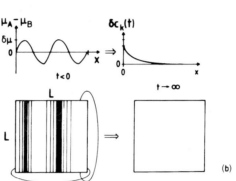

Fig. 4.24 An AB
binary alloy model for
the study of Onsager
coefficients. (a) A
linear gradient of the
chemical potential μ_A
(or μ_B respectively)
leading to steady-state
current. (b) Periodic
boundary variation of
the chemical potential
difference,
commensurate with
the linear dimension L
and leading to a
concentration wave $\delta(x)$
$= \delta c_k \exp(2\pi i x/\lambda)$.

allows us to measure currents and thus extract estimates for Onsager coefficients. (Note that a linear increase in the chemical potential with position is inconsistent with a static equilibrium in a box, because of the periodic boundary condition: particles leaving the box through the right wall reenter through the left wall.) For small enough $\delta\mu$ there is a linear relationship between chemical potential and the currents. Using the continuity equations together with the constitutive current expressions, we can extract coupled diffusion equations whose solutions yield decays which are governed by the Onsager coefficients. All three Onsager coefficients were successfully estimated for the non-interacting alloy (Kehr *et al.*, 1989). While the phenomenological description of diffusion in alloys as outlined above involves many unknown parameters, the obvious advantage of the simulation is that these parameters can be 'measured' in the simulation from their definition. Other scenarios may be studied by simulation. If a periodic variation of the chemical potential is created instead (see Fig. 4.24b), a concentration wave develops. Following the ideas of linear response theory, we 'shut off' this perturbation at $t = 0$, and simply watch the decay of the concentration with time. A decay proportional to $\exp(-D_{int}k^2t)$ where $k = 2\pi/\lambda$ allows us to determine the interdiffusion constant D_{int}.

Monte Carlo simulations were also used to study tracer diffusion in the binary alloy and no simple relationship was found to interdiffusion.

Diffusion can also be considered in interacting systems. Within the context of the Ising lattice gas model a particle can jump to a nn-vacancy site with probability

$$P(i \rightarrow l_i) = \exp(-\Delta E / k_B T), \tag{4.86}$$

where

$$\Delta E = \begin{cases} \varepsilon(l - z + 1) & \text{for repulsion } (\varepsilon < 0) \\ \varepsilon l & \text{for attraction } (\varepsilon > 0) \end{cases} \tag{4.87}$$

where z is the coordination number and l is the number of nn-particles in the initial state. Monte Carlo simulations were used to study both self-diffusion and collective diffusion as a function of the concentration of vacancies and of the state of order in the alloy (Kehr and Binder, 1984). Similarly, two-dimensional models of adsorbed monolayers can be considered and the self-diffusion and collective diffusion can be studied.

4.4.4 Hydrodynamic slowing down

The conservation of the concentration (or magnetization) during a simulation also has important consquences for the kinetics of fluctuations involving long length scales. If we consider some quantity A which has density ρ_A the appropriate continuity equation is

$$\frac{\partial \rho_A(x, t)}{\partial t} + \nabla j_A(x, t) = 0 \tag{4.88}$$

where j_A is a current density. Near equilibrium and for local changes of A, we may approximate the current by

$$\mathbf{j}_A(x, t) = D_{AA} \nabla a(x, t). \tag{4.89}$$

Taking the Fourier transform of Eqn. (4.80) and integrating we find

$$A(k, t) = A(k, \infty) + [A(k, 0) - A(k, \infty)] e^{-D_{AA} k^2 t}. \tag{4.90}$$

This equation exhibits 'hydrodynamic slowing down' with characteristic time $\tau_{AA}(k) = (D_{AA} k^2)^{-1}$. This argument justifies the result already discussed in Section 2.3.4. Thus, equilibrium will be approached quite slowly for all properties which describe long wavelength (i.e. small k) properties of the system.

4.5 MICROCANONICAL METHODS

4.5.1 Demon algorithm

In principle, a microcanonical method must work at perfectly constant energy. The demon algorithm first proposed by Creutz (1983) is not strictly microcanonical, but for large systems the difference becomes quite small. The procedure is quite simple. One begins by choosing some initial state. A 'demon' then proceeds through the lattice, attempting to flip each spin in turn and either collecting energy given off by a spin-flip or providing the energy needed to enable a spin-flip. The demon has a bag which can contain a maximum amount of energy, so that if the capacity of the bag is reached no spin-flip is allowed which gives off energy. On the other hand, if the bag is empty, no flip is possible that requires energy input. Thus, the energy in the bag E_D will vary with time, and the mean value of the energy stored in the bag can be used to estimate the mean value of the temperature during the course of the simulation,

$$K = \tfrac{1}{4}\ln(1 + 4\mathcal{J}/\langle E_D \rangle). \qquad (4.91)$$

If the bag is too big, the simulation deviates substantially from the microcanonical condition; if the bag is too small, it becomes unduly difficult to produce spin-flips. Note that once the initial state is chosen, the method becomes deterministic!

Problem 4.10 Simulate an $L = 10$ Ising square lattice using the microcanonical 'demon' method at two different values of energy E and estimate the temperatures. Carry out canonical ensemble simulations at these temperatures and compare the values of energy with your initial choices of E.

4.5.2 Dynamic ensemble

This method uses a standard Monte Carlo method for a system coupled to a suitably chosen finite bath (Hüller, 1993). We consider an N-particle system with energy E coupled to a finite reservoir which is an ideal gas with M degrees of freedom and kinetic energy k. One then studies the microcanonical ensemble of the total, coupled system with fixed total energy G. An analysis of detailed balance shows that the ratio of the transition probabilities between two states is then

$$\frac{W_{b \to a}}{W_{a \to b}} = (G - E_a)^{\frac{N-2}{N}}/(G - E_b)^{\frac{N-2}{N}} \approx e^{-\zeta(E_b - E_a)} \qquad (4.92)$$

where $\zeta = (N - 2)/2Nk_b$ where $k_b = (G - E_b)/N$ is the mean kinetic energy per particle in the bath. The only difference in the Monte Carlo method is that the effective temperature ζ is adjusted dynamically during the course of the simulation. Data are then obtained by computing the mean value of the energy on the spin system $\langle E \rangle$ and the mean value of the temperature from $\langle k_b \rangle$. This method becomes accurate in the limit of large system size. Plots of E vs. T then trace out the complete 'van der Waals loop' at a first order phase transition.

4.5.3 Q2R

The Q2R cellular automaton has been proposed as an alternative, microcanonical method for studying the Ising model. In a cellular automaton model the state of each spin in the system at each time step is determined completely by consideration of its near neighbors at the previous time step. The Q2R rule states that a spin is flipped if and only if half of its nearest neighbors are up and half down. Thus, the local (and global) energy change is zero. A starting spin configuration of a given energy must first be chosen and then the Q2R rule applied to all spins; this method is thus also deterministic after the initial state is chosen. Thermodynamic properties are generally well reproduced although the susceptibility below T_c is too low. (Other cellular automata will be discussed in Chapter 8.)

Problem 4.11 Simulate an $L = 10$, $q = 10$ Potts model square lattice using a microcanonical method and estimate the transition temperature. How does your answer compare with that obtained in Problem 4.8?

4.6 GENERAL REMARKS, CHOICE OF ENSEMBLE

We have already indicated how models may be studied in different ensembles by different methods. There are sometimes advantages in using one ensemble over the other. In some cases there may be computational advantages to choosing a particular ensemble, in other situations there may be a symmetry which can be exploited in one ensemble as opposed to the other. One of the simplest cases is the study of a phase diagram of a system with a tricritical point. Here there are both first order and second order transitions. As shown in Fig. 4.25 the phase boundaries look quite different when shown in the canonical and grand-canonical ensembles. Thus, for low 'density' (or magnetization in magnetic language) two phase transitions are encountered as the temperature is increased whereas if the 'field' is kept fixed as the temperature is swept only a single transition is found. Of course, to trace out the energy–field relation in the region where it is double valued, it is preferable to use a microcanonical ensemble (as was described in the previous section) or even other ensembles, e.g. a Gaussian ensemble (Challa and Hetherington, 1988).

4.7 STATICS AND DYNAMICS OF POLYMER MODELS ON LATTICES

4.7.1 Background

Real polymers are quite complex and their simulation is a daunting task (Binder, 1995). There are a number of physically realistic approximations which can be made, however, and these enable us to construct far simpler models which (hopefully) have fundamentally the same behavior. First we recognize that the bond lengths of polymers tend to be rather fixed as do bond angles. Thus, as a more computationally friendly model we may construct a 'polymer' which is made up of bonds which connect nearest neighbor

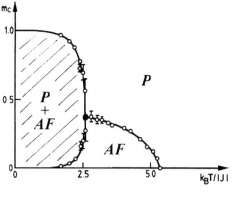

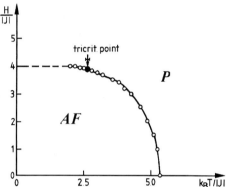

Fig. 4.25 Phase diagram for an Ising antiferromagnet with nearest and next-nearest neighbor couplings with a tricritical point: (top) grand canonical ensemble, the shaded area is a region of two-phase coexistence; (bottom) canonical ensemble.

sites (monomers) on a lattice and which obey an excluded volume constraint. The sites and bonds on the lattice do not represent individual atoms and molecular bonds but are rather the building blocks for a coarse-grained model. Even within this simplified view of the physical situation simulations can become quite complicated since the chains may wind up in very entangled states in which further movement is almost impossible.

4.7.2 Fixed length bond methods

The polymer model just described may be viewed as basically a form of self-avoiding-walk (SAW) which can be treated using Monte Carlo growth algorithms which have already been discussed (see Section 3.6.3). Another class of algorithms are dynamic in nature and allow random moves of parts of the polymer which do not allow any change in the length of a bond connecting two monomers. The range of possible configurations for a given polymer model can be explored using a variety of different 'dynamic' Monte Carlo algorithms which involve different kinds of move, three examples of which are shown in Fig. 4.26. In the generalized 'kink-jump' method single sites may be moved, obeying the restriction that no bond length changes. In the 'slithering snake' (reptation) method, a bond is removed from one end and then glued to the other end of the polymer in a randomly chosen orientation.

Fig. 4.26 Dynamic Monte Carlo algorithms for SAWs on a simple cubic lattice: (a) generalized Verdier–Stockmayer algorithm; (b) slithering snake algorithm; (c) pivot algorithm.

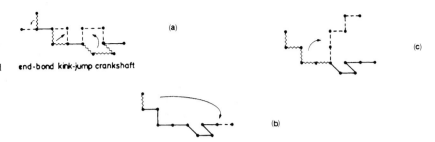

end-bond kink-jump crankshaft

In Fig. 4.26c we show the pivot ('wiggle') move in which a large part of the chain is rotated about a single site in the chain. (Obviously, not all moves reflect real, physical time development.) Different kinds of moves are useful for avoiding different kinds of 'trapped' configurations, and an intelligent choice of trial moves is essential in many cases. There are a large number of off-lattice models which are useful for studying more complex behavior, but these are beyond the scope of consideration here. More details about the methods shown in Fig. 4.26 can be found in Kremer and Binder (1988) and additional methods are discussed by Sokal (1995).

4.7.3 Bond fluctuation method

A very powerful 'dynamic' method which relaxes the rigid bond constraint slightly employs the 'bond fluctuation' model (Carmesin and Kremer, 1988). In this approach a monomer now occupies a nearest neighbor plaquette and attempts to move randomly by an amount which does not stretch or compress the bonds to its neighbors too much, and in the process to expand the range of configuration space which can be explored. Note that these moves may also allow some change in the bond angle as well as bond length. The excluded volume constraint is obeyed by not allowing overlap of monomer plaquettes. Examples of possible moves are shown in Fig. 4.27. At each step a randomly

---

Bond fluctuation Monte Carlo method

(1) Choose an initial state
(2) Randomly choose a monomer
(3) Randomly choose a 'plaquette' (from among the allowed possibilities) to which a move will be attempted
(4) Check the excluded volume and bond length restrictions; if these are violated return to step (2)
(5) Calculate the energy change ΔE which results if the move is accepted
(6) Generate a random number r such that $0 < r < 1$
(7) If $r < \exp(-\Delta E/k_B T)$, accept the move
(8) Choose another monomer and go to (3)

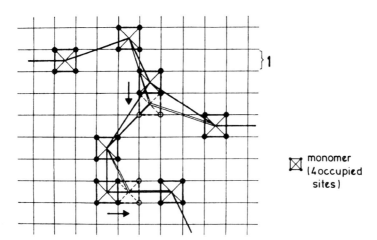

Fig. 4.27 Sample moves for the Bond fluctuation algorithm on a square lattice.

chosen monomer moves to a randomly chosen plaquette subject to excluded volume constraints as well as the limitations on bond length mentioned above. The bond fluctuation method can be effective in getting the system out of 'blocking' configurations and, as shown in Fig. 4.27, can also be applied to lattice model branched polymers.

4.7.4 Polymers in solutions of variable quality: θ-point, collapse transition, unmixing

So far the only interaction between monomers that are not nearest neighbors along the chain, is the (infinitely strong) repulsive excluded volume interaction. Obviously, this is an extremely simplified view of the actual interactions between the effective monomers that form a real macromolecule. Physically, this corresponds to the 'athermal' limit of a polymer chain in a good solvent: the solvent molecules do not show up explicitly in the simulation, they are just represented by the vacant sites of the lattice.

Given the fact that interactions between real molecules or atoms in fluids can be modeled rather well by the Lennard-Jones interaction, which is strongly repulsive at short distances and weakly attractive at somewhat longer distances, it is tempting to associate the above excluded volume interaction (incorporated both in the SAW and the bond fluctuation model) with the repulsive part of the Lennard-Jones interaction, and add an attractive energy which acts at somewhat longer distances, to represent the attractive part of the Lennard-Jones interaction. The simplest choice for the SAW model is to allow for an energy, ε, if a pair of monomers (which are not nearest neighbors along the chain) occupy nearest neighbor sites on the lattice. In fact, such models can be (and have been!) studied by simple sampling Monte Carlo methods as described in Chapter 3. To do this one simply has to weigh each generated SAW configuration with a weight proportional to the Boltzmann factor $\exp(n\varepsilon/k_B T)$, n being the number of such nearest neighbor contacts in each configuration. However, the problem of generating a sufficiently large statistical sample for long chains is now even worse than in the athermal case: we have seen that the success rate to construct a SAW from unbiased growth scales as $\exp(-\text{const.}N)$, for chains of N steps, and actually a very small fraction of these successfully generated walks will have a large Boltzmann weight. Therefore, for such problems, the 'dynamic' Monte Carlo methods treated in the present chapter are clearly preferred.

While in the case of the pure excluded volume interaction the acceptance probability is either one (if the excluded volume constraint is satisfied for the trial move) or zero (if it is not), we now have to compute for every trial move the change in energy $\Delta E = \Delta n\varepsilon$ due to the change Δn in the number of nearest neighbor contacts due to the move. This energy change has to be used in the acceptance probability according to the Metropolis method in the usual way, for all trial moves that satisfy the excluded volume constraint. This is completely analogous to the Monte Carlo simulation of the Ising model or other lattice models discussed in this book.

Of course, it is possible to choose interaction energies that are more complicated than just nearest neighbor. In fact, for the bond fluctuation model discussed above it is quite natural to choose an attractive interaction of somewhat longer range, since the length of an effective bond (remember that this length is in between 2 and $\sqrt{10}$ lattice spacings in $d = 3$ dimensions) already creates an intermediate length scale. One then wishes to define the range of the attractive interaction such that in a dense melt (where 50% or more of the available lattice sites are taken by the corners of the cubes representing the effective monomers) an effective monomer interacts with all nearest neighbor effective monomers that surround it. This consideration leads to the choice (e.g. Wilding et al., 1996) that effective monomers experience an energy ε if their distance r is in the range $2 \leq r \leq \sqrt{6}$ and zero else. In the bond fluctuation algorithm quoted above, the presence of some energy parameters such as ε was already assumed.

What physical problems can one describe with these models? Remember that one typically does not have in mind to simulate a macromolecule in vacuum but rather in dilute solution, so the vacant sites of the lattice represent the small solvent molecules, and hence ε really represents a difference in interactions $(\varepsilon_{\text{mm}} + \varepsilon_{\text{ss}})/2 - \varepsilon_{\text{ms}}$ where ε_{mm}, ε_{ss}, ε_{ms} stand for interactions between pairs of monomers (mm), solvent (ss) and monomer–solvent (ms), respectively. In this sense, the model is really a generalization of the ordinary lattice model for binary alloys (A, B), where one species (A) is now a much more complicated object, taking many lattice sites and exhibiting internal configurational degrees of freedom. Thus already the dilute limit is non-trivial, unlike the atomic binary mixture where both species (A, B) take a lattice site and only the concentrated mixture is of interest. Changing the parameter $\varepsilon/k_B T$ then amounts to changing the quality of the solvent: the larger $\varepsilon/k_B T$ the more the polymer coil contracts, and thus the mean square radius of gyration $\langle R_{\text{gyr}}^2 \rangle_{N,T}$ is a monotonically decreasing function when $\varepsilon/k_B T$ increases. Although this function is smooth and non-singular for any finite N, a singularity develops when the chain length N diverges: for all temperatures T exceeding the so-called 'theta temperature', θ, we then have the same scaling law as for the SAW, $\langle R^2_{\mathrm{gyr}} \rangle_{N,T} = A(T)N^{2\nu}$ with $\nu \approx 0.588$, only the amplitude factor $A(T)$ depends on temperature, while the exponent does not. However, for $T = \theta$ the macromolecule behaves like a simple random walk, $\langle R^2_{\mathrm{gyr}} \rangle_{N,T} = A'(\theta)N$ (ignoring logarithmic corrections), and for $T < \theta$ the chain configurations are compact, $\langle R^2_{\mathrm{gyr}} \rangle_{N,T} = A''(\theta)N^{2/3}$. This singular behavior of a single chain is called the 'collapse transition' (generalizations of this simple model also are devised for biopolymers, where one typically has a sequence formed from different kinds of monomers, such as proteins where the sequence carries the information about the genetic code).

Now we have to add a warning to the reader: just as power laws near a critical point are only observed sufficiently close, also the power laws quoted above are only seen for $N \to \infty$; in particularly close to θ one has to deal with 'crossover' problems: for a wide range of N for T slightly above θ the chain already behaves classically, $\langle R^2_{\mathrm{gyr}} \rangle \propto N$, and only for very large N does one have a chance to detect the correct asymptotic exponent. In fact, the θ-point can be related to tricritical points in ferromagnetic systems (DeGennes, 1979). Thus the Monte Carlo study of this problem is quite difficult and has a long history. Now it is possible to simulate chains typically for N of the order of 10^4, or even longer, and the behavior quoted above has been nicely verified, both for linear polymers and for star polymers (Zifferer, 1999). A combination of all three algorithms shown in Fig. 4.26 is used there.

The simulation of single chains is appropriate for polymer solutions only when the solution is so dilute that the probability that different chains interact is negligible. However, a very interesting problem results when only the concentration of monomers is very small (so most lattice sites are still vacant) but typically the different polymer coils already strongly penetrate each other. This case is called the 'semidilute' concentration regime (DeGennes, 1979). For good solvent conditions, excluded volume interactions are screened at large distances, and the gyration radius again scales classically, $\langle R^2_{\mathrm{gyr}} \rangle_{N,T} = A(T, \phi)N$, where ϕ is the volume fraction of occupied lattice sites. While the moves of types (a) and (b) in Fig. 4.26 are still applicable, the acceptance probability of pivot moves (type c) is extremely small, and hence this algorithm is no longer useful. In fact, the study of this problem is far less well developed than that of single polymer chains, and the development of better algorithms is still an active area of research (see e.g. the discussion of the configurational bias Monte Carlo algorithm in Chapter 6 below). Thus, only chain lengths up to a few hundred are accessible in such many-chain simulations.

When the solvent quality deteriorates, one encounters a critical point $T_c(N)$ such that for $T < T_c(N)$ the polymer solution separates into two phases: a very dilute phase ($\phi_I(T) \to 0$) of collapsed chains, and a semidilute phase ($\phi_{II}(T) \to 1$ as $T \to 0$) of chains that obey Gaussian statistics at larger distances. It has been a longstanding problem to understand how the critical concentration $\phi_c(N) (= \phi_I(T_c) = \phi_{II}(T_c))$ scales with chain length N, as well as how $T_c(N)$ merges with θ as $N \to \infty$, $\phi_c(N) \propto N^{-x}$, $\theta - T_c(N) \propto N^{-y}$, where x, y are some exponents (Wilding *et al.*, 1996). A study of this problem is carried out best in the grand-canonical ensemble (see Chapter 6), and near $T_c(N)$ one has to deal with finite size rounding of the transition, very similar to the finite size effects that we have encountered for the Ising model!

This problem of phase separation in polymer solutions is just one problem out of a whole class of many-chain problems, where the 'technology' of an efficient simulation of configurations of lattice models for polymer chains and the finite size scaling 'technology' to analyze critical phenomena and phase coexistence need to be combined in order to obtain most useful results. One other example, the phase diagram of 'equilibrium polymers', will now be described in more detail below.

4.7.5 Equilibrium polymers: a case study

Systems in which polymerization is believed to take place under conditions of chemical equilibrium between the polymers and their respective monomers are termed 'living polymers'. These are long linear-chain macromolecules that can break and recombine, e.g. liquid and polymer-like micelles. (In fact, in the chemistry community the phrase 'living polymers' is applied to radical initiated growth, or scission, that can occur only at one end of the polymer. In the model presented here, these processes can occur any place along the polymer chain. These systems are sometimes now referred to as 'equilibrium polymers'.) In order to study living polymers in solutions, one should model the system using the dilute $n \to 0$ magnet model (Wheeler and Pfeuty, 1981); however, theoretical solution presently exists only within the mean field approximation (Flory, 1953). For semiflexible chains Flory's model predicts a first order phase transition between a low temperature ordered state of stiff parallel rods and a high temperature disordered state due to disorientation of the chains.

Simulating the behavior of a system of living polymers is extremely difficult using a description which retains the integrity of chains as they move because the dynamics becomes quite slow except in very dilute solutions. An alternative model for living polymers, which is described in more detail elsewhere (Milchev, 1993), maps the system onto a model which can be treated more easily. Consider regular L^d hypercubic lattices with periodic boundary conditions and lattice sites which may either be empty or occupied by a (bifunctional) monomer with two strong (covalent) 'dangling' bonds, pointing along separate lattice directions. Monomers fuse when dangling bonds of nearest-neighbor monomers point toward one another, releasing energy $v > 0$ and forming the backbone of self-avoiding polymer chains (no crossing at vertices). Right-angle bends, which ensure the semiflexibility of such chains, are assigned an additional activation energy $\sigma > 0$ in order to include the inequivalence between rotational isomeric states (e.g. *trans* and *gauche*) found in real polymers. The third energetic parameter, w, from weak (van der

Waals) *inter* chain interactions, is responsible for the phase separation of the system into dense and sparse phases when T and/or μ are changed. w is thus the work for creation of empty lattice sites (holes) in the system. One can define $q = 7$ possible states, S_i, of a monomer i on a two–dimensional lattice (two straight 'stiff' junctions, $S_i = 1, 2$, four bends, $S_i = 3, \ldots, 6$, and a hole $S_i = 7$), and $q = 216$ monomer states in a simple cubic lattice. The advantage of this model is that it can be mapped onto an unusual q-state Potts model and the simulation can then be carried out using standard single spin-flip methods in this representation! The Hamiltonian for the model can be written:

$$\mathcal{H} = \sum_{i<j} \mathcal{F}_{ij} n(S_i) n(S_j) - \sum_i (\mu + \varepsilon) n(S_i), \qquad (4.93)$$

where $n(S_i) = 1$ for $i = 1, 2, \ldots, 6$, and $n(S_i) = 0$ (a hole) for $i = 7$ in two dimensions. Note that the interaction constant depends on the mutual position of the nearest neighbor monomer states, $\mathcal{F}_{ij} \neq \mathcal{F}_{ji}$. Thus, for example, $\mathcal{F}_{13} = -w$ whereas $\mathcal{F}_{31} = -v$. The local energies $\varepsilon_i = \sigma$ for the bends, and $\varepsilon_i = 0$ for the *trans* segments. The mapping to the different Potts states is shown in Fig. 4.28. The ground states of this model depend on the relative strengths of v, w and σ; long chains at low temperature are energetically favored only if $v/w > 1$. This model may then be simulated using single 'spin flip' methods which have already been discussed; thus the polymers may break apart or combine quite easily. (The resultant behavior will also give the correct static properties of a polydisperse solution of 'normal' polymers, but the time development will obviously be incorrect.) Even using the Potts model mapping, equilibration can be a problem for large systems so studies have been restricted to modest size lattices. An orientational order parameter must be computed: in two dimensions $\psi = \langle c_1 - c_2 \rangle$ (c_i is the concentration of segments in the ith state) where c_1 and c_2 are the fractions of stiff *trans* segments pointing horizontally and vertically on the square lattice. In $d = 3$ there are many more states than are shown in the figure, which is only for $d = 2$, and we do not list these explicitly here. In $d = 3$ then, the order parameter is defined as $\psi = \sqrt{(c_1 - c_2)^2 + (c_1 - c_8)^2 + (c_2 - c_8)^2}$, and c_1, c_2, c_8 are the fractions of *trans* bonds pointing in the x, y and z directions.

For two dimensions at $T = 0$, the lattice is completely empty below $\mu_c = -(v + w)$. Finite temperature phase transitions were found from the simulation data and, as an example, the resultant phase diagram for $v = 2.0$, $w =$

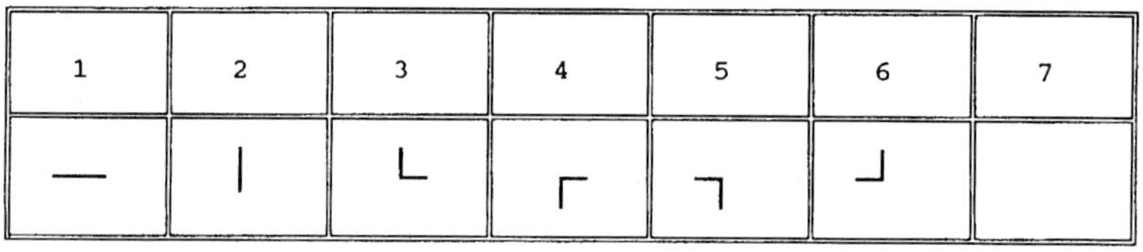

1	2	3	4	5	6	7
—	|	L	⌐	⌐	⌐	

Fig. 4.28 Different allowed monomer bond states and their Potts representation.

0.1 is shown in Fig. 4.28 for two different values of σ. In both cases the transition is first order at low temperatures, but above a tricritical point $T_t = 0.3$, it becomes second order. While for $\mu < \mu_c$ the density is quite high in both the ordered phase as well as the high temperature disordered phase, for $\mu < \mu_c$ the lattice is virtually empty below a temperature (the Lifshitz line) at which a rather steep (but finite) increase in θ is accompanied by pronounced maxima in the second derivatives of the thermodynamic potentials. A finite size scaling analysis along the second order portion of the boundary indicates critical behavior consistent with that of the two-dimensional Ising model. Fig. 4.29 shows the phase diagram in θ–T space; the first order portion of the phase boundary has opened up into a large coexistence region leaving only a relatively small area of the pure ordered phase. Figure 4.29c shows that as the chains become stiffer, T_c rises monotonically.

On a simple cubic lattice the groundstate is triply degenerate with parallel rods pointing along any of the three Cartesian axes. Moreover, a sort of a smectic ordered state with planes of differently oriented parallel rigid chains will be formed at low temperature if the *inter* chain interaction, w, between nearest neighbor monomers does not differentiate between pairs of rods which are parallel (in plane) or which cross at right angles when they belong to neighboring planes. Viewing these bonds as rough substitutes for the integral effect of longer range interactions, one could assume that the ws in both cases would differ so that in the former case (parallel rods) $w_{\parallel}$ is somewhat stronger than the latter one, $w_{\perp}$. Such an assumption leads to a groundstate consisting only of stiff chains, parallel to one of the three axes, whereby the order parameter in three dimensions attains a value of unity in the ordered state. A finite size scaling analysis of data for both $w_{\perp} \neq w_{\parallel}$ and $w_{\perp} = w_{\parallel}$ showed that the transition was first order.

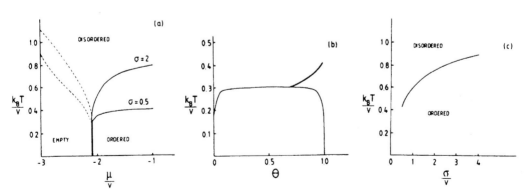

Fig. 4.29 Phase diagram of the two-dimensional system of living polymers for $v = 2.0$, $w = 0.1$: (a) T_c vs. chemical potential μ for two values of the rigidity parameter σ. The single line indicates a second order phase transition, the double line denotes a first order transition, and dots mark the Lifshitz line. (b) T_c as a function of coverage θ for $\sigma = 0.5$. (c) Variation of T_c with σ for $\mu = -1.4$. From Milchev and Landau (1995).

4.8 SOME ADVICE

We end this chapter by summarizing a few procedures which in our experience can be useful for reducing errors and making simulations studies more effective. These thoughts are quite general and widely applicable. While these 'rules' provide no 'money-back' guarantee that the results will be correct, they do provide a prudent guideline of steps to follow.

(1) In the very beginning, *think*!

What problem do you really want to solve and what method and strategy is best suited to the study. You may not always choose the best approach to begin with, but a little thought may reduce the number of false starts.

(2) In the beginning think small!

Work with small lattices and short runs. This is useful for obtaining rapid turnaround of results and for checking the correctness of a program. This also allows us to search rather rapidly through a wide range of parameter space to determine ranges with physically interesting behavior.

(3) Test the random number generator!

Find some limiting cases where accurate, or exact values of certain properties can be calculated, and compare your results of your algorithm with different random number sequences and/or different random number generators.

(4) Look at systematic variations with system size and run length!

Use a wide range of sizes and run lengths and then use scaling forms to analyze data.

(5) Calculate error bars!

Search for and estimate both statistical and systematic errors. This enables both you and other researchers to evaluate the correctness of the conclusions which are drawn from the data.

(6) Make a few very long runs!

Do this to ensure that there is not some hidden time scale which is much longer than anticipated.

REFERENCES

Baxter, R. J. (1972), Ann. Phys. (N.Y.) **70**, 193.

Baxter, R. J. and Wu, F. Y. (1973), Phys. Rev. Lett. **31**, 1294.

Binder, K. (1981), Z. Phys. B **43**, 119.

Binder, K. (1983), in *Phase Transitions and Critical Phenomena*, vol. 8, eds. C. Domb and J. L. Lebowitz (Academic Press, London) p. 1.

Binder, K. (1987), Rep. Prog. Phys. **50**, 783.

Binder, K. (1992), in *Computational Methods in Field Theory*, eds. C. B. Lang and H. Gausterer (Springer, Berlin).

Binder, K. (ed.) (1995), *Monte Carlo and Molecular Dynamics Simulations in*

Polymer Science (Oxford University Press, New York).

Binder, K. and Hohenberg, P. C. (1974), Phys. Rev. B **9**, 2194.

Binder, K. and Landau, D. P. (1984), Phys. Rev. B **30**, 1477.

Binder, K. and Müller-Krumbhaar, H. (1973), Phys. Rev. B **7**, 3297.

Binder, K. and Young, A. P. (1986), Rev. Mod. Phys. **58**, 801.

Borgs, C. and Kotecký, R. (1990), J. Stat. Phys. **61**, 79.

Caillol, J. M. (1993), J. Chem. Phys. **99**, 8953.

Carmesin, I. and Kremer, K. (1988), Macromolecules **21**, 2878.

Challa, M. S. S. and Hetherington, J. H. (1988), in *Computer Simulation Studies in Condensed Matter Physics I*, eds. D. P. Landau, K. K. Mon and H.-B. Schüttler (Springer, Heidelberg).

Challa, M. S. S. and Landau, D. P. (1986), Phys. Rev. B **33**, 437.

Challa, M. S. S., Landau, D. P., and Binder, K. (1986), Phys. Rev. B **34**, 1841.

Creutz, M. (1983), Phys. Rev. Lett. **50**, 1411.

de Gennes, P. G. (1979), in *Scaling Concepts in Polymer Physics* (Cornell University Press, Ithaca) Chapter 1.

Ferrenberg, A. M. and Landau, D. P. (1991), Phys. Rev. B **44**, 5081.

Ferrenberg, A. M., Landau, D. P., and Binder, K. (1991), J. Stat. Phys. **63**, 867.

Fisher, M. E. (1971), in *Critical Phenomena*, ed. M. S. Green (Academic Press, London) p. 1.

Flory, P. J. (1953), *Principles of Polymer Chemistry* (Cornell University Press, Ithaca, New York).

Glauber, R. J. (1963), J. Math. Phys. **4**, 294.

Gompper, G. and Goos, J. (1995), in *Annual Reviews of Computational Physics II*, ed. D. Stauffer (World Scientific, Singapore) p. 101.

Graim, T. and Landau, D. P. (1981), Phys. Rev. B **24**, 5156.

Gunton, J. D., San Miguel, M., and Sahni, P. S. (1983), in *Phase Transitions and Critical Phenomena*, vol. 8, eds. C. Domb and J. L. Lebowitz (Academic Press, London) p. 267.

Hohenberg, P. C. and Halperin, B. I. (1977), Rev. Mod. Phys. **49**, 435.

Hüller, A. (1993), Z. Phys. B **90**, 207.

Ito, N. (1993), Physica A **196**, 591.

Janssen, H. K., Schaub, B., and Schmittmann, B. (1989), Z. Phys. B **73**, 539.

Jasnow, D. (1984), Rep. Prog. Phys. **47**, 1059.

Kawamura, H. and Kikuchi, M. (1993), Phys. Rev. B **47**, 1134.

Kawasaki, K. (1972), in *Phase Transitions and Critical Phenomena*, vol. 2, eds. C. Domb and M. S. Green (Academic Press, London).

Kehr, K. W. and Binder, K. (1984), in *Applications of the Monte Carlo Method in Statistical Physics*, ed. K. Binder (Springer, Heidelberg).

Kehr, K. W., Reulein, S., and Binder, K. (1989), Phys. Rev. B **39**, 4891.

Kikuchi, M. and Ito, N. (1993), J. Phys. Soc. Japan **62**, 3052.

Koch, W. and Dohm, V. (1998), Phys. Rev. E **58**, 1179.

Koch, W., Dohm, V., and Stauffer, D. (1996), Phys. Rev. Lett. **77**, 1789.

Kremer, K. and Binder, K. (1988), Computer Phys. Rep. 7, 261.

Landau, D. P. (1976), Phys. Rev. B **13**, 2997.

Landau, D. P. (1996), in *Monte Carlo and Molecular Dynamics of Condensed Matter Systems*, eds. K. Binder and G. Ciccotti (Societa Italiana di Fisica, Bologna).

Landau, D. P. and Binder, K. (1985), Phys. Rev. B **31**, 5946.

Landau, D. P. and Binder, K. (1990), Phys. Rev. B **41**, 4633.

Landau, D. P., Tang, S., and Wansleben, S. (1988), J. de Physique **49**, C8-1525.

Li, Z. B., Ritschel, U., and Zhang, B. (1994), J. Phys. A: Math. Gen. **27**, L837.

Liu, A. J. and Fisher, M. E. (1990), J. Stat. Phys. **58**, 431.

Metropolis, N., Rosenbluth, A. W., Rosenbluth, M. N., Teller, A. M., and Teller, E. (1953), J. Chem Phys. **21**, 1087.

Milchev, A. (1993), Polymer **34**, 362.

Milchev, A. and Landau, D. P. (1995), Phys. Rev. E **52**, 6431.

Milchev, A., Binder, K., and Heermann, D. W. (1986), Z. Phys. B **63**, 527.

Müller-Krumbhaar, H. and Binder, K. (1973), J. Stat. Phys. **8**, 1.

Nightingale, M. P. and Blöte, H. W. J. (1998), Phys. Rev. Lett. **80**, 1007.

Novotny, M. A. and Landau, D. P. (1981), Phys. Rev. B **24**, 1468.

Onsager, L. (1944), Phys. Rev. **65**, 117.

Parry, A. D. and Evans, R. (1992), Physica A **181**, 250.

Potts, R. B. (1952), Proc. Cambridge Philos. Soc. **48**, 106.

Privman, V. (1990) (ed.), *Finite Size Scaling and Numerical Simulation of Statistical Systems* (World Scientific, Singapore).

Privman, V., Hohenberg, C., and Aharony, A. (1991), in *Phase Transitions and Critical Phenomena*, Vol. 14, eds. C. Domb and J. L. Lebowitz (Academic Press, London).

Schmid, F. and Binder, K. (1992), Phys. Rev. B **46**, 13553; *ibid*, 13565.

Selke, W. (1992), in *Phase Transitions and Critical Phenomena*, Vol. 15, eds. C. Domb and J. L. Lebowitz (Academic Press, London) p. 1.

Sokal, A. D. (1995), in *Monte Carlo and Molecular Dynamics Simulations in Polymer Science*, ed. K. Binder (Oxford University Press, New York) Chapter 2.

Stauffer, D. (1997), Physica A **244**, 344.

Stoll, E., Binder, K., and Schneider, T. (1973), Phys. Rev. B **8**, 3266.

Wang, J.-S. and Gan, C. K. (1998), Phys. Rev. E **57**, 6548.

Wansleben, S. and Landau, D. P. (1991), Phys. Rev. B **43**, 6006.

Werner, A., Schmid, F., Müller, M., and Binder, K. (1997), J. Chem. Phys. **107**, 8175.

Wheeler J. C. and Pfeuty, P. (1981), Phys. Rev. A **24**, 1050.

Wilding, N. B. (1995), in *Computer Simulation Studies in Condensed Matter Physics VIII*, ed. D. P. Landau, K. K. Mon. and H.-B. Schüttler (Springer, Heidelberg).

Wilding, N. B. and Bruce, A. D. (1992), J. Phys. Condens. Matter **4**, 3087.

Wilding, N. B., Müller, M., and Binder, K. (1996), J. Chem. Phys. **105**, 802.

Yaldram, K. and Binder, K. (1991), J. Stat. Phys. **62**, 161.

Young, A. P. and Kawashima, N. (1996), Int. J. Mod. Phys. C **7**, 327.

Zifferer, G. (1999), Macromol. Theory Simul. **8**, 433.

5 More on importance sampling Monte Carlo methods for lattice systems

5.1 CLUSTER FLIPPING METHODS

5.1.1 Fortuin–Kasteleyn theorem

Advances in simulational methods sometimes have their origin in unusual places; such is the case with an entire class of methods which attempt to beat critical slowing down in spin models on lattices by flipping correlated clusters of spins in an intelligent way instead of simply attempting single spin-flips. The first steps were taken by Fortuin and Kasteleyn (Kasteleyn and Fortuin, 1969; Fortuin and Kasteleyn, 1972) who showed that it was possible to map a ferromagnetic Potts model onto a corresponding percolation model. The reason that this observation is so important is that in the percolation problem states are produced by throwing down particles, or bonds, in an uncorrelated fashion; hence there is *no critical slowing down*. In contrast, as we have already mentioned, the q-state Potts model when treated using standard Monte Carlo methods suffers from slowing down. (Even for large q where the transition is first order, the time scales can become quite long.) The Fortuin–Kasteleyn transformation thus allows us to map a problem with slow critical relaxation into one where such effects are largely absent. (As we shall see, not all slowing down is eliminated, but the problem is reduced quite dramatically!)

The partition function of the q-state Potts model (see Eqn. (2.38)) is

$$Z = \sum_{\{\sigma_i\}} e^{-K \sum_{i,j} (\delta_{\sigma_i \sigma_j} - 1)}, \tag{5.1}$$

where $K = J/k_B T$ and the sum over $\{\sigma_i\}$ is over all states of the system. The transformation replaces each pair of interacting Potts spins on the lattice by a bond on an equivalent lattice with probability

$$p = 1 - e^{-K \delta_{\sigma_i \sigma_j}}. \tag{5.2}$$

This means, of course, that there is only a non-zero probability of bonds being drawn if the pair of spins on the original lattice is in the same state. This process must be carried out for all pairs of spins, leaving behind a lattice with bonds which connect some sites and forming a set of clusters with different sizes and shapes. Note that all spins in each cluster must have the same value. The spins may then be integrated out (leaving a factor of q

behind for each cluster) and for the N_c clusters which remain (including single site clusters) the resultant partition function is

$$Z = \sum_{\text{bonds}} p^b (1-p)^{(N_b - b)} q^{N_c}, \qquad (5.3)$$

where b is the number of bonds and N_b is the total number of possible bonds. The quantity $(1-p)$ is simply the probability that no bond exists between a pair of sites. Thus, the Potts and percolation problems are equivalent. This equivalence was first exploited by Sweeny (1983) who generated graph configurations directly for the weighted percolation problem and showed that this was a more efficient approach than using the Metropolis method. In the following two sub-sections we shall demonstrate two particularly simple, different ways in which this equivalence may be exploited to devise new Monte Carlo methods which work 'directly' with the spin systems.

5.1.2 Swendsen–Wang method

The first use of the Fortuin–Kasteleyn transformation in Monte Carlo simulations was by Swendsen and Wang (1987); and although this is seldom the most efficient method, it remains an important tool. Just as in the Metropolis method, we may begin with any sort of an initial spin configuration. We then proceed through the lattice, placing bonds between each pair of spins with the probability given by Eqn. (5.2). A Hoshen–Kopelman method (see Section 3.4) is used to identify all clusters of sites which are produced by a connected network of bonds. Each cluster is then randomly assigned a new spin value, using a random number, i.e. each site in a cluster must have the same new spin value. The bonds are 'erased' and a new spin configuration is produced. See Fig. 5.1 for a schematic representation of the implementation of this algorithm. Since the probability of placing a bond between pairs of sites depends on temperature, it is clear that the resultant cluster distributions will vary dramatically with temperature. At very high temperature the clusters will tend to be quite small. At very low temperature virtually all sites with nearest neighbors in the same state will wind up in the same cluster and there will be a tendency for the system to oscillate back and forth between quite similar structures. Near a critical point, however, a quite rich array of clusters is produced and the net result is that each configuration differs substantially from its predecessor; hence, critical slowing down is reduced! In addition to the above intuitive argument, the reduction in characteristic time scales has been measured directly. It is thus known that the dynamic critical exponent z is reduced from a value of just over 2 for Metropolis single-site spin-flipping to a value of about 0 (i.e. log) in two dimensions and ~ 0.5 in three dimensions (Wang, 1990). Please don't forget, however, that the overall performance of the algorithm also depends strongly on the complexity of the code which is usually much greater than for single spin-flip methods. Hence, for small lattices the Swendsen–Wang technique may not

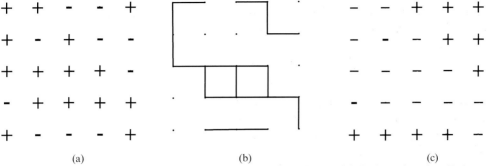

Fig. 5.1 Schematic view of the Swendsen–Wang algorithm for an Ising model: (a) original spin configuration; (b) clusters formed; (c) 'decorated' clusters.

Swendsen–Wang algorithm for a q-state Potts model

(1) Choose a spin
(2) Calculate $p = 1 - e^{-K\delta_{\sigma_i \sigma_j}}$ for each nearest neighbor
(3) If $p < 1$, generate a random number $0 < rng < 1$;
   If $rng < p$ place a bond between sites i and j
(4) Choose the next spin and go to (2) until all bonds have been considered
(5) Apply the Hoshen–Kopelman algorithm to identify all clusters
(6) Choose a cluster
(7) Generate a random integer $1 \le R_i \le q$
(8) Assign $\sigma_i = R_i$ to all spins in the cluster
(9) Choose another cluster and go to (7)
(10) When all clusters have been considered, go to (1)

---

offer much advantage, or may actually be slower in real time!, but for sufficiently large lattices it will eventually become more efficient.

This method may be extended to more complicated systems if one gives a little thought to modification. Magnetic fields can be included using two equivalent methods: either a 'ghost spin' is added which interacts with every spin in the system with a coupling equal to the magnetic field, or each cluster is treated as a single spin in a magnetic field whose strength is equal to the product of the field times the size of the cluster. If the interactions in an Ising model are antiferromagnetic instead of ferromagnetic, one simply places 'antibonds' between antiparallel spins with probability

$$p = 1 - e^{-|K|} \tag{5.4}$$

and proceeds as before. A further extension is to antiferromagnetic q-state Potts models for which the groundstate is multiply degenerate (see Wang, 1989). Two different spin values are randomly chosen and all spins which have different values are frozen. The spins which are still free are then simulated with the Swendsen–Wang algorithm with the frozen spins playing the role of quenched, non-interacting impurities. Two new spin values are chosen and the process is repeated. This method can also be applied to spin glass models but does not bring an improvement in performance due to the strong frustration.

The connection between cluster configurations and spin configuration raises a number of interesting issues which have been studied in detail by De Meo *et al.* (1990) for the Ising ferromagnet. In spite of the initial belief that the 'geometric clusters' formed by simply connecting *all* like spins in a given configuration could describe the Ising transition, it is clear that the actual 'physical clusters' which can be used for theoretical descriptions in terms of cluster theories are different. The Swendsen–Wang algorithm quite naturally selects only portions of a geometric cluster in creating new configurations. It is possible, however, to describe the thermal properties of a system in terms of the cluster properties, so one question becomes: just how well do the two agree? For the order parameter M the estimate in terms of clusters is given by the sum over all clusters of like spin direction. In contrast, the percolation probability P_∞ is determined only by the largest cluster. In a finite system the two may thus be expected to be different and indeed, as Fig. 5.2 shows, the finite size behaviors of the order parameter M and the percolation probability P_∞ are not the same. They also showed that for large lattices and $p < p_c$ in d-dimensions

$$\langle M \rangle \propto L^{-d/2}, \qquad L \to \infty, \qquad (5.5a)$$

$$\langle P_\infty \rangle \propto L^{-d} \log L, \qquad L \to \infty. \qquad (5.5b)$$

Related differences are present for the fluctuation quantities such as specific heat and susceptibility for which one has to separate out contributions from the clusters other than the largest one and those in the size of the largest cluster.

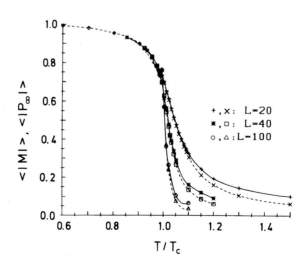

Fig. 5.2 Magnetization (—) and percolation probability (- -) plotted vs. reduced temperature for $L \times L$ Ising models studied using the Swendsen–Wang algorithm. After De Meo *et al.* (1990).

Problem 5.1 Perform a Swendsen–Wang simulation of a 32×32 Ising square lattice with periodic boundary conditions at $T = 2.27 \, J/k_B$ and $T = 3.0 \, J/k_B$. Determine the correlation times for the internal energy and compare the answers with the corresponding results for a Metropolis simulation at these temperatures. Comment on your findings.

5.1.3 Wolff method

One obvious shortcoming of the Swendsen–Wang approach is that significant effort is expended in dealing with small clusters as well as large ones. These small clusters do not contribute to the critical slowing down, so their consideration does not accelerate the algorithm. In order to partially eliminate this constraint, Wolff (Wolff, 1989a) proposed an alternative algorithm based on the Fortuin–Kasteleyn theorem in which single clusters are grown and flipped sequentially; the resultant performance generally exceeds that of the Swendsen–Wang method. The algorithm begins with the (random) choice of a single site. Bonds are then drawn to all nearest neighbors which are in the same state with probability

$$p = 1 - e^{-K}. \tag{5.6}$$

One then moves to all sites in turn which have been connected to the initial site and places bonds between them and any of their nearest neighbors which are in the same state with probability given by Eqn. (5.6). The process continues until no new bonds are formed and the entire cluster of connected sites is then flipped. Another initial site is chosen and the process is then repeated. The Wolff dynamics has a smaller prefactor and smaller dynamic exponent than does the Swendsen–Wang method. Of course the measurement of Monte Carlo time is more complicated since a different number of spins is altered by each cluster flip. The generally accepted method of converting to MCS/site is to normalize the number of cluster flips by the mean fraction of sites $\langle c \rangle$ flipped at each step. The Monte Carlo time then becomes

---

Wolff cluster flipping method for the Ising model

(1) Randomly choose a site
(2) Draw bonds to all nearest neighbors with probability $p = 1 - e^{-K \delta_{\sigma_i \sigma_j}}$
(3) If bonds have been drawn to any nearest neighbor site j, draw bonds to all nearest neighbors k of site j with probability $p = 1 - e^{-K \delta_{\sigma_j \sigma_k}}$
(4) Repeat step (3) until no more new bonds are created
(5) Flip all spins in the cluster
(6) Go to (1)

well-defined only after enough flips have occurred so that $\langle c \rangle$ is well defined. Later in this chapter we shall see just how important the Wolff algorithm can be for testing random number generators.

Problem 5.2 Perform a Wolff simulation of a 32 × 32 Ising square lattice with periodic boundary conditions at $T = 2.27\,J/k_B$ and $T = 3.0\,J/k_B$. Determine the correlation times for the internal energy and compare the answers with the corresponding results for a Metropolis simulation at these temperatures. Comment on your findings.

5.1.4 'Improved estimators'

In general, it may be possible to find multiple ways to calculate the same physical property of the system, and it may also turn out that the fluctuations in one estimator cancel more than for another estimator. (In earlier chapters we saw that the specific heat could be determined as a numerical derivative of the internal energy or from the fluctuations. The zero field susceptibility can be computed from the fluctuation of the order parameter or from the sum of the site–site correlation functions.) Since individual clusters are independent for the cluster flipping methods just discussed, for some quantities which can be calculated using cluster properties, 'noise reduction' occurs. It is then convenient to express various quantities in terms of clusters and use these expressions to answer the thermodynamic questions of interest (Sweeny, 1983; Wolff, 1988, 1990). Thus, for example, the susceptibility for $O(N)$ models is given by the mean cluster size, i.e.

$$\chi = \beta \langle |C| \rangle \tag{5.7}$$

where $|C|$ is the size of a cluster. The statistical error in the cluster definition of the suceptibility is smaller than that obtained using the fluctuations in the order parameter since the fluctuations due to the very small clusters cancel out. As discussed in the first section, however, for finite systems the behavior is not exactly the same as the true susceptibility, but in the thermodynamic limit it yields the same behavior. An improved estimator for the correlation function of the non-linear sigma model also yields substantial reduction in statistical error (Hasenbusch, 1995) and this property can be used for the classical spin systems that will be discussed shortly. The conclusion to which one might reasonably come is that not only the simulation method but also the method of analyzing the data needs careful consideration. We shall see in Chapter 7 just how important this consideration can be.

5.2 SPECIALIZED COMPUTATIONAL TECHNIQUES

5.2.1 Expanded ensemble methods

In many cases it is preferable to work in other ensembles, e.g. to include the temperature T in the set of dynamic variables (i.e. in the Markov process a random walk is also carried out over a range of temperatures). These methods will be treated in some detail in the next chapter. In the remainder of this section we shall concentrate on specialized techniques that apply primarily to spin systems.

5.2.2 Multispin coding

Multispin coding is a name given to a variety of very closely related algorithms which pack multiple spins into a single computer word, and then, through the use of a control word, carry out the spin–flip acceptance or non-acceptance for all spins in the word simultaneously (Zorn et al. (1981); Wansleben (1987)). The goal is to reduce both storage and cpu times, and the performance of multispin coding is very strongly machine dependent! Since all spins in a word will be considered in a single step, it is essential that they do not interact with each other. The checkerboard decomposition, described in Section 4.2.1, was developed explicitly for the purpose of implementing Monte Carlo on vector computers, and the use of a checkerboard decomposition is necessary for multispin coding on any computer. First, spins on a single sublattice are packed into 'multispin storage words' *is*. For an Ising model 'n' spins may be packed into a single word, where the word length is 'n' bits. For a q-state Potts model or other discrete state models, fewer spins may be packed into each word, depending on the number of bits needed to represent the possible spin states. The flipping probabilities are computed for each spin and compared with a random number creating a 'multispin flip word' *iscr* and spin flipping is then carried out by the exclusive or operation $is = iscr.XOR.is$. Sublattices are alternated in turn, and the resultant algorithm may yield substantial enhancement of performance.

In a variation of this method, which we refer to as 'multilattice coding', the same site from multiple, independent lattices is packed into a single word. Thus, for an Ising model 'n' lattices may be packed into a word of 'n' bits. Each lattice may be for instance at a different temperature. The advantage of this technique is that it offers the possibility of rapid production of data for 'n' different, independent systems and hence the possibility of calculating error bars based upon 'n' different runs. Since there is only one spin per system per word, there is no saving in memory.

5.2.3 *N*-fold way and extensions

The algorithms which we have discussed so far have been time-step driven methods, i.e. they were based on following the development of the system at each tick of some fictitious clock. At very low temperatures the flipping probability becomes quite small and virtually nothing happens for a long time. In order to avoid this wasteful procedure Bortz *et al.* (1975) introduced an event driven algorithm (the '*N*-fold way') in which a flip occurs at each step of the algorithm and one then calculates how many ticks of the clock would have elapsed if one had done it the 'old way'. They applied their method to the two-dimensional Ising model and we shall describe it now in terms of this system even though it can be applied to other discrete spin systems as well.

We begin by recognizing that there are only a small number of possible local environments which a spin can possibly have and consequently a limited number of different flipping probabilities. One thus collects all spins in the system into lists, in which each member has the identical energetic local environment. For an Ising $S = 1/2$ square lattice, for a spin with $\sigma = +1$ there are only 5 possible arrangements of nearest neighbors with different energies, i.e. the number of neighbors which also have $\sigma = 1$ may only be 4, 3, 2, 1, or 0. The same number of possibilities exist for a spin $\sigma = -1$, so every spin in the system can belong to one of only 10 classes. (If next-nearest neighbor interactions are present or the system is three-dimensional the number of classes will be different, but in all cases it will be some modest size integer N. Hence the name N-fold way.) The total probability of any spin of class l being flipped in a given step is

$$p_l = n_l e^{-\Delta E_l / k_B T}, \tag{5.8}$$

where n_l is the number of spins which are in class l. The integrated probability of 'some' event occurring in a given step for the first M classes is simply

$$Q_M = \sum_{l \leq M} p_l. \tag{5.9}$$

Then Q_N is the total probability for all N classes. The procedure is then to generate a random number $0 < rn < Q_N$ to determine the class from which the next spin to be overturned will come, i.e. class M is chosen if $Q_{M-1} < rn < Q_M$. Once the class has been chosen, another random number must be chosen to pick a spin from among those in the class. Finally, a third random number will be used to determine how much time has elapsed before this event has taken place, and we will discuss this part of the algorithm in a minute. First, we want to say a few words about bookkeeping. Each time a spin is flipped, it changes class. The site must then be removed from the list belonging to its original class and added to the new list corresponding to its new class. In addition, all of its (interacting) near neighbors change class. The key to an efficient N-fold way algorithm is thus an effective way of maintaining and updating the lists.

In order to determine the 'lifetime' of a state we first consider the probability that the system is in state $\{\sigma\}$ at time t and then undergoes a transition between time t and time $t + \Delta t$:

$$\Delta P(t) = -P(t)\frac{Q_l}{\tau}\Delta t, \tag{5.10}$$

where τ is the time needed to execute a flip. The probability of a flip of a spin in any class is then

$$P(\Delta t) = \exp\left(-\frac{Q_l}{\tau}\Delta t\right). \tag{5.11}$$

Treating this as a stochastic process, we can generate a random number R between 0 and 1, and inverting Eqn. (5.11), we find that the 'lifetime' of the state before flipping occurs becomes

$$\Delta t = -\frac{\tau}{Q_N}\ln R. \tag{5.12}$$

The thermodynamic averages of properties of the system are then calculated by taking the lifetime weighted average over the different states which are generated. The N-fold way is rather complicated to implement and each spin-flip takes a considerable amount of cpu time; however, at low temperatures, the net gain in performance can be orders of magnitude.

A generalization of the N-fold way algorithm is the technique of 'absorbing Markov chains', or MCAMC, (Novotny, 1995a) which offers substantial additional advantage in looking at magnetization switching in nanoscale ferromagnets and related problems. At low temperatures a strongly magnetized ferromagnet will not immediately reverse when a magnetic field is applied in the opposite direction because the nucleation barrier to the formation of a cluster of overturned spins is large. In a Monte Carlo simulation the same problem occurs and very long times are needed to follow the magnetization reversal process using standard techniques. The MCAMC approach extends the N-fold way algorithm to allow the simultaneous flipping of more than one spin to enhance the speed at which a nucleation cluster is formed; the 'level' of the method determines how many spins may be overturned in a single step. The level 1 MCAMC is essentially a discrete time version of the N-fold way (Novotny, 1995b) and is best used for an initial state in which all spins are up, i.e. for class 1 spins. A random number R is picked and then the lifetime m of the state is determined from $p_o^m < R < p_o^{m-1}$ where $p_o = 1 - p_1$. A spin is then randomly chosen and overturned. Level 2 MCAMC offers a decided advantage in the case that the nucleation cluster size is at least two, since it avoids the tendency to flip back those spins that have just been overturned. The level 2 MCMAC begins with a fully magnetized state and overturns two spins; these may either be nearest neighbors of each other or may be more distant neighbors which do not interact. Then one must define a

transient submatrix T which describes the single time step transition probabilities, i.e. for overturning one spin to reach a transient (intermediate) state,
and the recurrent submatrix R which gives the transition probabilities from
the transient to the absorbing (final) states. Again a random number R is
chosen and the lifetime of the state is determined by $\mathbf{v}^T T^m \mathbf{e} < R < \mathbf{v}^T T^{m-1} \mathbf{e}$
where $\mathbf{v}$ is the vector describing the initial state and $\mathbf{e}$ is the vector with all
elements equal to one. Another random number is then generated to decide
which spins will actually flip. Following generation of the 'initial cluster' as
just described, the N-fold way may then be used to continue. This method
may be systematically extended to higher order when the size of the nucleation cluster is larger so that the process of overturning a cluster is 'seeded'. It
is also possible to use the concept of spin classes to devise another algorithm
that can bridge the disparate time and length scales desired in Monte Carlo
simulations (Kolesik et al., 1998).

Problem 5.3 Perform an N-fold way of a 32×32 Ising square lattice with
periodic boundary conditions at $T = 1.5 J/k_B$. Determine the results for the
internal energy and the correlation time for the internal energy and compare the answers with the corresponding results for a Metropolis simulation
at this temperature. Repeat this comparison for $T = 0.5 J/k_B$.

5.2.4 Hybrid algorithms

In this section we consider methods which employ a combination of different
algorithms. The goal of this approach is to take advantage of the characteristics of each component algorithm to produce a method which is superior to
each individually. Microcanonical algorithms generate new states very
rapidly, but all states are confined to a constant energy surface (see e.g.
Creutz, 1980). By mixing Metropolis and microcanonical algorithms, one
produces a technique which is ergodic and canonical. Also the mixing of
Monte Carlo and molecular dynamics algorithms goes under the name
'hybrid Monte Carlo' but this will be discussed in Chapter 12, Section 12.2.4.

5.2.5 Multigrid algorithms

Multigrid methods are an alternative approach to the reduction of critical
slowing down. 'Blocks' of spins of various sizes are considered at different
time steps and all the spins within a given block are either flipped or not in a
sort of coarse-graining procedure. The change in block size is done in a
systematic fashion, and examples are shown in Fig. 5.3. While multigrid
MC (Kandel et al., 1988; 1989) can be shown to eliminate critical slowing
down perfectly for continuous spin models where the single-site probability is
a Gaussian, the method is already less successful for cases where the single
site probability is a ϕ^4 model (Goodman and Sokal, 1986) or for models with
discrete spins. Thus we do not describe this method further here.

Fig. 5.3 Schematic
description of a
multigrid Monte Carlo
cycle. The degree of
blocking is given by n
so that the size of the
'blockspin' lattice is
L/b^n where the size of
the blocking is
denoted by b.

5.2.6 Monte Carlo on vector computers

The use of vector computers for Monte Carlo calculations has been immensely successful; and even though everyone anticipates the dominance of parallel computing in the future, in many cases vector computing is still the most efficient and user friendly computing tool. Compilers on vector computers tend to be quite mature and rather efficient code can thus be produced without enormous user effort. The basic idea of vector computing is to speed up computation by arranging the problem so that essentially the same operation can be performed on an entire vector of variables which is loaded into the 'vector pipe' at the start. This necessitates program construction which insures that all elements of the vector are independent and that the change of one does not affect any other. For a Monte Carlo calculation on a simple lattice model, the checkerboard decomposition discussed in Chapter 4 achieves this goal. For more details on implementation of Monte Carlo programs and application examples we refer to a separate review (Landau, 1992). Certainly a familiarity with vector computing provides, at the very least, a better understanding of the issues raised in over a decade of the literature.

5.2.7 Monte Carlo on parallel computers

One of the most effective uses of parallel architectures is to simply perform independent Monte Carlo simulations of a system under different conditions on different processors. This approach, called 'trivial parallelism', is obviously not the goal of designers of these machines, but is often the most effective from the user point of view. For very large problems, parallel architectures offer the hope of speeding up the simulation dramatically so that data are produced within a reasonable turnaround time. Broadly speaking, parallel algorithms may be of two different types. The work to be done on a system may be spread among multiple processors, or the system may be decomposed into different parts and each processor may be assigned to work on a different part of the system. This latter approach is almost always more effective if a substantial number of processors is available. Simple lattice systems may be split up into squares or strips. For systems with continuous positional degrees of freedom, one may either assign a fixed region of space to a processor or a fixed set of particles. The determination of which approach is more efficient depends on the characteristics of the problem at hand. For example, for systems with very strong density fluctuations, a rigid spatial decomposition of the problem may result in some processors having the responsibility for many particles and others have no work to do within a given time interval. One particularly important consideration in the development of any type of parallel program is the relative importance of communication time and computation time. In the case of geometric parallelization, i.e. decomposition of a system into strips, if the size of the individual regions is too small, the time used to communicate information between processors may not be small compared to the time needed on each processor for computation. In such a case, the performance may actually get *worse* as processors are added (Heermann and Burkitt, 1990). For systems with long range interactions (e.g. spin systems with exchange constants which decay with distance according to a power law) most of the computational effort goes into the calculation of energy changes, and then the communication overhead is much less of a problem.

5.3 CLASSICAL SPIN MODELS

5.3.1 Introduction

There are many important lattice models in statistical mechanics which do not have discrete degrees of freedom but rather variables which vary continuously. Just as the Ising model is the 'standard' example of a discrete model, the classical Heisenberg model is the most common prototype of a model with continuous degrees of freedom. A more general model which includes the Heisenberg system as a special case involves classical spin vectors S_i of unit length which interact via a Hamiltonian given by

$$\mathcal{H} = -\mathcal{J}\sum_{i,j}S_i \cdot S_j - D\sum_i S_i^2, \qquad |S_i| = 1, \qquad (5.13)$$

where the first term is the Heisenberg exchange interaction and the second term represents single ion anisotropy. This Hamiltonian describes many physically interesting magnetic systems, and even examples of $D = 0$ have been experimentally verified for magnetic ions with large effective spin values, e.g. $RbMnF_3$. We must remember, of course, that at sufficiently low temperatures the classical Hamiltonian cannot be correct since it neglects quantum mechanical effects (see Chapter 8). In the remaining parts of this section we shall consider methods which can be used to simulate the Heisenberg model and its anisotropic variants.

5.3.2 Simple spin-flip method

The Metropolis method can be used for Monte Carlo simulations of classical spin vectors if we allow a spin to 'tilt' towards some new direction instead of simply flipping as in an Ising model. In the simplest approach, some new, random direction is chosen and the energy change which would result if this

new spin orientation is kept is then calculated. The usual Metropolis prescription is then used to determine, by comparison with a random number
generated uniformly in the interval [0, 1], whether or not this new direction is
accepted, i.e. the transition rate is

$$W_{n \to m} = \tau_{\rm o}^{-1} \exp(-\Delta E/k_{\rm B} T), \qquad \Delta E > 0 \qquad (5.14a)$$
$$= \tau_{\rm o}^{-1}, \qquad \Delta E < 0 \qquad (5.14b)$$

where ΔE is simply the difference between the initial and trial state. When
beginning such a simulation one must first make a decision about whether
information about the spins will be kept by keeping track of Cartesian spin
components or by keeping track of angles in spherical coordinates. The
manipulation of spins in angular coordinates is usually quite time consuming,
and it is generally more efficient to use Cartesian coordinates. (One price
which one must then pay is that the spin length is no longer fixed to remain
exactly equal to unity.) A new spin direction can then be chosen by randomly
choosing new spin components and normalizing the total spin length to unity.
The simplest way to accomplish this is to generate a new random number in
the interval [0, 1] for each component and then subtract 0.5 to produce
components such that $-0.5 < S_\alpha < 0.5$; by normalizing by the length of
the spin one obtains a new spin of length unity. If this procedure is used,
however, the spins are not part of a uniform distribution of directions since
they are more likely to point towards the corners of a unit cube than in other
directions. Instead, one must first discard any new spin which has a length
greater than 0.5 before renormalization, but if this is done, the new spins will
be uniformly distributed on the unit sphere. An interesting alternative procedure was suggested by Marsaglia (1972). Two random numbers r_1 and r_2
are chosen from a uniform distribution to produce a vector with two components $\zeta_1 = 1 - 2r_1$ and $\zeta_2 = 1 - 2r_2$. The length of the vector is determined by $\zeta^2 = \zeta_1^2 + \zeta_2^2$ and if $\zeta^2 > 1$ a new spin vector is then computed with
components

$$S_x = 2\zeta_1(1 - \zeta^2)^{1/2}, \qquad S_y = 2\zeta_2(1 - \zeta^2)^{1/2}, \qquad S_z = 1 - 2\zeta^2. \quad (5.15)$$

Note that this procedure is *not* simply the generation of points randomly in
the unit circle and then projecting them onto the unit sphere since this would
not produce a uniform distribution. Any of the methods for producing new
trial spin configurations require multiple random numbers, moreover the
continuous variation of possible energies eliminates the possibility of building
a table of probabilities. Thus, continuous spin models are much more time
consuming to simulate. (A trick which can be used is to approximate the
possible spin directions by a discrete distribution, e.g. for a two-dimensional
XY-model one could use an n-state clock model with the spins confined to
point in one of n different equally spaced directions. The discreteness which
results would then allow table building, however, it may also modify the
behavior. At low temperatures, for example, the effective anisotropy introduces a gap into the excitation spectrum which is not present in the original

model. In two-dimensional models the nature of the phase transitions is also modified. Thus, even though such approaches may improve performance, they must be treated with caution.)

One additional feature that needs to be discussed is the choice of an order parameter. These systems now have order parameters with multiple components, and the nature of the Hamiltonian detemines just which components are important. In the case of single ion anisotropy ordering will occur only along the z-direction so this component must be kept track of separately from the other components. In the fully isotropic case, all components are equivalent. The order parameter is then invariant under global rotation so it is the magnitude of the order parameter which matters, i.e.

$$m = \sqrt{M_x^2 + M_y^2 + M_z^2}, \quad \text{where} \quad M_\alpha = \frac{1}{N}\sum_i S_{i\alpha}. \tag{5.16}$$

In this case the order parameter can never be zero, even above T_c so finite size effects are always quite pronounced. The usual fluctuation definition of the susceptibility is also no longer valid although it can be used as an 'effective' susceptibility. Above T_c the best estimate for the true susceptibility is simply

$$\chi = \frac{N}{k_{\rm B}T}\langle m^2\rangle \tag{5.17}$$

since $\langle m\rangle$ will be zero in the thermodynamic limit.

Problem 5.4 Perform a Metropolis simulation of a $4 \times 4 \times 4$ classical Heisenberg model on a simple cubic lattice with periodic boundary conditions at $T = 2.0\,J/k_{\rm B}$ and $T = 4.0\,J/k_{\rm B}$. Determine the internal energy and order parameter. Comment on your findings.

5.3.3 Heatbath method

A variation of this method which was first suggested for application to lattice gauge theories (see Chapter 11) corresponds to touching each spin in turn (selected either in order or randomly) to a 'heat bath' (Creutz, 1980). Instead of allowing the change in energy to determine the 'new' spin configuration, one can simply randomly select a new spin direction and then compare a random number rn with the Boltzmann probability of the trial configuration, i.e. accept the new configuration if $rn < \exp(-E'/k_{\rm B}T)$ where E' is the energy of the trial state. This method is most useful in circumstances where the Metropolis-like approach described above has a very low acceptance rate. In simulations of lattice gauge models the determination of the energy of a given state may be very time consuming so one may repeat the heatbath process many times, with the same new trial state, and use the collection of configurations which result for the statistical averaging. The entire process must be repeated many times, and after equilibration has occurred, many Monte Carlo steps must be made to obtain good statistical averaging. The heatbath method may also be used for Ising model simulations for which there are only two different states for each spin. Here the spin may be set equal to $+1$ with probability p_i and equal to -1 with probability $1 - p_i$ where

$$p_i = \frac{e^{2\beta \sum_{nn} \sigma_j}}{1 + e^{2\beta \sum_{nn} \sigma_j}}. \tag{5.18}$$

This may be easily implemented by generating a random number rn and setting

$$\sigma_i' = \text{sign}(p_i - rn). \tag{5.19}$$

Note that the probability of a spin being up or down is the same for Glauber dynamics, however the implementation is different since

$$\sigma_i' = \text{sign}(rn - (1 - p_i)) \qquad \text{if } \sigma = +1, \tag{5.20a}$$
$$\sigma_i' = \text{sign}(p_i - rn) \qquad \text{if } \sigma_i = -1. \tag{5.20b}$$

This means that the random numbers are used differently and the actual sequence of states will be different (Herrmann, 1990).

5.3.4 Low temperature techniques

5.3.4.1 Sampling

In classical spin systems there is no gap in the excitation spectrum. Very low energy excitations dominate at low temperatures, but a random choice of new spin direction will generally produce a large energy change and is thus unlikely to be accepted. The acceptance rate can be increased by restricting the new spin-flip attempts to a small cone about the initial position. If the cone is made too narrow, however, the changes are so small that the system again evolves quite slowly. Hence some initial trials followed by an intelligent choice of the angle for the cone of maximum displacement must be made.

5.3.4.2 Interpretation

At low temperatures the excitations are spin waves which can be most readily explained by a harmonic analysis in reciprocal (momentum) space. For small lattices, however, the reciprocal space is quite coarse grained and the number of momentum points $\mathbf{q}$ is limited. Thus, finite size effects can become important, not because of any critical behavior but because of the restrictions on the number of modes.

5.3.5 Over-relaxation methods

Strictly speaking over-relaxation (Brown and Woch, 1987; Creutz, 1987) techniques are deterministic, but they are of great value when used in combination with other, stochastic approaches. The effective interaction field for a spin is determined by examining all neighbors to which it is coupled. The spin is then precessed about this interaction field by an angle θ, making use of the equation of motion

$$\dot{\mathbf{S}} = -\mathbf{S} \times \mathbf{H}_{\text{eff}}. \tag{5.21}$$

This process is microcanonical since the energy is a constant of the motion, but for large values of θ it can enhance decorrelation. If a checkerboard algorithm is used, every spin on a single sublattice may be considered, and then each spin on the next sublattice treated. This algorithm is deterministic, but when used together with a stochastic technique, e.g. Metropolis, the resultant states are drawn from a canonical ensemble. This method is quite efficient and vectorizes extremely well.

5.3.6 Wolff embedding trick and cluster flipping

At first glance the cluster flipping methods which have been described earlier would seem to be restricted to systems with discrete states, but Wolff (1989a,b) has also shown how these methods can be applied to general $O(n)$ models. This approach, known as the embedding trick, turns the original uniform interaction classical spin model into an Ising model with inhomogeneous couplings. It proceeds in the following manner. First a direction $\hat{n}$ is chosen randomly in space. The spins are then projected onto that direction to form an Ising model with interactions between pairs which depend on the projections of each spin. In principle the Metropolis method could then be used to flip spins, but it is clearly more effective to use a cluster flipping method. If the single cluster (Wolff) flipping algorithm is to be used, bonds are added between nearest neighbor sites with probability

$$p = 1 - \exp\{\min[0, 2\beta \mathcal{J}(\hat{n} \cdot S_i)(\hat{n} \cdot S_j)]\} \tag{5.22}$$

to form a connected cluster of sites in the same way as for a simple Ising model. The components parallel to $\hat{n}$ are then reversed for every spin in the cluster to yield a new spin configuration. Note that in this case the projection need only be carried out for those spins which have a chance to join the cluster to be flipped. A new direction is randomly chosen in space and the process is repeated. Data are collected in the usual way. (See also Section 5.1.3 in this chapter for a quick review of the cluster flipping technique for the Ising model.)

We wish to emphasize that this trick is not just of academic interest since it has already been used to extend the studies of critical phenomena in classical spin systems well beyond what was previously possible. For example, a very successful investigation of the three-dimensional classical, Heisenberg model has been made using the embedding trick Wolff flips together with histogram reweighting (see Chapter 7) and a finite size scaling analysis to determine the critical temperatures on several lattices with quite high precision (Chen *et al.*, 1993). Lattices as large as $40 \times 40 \times 40$ with periodic boundary conditions were simulated with the results: $\mathcal{J}/k_{\text{B}}T_{\text{c}} = 0.693\,035(37)$ (body centered cubic lattice with two atoms per unit cell) and $J/k_B T_c = 0.486\,798(12)$ (simple cubic lattice). The critical exponents were found with high precision and the values agreed quite closely for the two lattices in full support of our ideas of universality. A Wolff cluster study of this system by Holm and Janke (1993) yielded similar results but with less resolution. Whereas these lattice sizes are still much smaller than those which are accessible for the Ising model, they represent a dramatic improvement over what could be treated by Metropolis sampling. Other systems have been studied with this method as well. The three-dimensional XY-model (plane rotator) was studied by Hasenbusch and Meyer (1990) using the Swendsen–Wang cluster update method together with the embedding trick and improved estimators. They found a critical coupling of $J/k_B T_c = 0.454\,21(8)$ and obtained estimates for static and dynamic critical exponents from finite size scaling. All of the above mentioned studies indicate that the combination of several methods, for both simulation and analysis, can indeed be quite powerful.

Problem 5.5 Using the embedding trick perform a Wolff cluster simulation of a $4 \times 4 \times 4$ classical Heisenberg model on a simple cubic lattice with periodic boundary conditions at $T = 2.0\,J/k_B$. Determine the internal energy and order parameter and compare the results with those of Problem 5.4.

5.3.7 Hybrid methods

Often it is advisable to combine different updating schemes into a single, more complicated scheme that is more efficient in destroying correlations between subsequently generated states on all length scales. Thus, it is straightforwardly possible to mix ordinary Metropolis or heatbath sweeps through the lattice with Wolff cluster flips, etc. A very successful study of the two-dimensional classical, XY-model (three component) used a mixture of Metropolis, over-relaxation and embedding trick Wolff flips together with a finite size scaling analysis to determine the Kosterlitz–Thouless temperature to much higher precision than had previously been possible (Evertz and Landau, 1996), $J/k_B T_{KT} = 0.700(5)$.

Another technique which is actually termed 'hybrid Monte Carlo' has been used in lattice gauge theories (see Chapter 11) but is also straightforward to implement for classical spin systems. Instead of choosing the trial configuration by random change of a single spin (or link for lattice gauge models) one can instead produce a trial state by changing all spins by a small amount determined from the canonical equations of motion. (Such time integration methods will be discussed in Chapter 12. As a note, we add that a symplectic integrator is best chosen to insure detailed balance. For lattice gauge models it may be necessary to introduce fictitious momenta in order to accomplish this.) The acceptance or rejection of the new trial state can then be made via standard Metropolis.

5.3.8 Monte Carlo dynamics vs. equation of motion dynamics

In the previous sections we have discussed a number of techniques which allow us to 'speed up' the Monte Carlo sampling through phase space through the intelligent choice of 'step' size and direction. For some systems such changes can be made with impunity since the time development of the system being modeled is stochastic. In some cases systems have true dynamics which are described by Poisson's equations if they are classical or by the commutator if they are quantum, i.e.

$$\dot{S}_i = -\frac{i}{\hbar}[\mathcal{H}, S_i], \tag{5.23}$$

where $\mathcal{H}$ is the Hamiltonian and S_i the operator in question. Equation (5.23) represents an equation of motion and takes the system along a deterministic path through phase space. This path has physical significance and the associated time is true time. In contrast, the Monte Carlo method is strongly dependent on the (arbitrary) transition rate which is chosen. For the Ising model, Eqn. (5.23) yields no equations of motion since the commutator is zero. The Ising model thus has only stochastic 'dynamics', i.e. kinetics. The time-dependent behavior of the Heisenberg model may be studied either by Monte Carlo kinetics or by integrating deterministic equations of motion obtained through Eqn. (5.30); the time-dependent critical behavior will be different in the two cases (Landau, 1994; Landau and Krech, 1999).

5.3.9 Topological excitations and solitons

In most situations discussed so far, deviations from the groundstate are produced by spin-flips or by periodic, spin-wave excitations. In some cases, other kinds of excitations have fundamental importance. In the two-dimensional XY-model, in addition to spin waves, topological excitations known as vortices play a crucial role. The vortex cores can be located by following the spin directions around an elementary plaquette and summing the differences in the relative spin angles with regard to some fixed direction. If the sum is 2π a vortex is present, if the sum is -2π an antivortex is present, and if the sum is 0 then no topological excitation is centered on the plaquette. Both spin waves and vortices are portrayed schematically in Fig. 5.4. At low temperatures a few tightly bound vortex–antivortex pairs are present in the two-dimensional XY-model, and as the temperature is increased the pairs unbind, signaling a special kind of phase transition. A Monte Carlo simulation does not manipulate the vortices directly since it is the spin degrees of freedom which are sampled, but the vortex behavior can be monitored along with the thermodynamic properties.

There are also slightly more complex systems which show a combination of order parameter fluctuations as well as topological excitations. As a simple 'case study' example we consider the two-dimensional Heisenberg antiferromagnet with exchange anisotropy in a magnetic field,

Fig. 5.4 Schematic
view of excitations in
classical spin models:
(a) spin waves; (b)
vortices in a two–
dimensional plane; (c)
solitons in a one-
dimensional lattice
with a symmetry
breaking field.

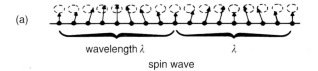

(a)

wavelength λ   λ

spin wave

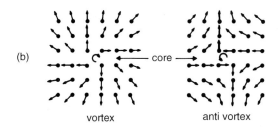

(b)   — core —

vortex   anti vortex

(c)

soliton

$$\mathcal{H} = J \sum_{\langle i,j \rangle} [(1 - \Delta)(s_{ix}s_{iy} + s_{iy}s_{jy}) + s_{iz}s_{jz}] + H_{\parallel} \sum_i s_{iz}, \tag{5.24}$$

where $J > 0$ is the antiferromagnetic nearest neighbor exchange parameter, Δ describes the exchange anisotropy, and $H_{\parallel}$ is an applied magnetic field in the z-direction. Data were obtained for this model using quite simple Monte Carlo methods by Landau and Binder (1981) either by varying the temperature at fixed field strength or by sweeping the field at constant temperature. $L \times L$ lattices with periodic boundaries were simulated using the Metropolis method. From a combination of data on order parameters, magnetization, internal energy, susceptibility, and specific heat, a phase diagram was extracted in $H_{\parallel}$–T space. This diagram, see Fig. 5.5, shows that in low fields below a field-dependent critical temperature, there is an Ising transition to a state in which the system shows antiferromagnetic order along the field direction. At high fields the z-components of the spins are aligned (but disordered) and only the x- and y-spin components are free to order. This is the so-called 'spin-flop' (SF) phase. However, since the symmetry is then the same as for a two-dimensional XY-model, we expect a Kosterlitz–Thouless transition with the formation of bound, topological excitations as the temperature is lowered. In three dimensions the upper and lower phase boundaries would meet at a Heisenberg-like bicritical point at some finite temperature, but in two dimensions the Heisenberg model does not order at any finite temperature so we would expect on theoretical grounds that they would meet at $T = 0$. The simulations show that these two phase boundaries

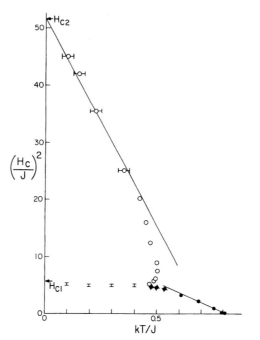

Fig. 5.5 Phase diagram for the two–dimensional anisotropic Heisenberg model (see Eqn. (5.24)). From Landau and Binder (1981).

come very close together, but it is not possible to determine whether or not they merge at some non-zero temperature. In the 'XY-like' phase, bound vortex–antivortex pairs are seen at low temperatures; in addition to increasing in density as the temperature is elevated, they begin to unbind, as is shown in Fig. 5.6a. The measured density is consistent with a non–zero excitation energy and the value of the 'gap' varies systematically with the applied field (see Fig. 5.6b). Of course, with modern computers and techniques one could obtain far better data on larger systems, but even these results which require quite modest computer resources clearly reveal the essential physics of the problem.

Another very intriguing situation is found in one–dimensional XY-models with a symmetry breaking field. In the simplest possible case there may be topological excitations in which the spins go through a 2π-twist as we move along the chain direction. This may be observed by simply monitoring the angular position of successive spins. These excitations are known as solitons, or more properly speaking solitary waves, and may exist in a variety of forms in magnetic models. (See Fig.5.4 for a schematic representation of a soliton excitation.) For example, in an antiferromagnet each sublattice may rotate through π to form a new kind of soliton. It is also possible for one sublattice to rotate through π and the other sublattice to rotate through $-\pi$. In a third variant, one sublattice is unchanged and the other rotates through 2π. All of these types of solitons have been observed in Monte Carlo simulations.

Fig. 5.6 Topological excitations in the two–dimensional anisotropic Heisenberg model. (a) 'Snapshots' of vortex behavior in the SF state for $L = 40$, $H_\parallel / \mathcal{J} = 4.0$. Open and closed circles represent vortices and antivortices, respectively. (b) Vortex-pair density in the SF state. The energy (in units of $\mathcal{J}$) needed to create a vortex–antivortex pair is 2μ. From Landau and Binder (1981).

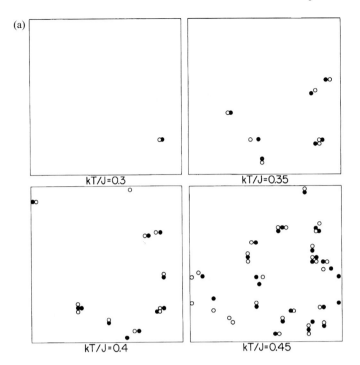

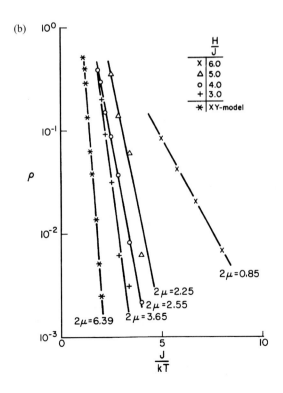

5.4 SYSTEMS WITH QUENCHED RANDOMNESS

5.4.1 General comments: averaging in random systems

By quenched randomness we imply that the model Hamiltonian of interest depends on random variables other than the degrees of freedom which are considered in the thermal average, and these random variables are kept fixed in one physical realization of the system. For example, consider a magnetic binary alloy $A_x B_{1-x}$, where a crystal is grown from a melt containing a fraction x of A-atoms and a fraction $1 - x$ of B-atoms. Assuming that both species carry Ising spins $S_i = \pm 1$, it is nevertheless natural to assume that the exchange constants $\mathcal{J}_{ij}$ depend on the type of pair that is considered: $\mathcal{J}_{AA}, \mathcal{J}_{AB}$ or $\mathcal{J}_{BB}$, respectively. Denoting the occupation variable $c_i = 1$ if site i is taken by an A-atom, $c_i = 0$ if it is taken by a B-atom, one would arrive at the Hamiltonian (assuming nearest neighbor exchange only)

$$
\mathcal{H}\{S_i, c_i\} = - \sum_{\langle i,j \rangle} \Big\{ c_i c_j \mathcal{J}_{AA} + \big[c_i(1 - c_j) + c_j(1 - c_i) \big] \mathcal{J}_{AB}
$$

$$
+ (1 - c_i)(1 - c_j) \mathcal{J}_{BB} \Big\} S_i S_j.
$$

(5.25)

Of course, this model includes the dilution of a magnetic crystal by a non-magnetic species as a special case (then $\mathcal{J}_{AB} = \mathcal{J}_{BB} = 0$). While the configurations of the spins $\{S_i\}$ in all averages are weighted with the Boltzmann factor $\exp[-\mathcal{H}\{S_i, c_i\}/k_B T]$ in all averages, the configurations of the $\{c_i\}$ are not assumed to occur with a weight given by the Boltzmann factor, but rather with a predetermined distribution $P\{c_i\}$. Depending on the history of sample preparation in the laboratory, one may wish to choose the c_i completely at random, but consistent with the chosen concentration x, or with some built-in correlations reflecting 'chemical' short range order. In any case, an average of some observable $A\{S_i, c_i\}$ (e.g. the magnetization M of the crystal) then becomes

$$
\Big[\langle A\{S_{i_j}, c_i\} \rangle_T \Big]_{av} = \int d\{c_i\} P\{c_i\} \frac{1}{Z\{c_i\}} \underset{\{s_i\}}{\mathrm{Tr}}\, A\{S_i, c_i\} \exp[-\mathcal{H}\{S_i, c_i\}/k_B T].
$$

(5.26)

Thus one sees that there is a double average that needs to be carried out: for a fixed realization $\{c_i\}$, one computes the thermal average as usual, and then this average is averaged once more with $P\{c_i\}$. While the thermal averaging is done with the usual Metropolis importance sampling, the disorder average $[\ldots]_{av} = \int d\{c_i\} P\{c_i\} \ldots$ can immediately be realized by simple sampling.

In principle, this problem is hence straightforwardly suitable for Monte Carlo simulation. However, the question arises how large the sample has to be for the averaging with $P\{c_i\}$ over the configurations $\{c_i\}$ of the quenched disorder variables. In an experiment, typically measurements are carried out for a single probe, there is no need to repeat the experiment for a large number of samples, the observable quantities are 'self-averaging'. One would expect that a similar self-averaging property would also apply to simulations, if very large systems away from any phase transition are studied, and then simulation of a single (or a few) realizations of the $\{c_i\}$ would suffice. However, the situation is rather different in the case of a finite size scaling analysis, where one considers systems of finite linear dimension L right at the critical temperature T_c of the model: the fluctuations from one sample $\{c_i\}$ to the next one cause a significant sample-to-sample fluctuation of the effective pseudo-critical temperature $T_c(L)$ of the system (defined e.g. by the max-imum of the specific heat or the maximum slope of the fourth order cumu-lant, etc.). This sample-to-sample fluctuation of $T_c(L)$ causes a lack of self-averaging for certain quantities (typically for the order parameter and its susceptibility) at T_c. This lack of self-averaging shows up when one considers ratios such as (Wiseman and Domany, 1995)

$$R_A \equiv \left[(\langle A \rangle_T - [\langle A \rangle_T]_{av})^2\right]_{av} / \left([\langle A \rangle_T]_{av}\right)^2. \tag{5.27}$$

Lack of self-averaging implies that (ξ is the correlation length)

$$R_A \to C_A \qquad \text{if } L/\xi \to 0 \text{ (i.e. for } T = T_c) \tag{5.28}$$

while away from T_c there is self-averaging, ratios such as R_A decay for $L \to \infty$ inversely proportional to the volume,

$$R_A \propto (\xi/L)^d \qquad \text{if } L \gg \xi. \tag{5.29}$$

The lack of self-averaging implies that a sample of the order $n \approx 10^4$ realiza-tions is desirable, in order to get the relative error of the disorder average at T_c $[\langle A \rangle_{T_c}]_{av}$ down to 1% or less. This consideration already shows that the Monte Carlo study of phase transitions in random systems may be very computer time consuming. Of course, sometimes a relative error of 10% may seem acceptable, and then only a sample of $n \approx 10^2$ realizations is required.

In addition, one has to be careful in the precise manner in which the disorder averaging is carried out. Suppose we consider the case $c = 0.5$ for the $A_x B_{1-x}$ alloy. We can generate a sample $\{c_i\}$ by drawing a random uniformly distributed number η_i with $0 \le \eta_i < 1$ for each lattice site, and choosing $c_i = 1$ if $\eta_i > x$ and otherwise setting $c_i = 0$. However, for a crystal with $N = L^d$ sites the average composition will differ from $x = 0.5$ also by a random deviation of order $1/\sqrt{N}$. Since often dependence of the critical temperature $T_c(x)$ on concentration x is rather strong, this sample-to-sample variation of the concentration may contribute substantially to the sample-to-sample fluctuation of the pseudo-critical temperature $T_c(L)$. However, this problem is avoided if one simply selects $N_x = xN$ lattice sites at random, setting $c_i = 1$ at each of these sites and otherwise putting $c_i = 0$. Then the concentration of every sample is strictly equal to x, and the sample-to-sample fluctuation of the concentration is suppressed. It turns out that the 'universal' numbers C_A defined above, that characterize the lack of self-averaging at T_c in a random system, do differ for these two choices (Wiseman and Domany, 1998). In a sense these two choices to fix the concentration of the random alloy correspond to the canonical and semi-grand canonical ensemble of statistical mechanics. If we were to treat the disorder as 'annealed' rather than 'quenched' for annealed disorder, the average would simply be

$$\langle A\{S_i, c_i\}\rangle_T = \frac{1}{Z} \operatorname*{Tr}_{\{s_i c_i\}} A\{S_i, c_i\} \exp(-\mathcal{H}\{S_i, c_i\}/k_B T), \qquad (5.30)$$

i.e. in the trace the two types of variables $\{S_i\}$, $\{c_i\}$ are now both included, and treated on an equal footing – so the local concentration on the lattice site also exhibits thermal fluctuations. (e.g. due to interdiffusion of the species A, B in the crystal), unlike the quenched case. In the semi-grand canonical ensemble of alloys, the chemical potential difference $\Delta\mu = \mu_A - \mu_B$ between the species is the independent thermodynamic variable, and then the concentration undergoes thermal fluctuations, while in the canonical ensemble x is the independent thermodynamic variable and hence strictly non-fluctuating (thermal fluctuations then occur in the conjugate variable $\Delta\mu$, but this variable often is not even recorded in a simulation). These distinctions between the various thermodynamic ensembles naturally have analogs for the calculation of quenched averages, since one can consider quenched averaging as an averaging of the disorder variables ($\{c_i\}$ in our example) as a thermal averaging at a different (higher) temperature: for a completely random selection of lattice sites, we average at infinite temperature, but we can introduce some correlations in the occupancy of lattice sites by defining

$$P\{c_i\} = \frac{1}{Z_0} \exp(-\mathcal{H}_c\{c_i\}/k_B T_0), \qquad (5.31)$$

where $\mathcal{H}_c$ is some model Hamiltonian describing the 'crystallographic' interaction between the species A, B, and one assumes that at the temperature T_0 ($\gg T$) the $\{c_i\}$ are still in full thermal equilibrium, before one quenches in the configurations of the $\{c_i\}$ thus generated by sudden cooling from T_0 to T, where the $\{c_i\}$ are forbidden to relax. Obviously, these considerations are motivated by the actual experimental procedures, but they also clarify that the different ensembles with which the averaging at T_0 is performed lead to different ensembles for carrying out quenched averages. In most cases one considers uncorrelated disorder, i.e. $1/T_0 \to 0$, but these considerations apply in this limit as well.

One important aspect about quenched averaging is that the distribution $P(A)$ generated in this way ($[\langle A\{S_i, c_i\}\rangle]_{av} = \int dA\, P(A)A$) typically is not symmetric around its average, mean value and most probable value may differ appreciably. Consider, for instance, the magnetization for the above model Hamiltonian at a temperature slightly above the average value of $T_c(L)$: those samples for which $T_c(L) > T$ due to the sample-to-sample fluctuation of $T_c(L)$ will have a large magnetization, while those samples where $T_c(L)$ deviates in the other direction will have a very small magnetization. This asymmetry of the distribution creates problems if one calculates quantities which have very small averages, e.g. spin correlations $[\langle S_i S_j \rangle_T]_{av}$ with large distances $\mathbf{r}_i - \mathbf{r}_j$ between the sites i, j.

An even more subtle effect may occur due to extremely rare fluctuations. Consider e.g. the case of simple dilution in the above model Hamiltonian, where $\mathcal{J}_{AB} = \mathcal{J}_{BB} = 0, \mathcal{J}_{AA} \equiv \mathcal{J}$. Then for $x < 1$ the critical temperature T_c (x) will be clearly less than $T_c(1)$. However, the probability is non-zero (albeit extremely small) that somewhere in the system we find a large compact region free of dilution sites. This region will effectively order already at $T_c(1)$, in a still disordered environment. A mathematical consideration of this problem shows that there is a whole temperature region $T_c(x) < T < T_c(1)$ where very weak singularities are already present. ('Griffiths singularities', Griffiths (1969)). One also expects that these singularities cause some anomalous tails in dynamic correlation functions at long times, but due to problems of sampling such very small correlations accurately enough this problem is not yet so well understood.

Monte Carlo simulation of systems with quenched disorder is a difficult task; due to the need of carrying out the double averaging procedure over both thermal disorder and quenched disorder the demand for computer resources is huge and the judgement of the accuracy is subtle, particularly due to metastability and slow relaxation at low temperatures. Many problems are still incompletely understood. In the following, we mention two types of problems more explicitly, but only on the level of rather introductory comments. For up-to-date reviews of the state of the art in this field, we refer to Young (1998).

5.4.2 Random fields and random bonds

The presence of certain kinds of randomness leads to some of the most complex behavior in statistical physics and occurs in several different kinds of deceptively simple models (see Young, 1996). (In this section we shall not discuss the case of spin glasses at all since these will be treated separately.) If a simple Ising ferromagnet is subjected to a magnetic field which is randomly up or down, what happens to the phase transition? This quite straightforward question is surprising difficult to answer. Imry and Ma (1975) examined the question of whether or not the groundstate would be ordered by considering the competition between the surface energy that would be needed by producing a domain of overturned spins and the Zeeman energy that would be gained. They concluded that for lattice dimension $d \leq d_l = 2$ an ordered state would be unstable against the formation of domains. (For continuous spins, $d_l = 4$.) These, and other random field models, have been simulated extensively; but the long correlation times and the need to average over different realizations of the random field have produced data which have been interpreted in different ways, including the presence of first order transitions for at least a portion of the phase diagram and a new 2-exponent hyperscaling relation. At this time there is still a pressing need for a dramatically improved algorithm to allow the unambiguous determination of the nature of the phase diagram.

For the case of random bond models in the absence of an applied field the situation is equally intriguing. Two separate kinds of problems have already been examined, although the descriptions are by no means complete. For q-state Potts models with large q the transition is known to be first order. A somewhat surprising prediction was made by Hui and Berker (1989) that the presence of two different strength ferromagnetic bonds would change the order of the transition to second order. This behavior was indeed observed by Chen *et al.* (1995) who used a 'multihit' Swendsen–Wang algorithm and histogram reweighting (see Chapter 7) to study the phase transition in the two-dimensional $q = 8$ Potts model with exactly 50% of weak bonds randomly spread throughout the lattice, a fraction for which the exact transition temperature is known. Their finite size analysis yielded critical exponents which were consistent with two-dimensional Ising values. Although there are now more refined predictions (Cardy and Jacobsen, 1997) that the exponents are not quite Ising-like, there is still no broad understanding of the effect of different kinds of randomness on first order transitions. If the transition is second order in the absence of any randomness there may again be several kinds of phenomena which result. If the randomly dispersed bonds are of zero strength, one can study the nature of the critical behavior, both for small dilution as well as as the percolation threshhold is approached. Extensive Monte Carlo simulations of the bond impure two-dimensional Ising model have suggested that the critical behavior is modified by logarithmic corrections (Selke *et al.*, 1994). Random antiferromagnetic bonds can also lead to frustration, although this does not necessarily destroy the transition if the percentage of bonds is below a critical value.

5.4.3 Spin glasses and optimization by simulated annealing

Spin glasses are disordered magnetic systems, where the interactions are 'frustrated' such that no ground state spin configuration can be found that is satisfactory for all the bonds (Binder and Young, 1986). Experimentally, such quenched disorder in the exchange constants is found in many strongly diluted magnets, e.g. a small percentage of (magnetic) Mn ions in a random Cu–Mn alloy interact with the Ruderman–Kittel indirect exchange which oscillates with distance as $\mathcal{J}_{ij} \propto \cos(k_F |\mathbf{r}_i - \mathbf{r}_j|)/|\mathbf{r}_i - \mathbf{r}_j|^3$, where the Fermi wavelength $2\pi/k_F$ is in general incommensurate with the lattice spacing. Since in such a dilute alloy the distances $|\mathbf{r}_i - \mathbf{r}_j|$ between the Mn ions are random, both ferro- and antiferromagnetic $\mathcal{J}_{ij}$ occur approximately with equal probability. Qualitatively, we may model such systems as Ising models with a Gaussian distribution $P(\mathcal{J}_{ij})$, see Eqn. (4.62), or by the even simpler choice of taking $\mathcal{J}_{ij} = \pm \mathcal{J}$ at random with equal probability as shown in Eqn. (4.63). A plaquette of four bonds on a square with three $+\mathcal{J}$ and one $-\mathcal{J}$ is enough to demonstrate the frustration effect: it is an easy exercise for the reader to show that such an isolated plaquette that is frustrated (i.e. sign $(\mathcal{J}_{ij}\mathcal{J}_{jk}\mathcal{J}_{kl}\mathcal{J}_{li}) = -1$)

has an energy $-2J$ and a 6-fold degenerate ground state, while for an unfrustrated plaquette the energy is $-4J$ and the degeneracy only 2-fold. An example of frustration, as well as a schematic 'energy landscape' for a frustrated system, is shown in Fig. 5.7. Note that in reality phase space is multidimensional, not one-dimensional, and finding low lying minima as well as optimal paths over low lying saddle points is still quite a challenge for simulations. An approach for tackling this challenge can be based on 'multicanonical sampling' (Berg and Neuhaus, 1991, 1992), as will be described in Section 7.5.3.

The experimental hallmark of spin glasses is a cusp (or kink) in the zero field static susceptibility and while mean field theory for an infinite range model (Edwards and Anderson, 1975) shows such a behavior, the properties of more realistic spin glasses have been controversial for a long time. As has already been emphasized above, for systems with such quenched disorder, a double averaging is necessary, $[\langle\ldots\rangle_T]_{av}$, i.e. the thermal average has to be carried out first, and an average over the random bond configuration (according to the above probability $P(J_{ij})$) afterwards. Analytic techniques yield only rather scarce results for this problem, and hence Monte Carlo simulations are most valuable.

However, Monte Carlo simulations of spin glasses are also very difficult to perform due to slow relaxation caused by the existence of many states with low lying energy. Thus, when one tries to estimate the susceptibility χ in the limit $H \to 0$, the symmetry $P(J_{ij}) = P(-J_{ij})$ implies that $[\langle S_i S_j\rangle]_{av} = \delta_{ij}$ and hence

$$\chi = \frac{1}{k_B T}\left(\sum_j [\langle S_i S_j\rangle_T - \langle S_i\rangle_T \langle S_j\rangle_T]_{av}\right) = \frac{1}{k_B T}(1 - q), \qquad (5.32)$$

where

$$q = [\langle S_i\rangle_T^2]_{av},$$

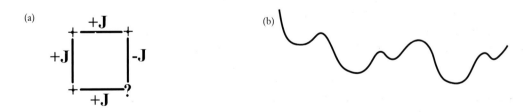

Fig. 5.7 (a) Frustrated plaquette in an Ising model with three $+J$ and one $-J$ bonds; (b) schematic view of an 'energy landscape' in a spin glass.

i.e. the cusp would result from onset of a spin glass order parameter q below the freezing temperature T_f. In the Monte Carlo simulation, the thermal averaging $\langle \ldots \rangle_T$ is replaced by time averaging, and hence (Binder, 1977)

$$\chi = \frac{1}{k_B T}(1 - q(t)), \tag{5.33}$$

where

$$q(t) = \left[\left(\frac{1}{t}\int_0^t S_i(t')\, dt' \right)^2 \right] = \left[\frac{2}{t}\int_0^t \left(1 - \frac{t'}{t}\right)\langle S_i(0)S_i(t')\rangle\, dt' \right].$$

This argument shows that an apparent (weakly time-dependent) spin glass order parameter $q(t)$ may arise if the spin autocorrelation function has not decayed during the observation time t. Thus Monte Carlo runs which are too short may show a cusp in χ as an observation-time effect, even if there is no transition at non-zero temperature in the static limit. This in fact is the explanation of 'cusps' found for χ in two-dimensional spin glasses (Binder and Schröder, 1976). It took great effort with dedicated machines (a special purpose processor for spin glass simulations was built by Ogielski (1985) at AT&T Bell Laboratories) or other advanced specialized computers, e.g. the 'distributed array processor' (DAP), to show that $T_f = 0$ for $d = 2$ but $T_f \approx 1.2\mathcal{J}$ for the $\pm\mathcal{J}$-model in $d = 3$. Again the cumulant intersection method, generalized to spin glasses (Bhatt and Young, 1985), turned out to be extremely useful: one considers the quantity

$$g_L(r) = \tfrac{1}{2}\left(3 - [\langle q^4\rangle_T]_{av}/[\langle q^2\rangle_T]_{av}\right), \tag{5.34}$$

the $\langle q^k\rangle$ being the moments of the distribution of the spin glass parameter. The fact that the curves for $q_L(T)$ for various L merge at T_f is evidence for the existence of the transition (Fig. 5.8). No analytic method has yielded results competitive with Monte Carlo for this problem. (Note, however, that more recent work (Kawashima and Young, 1996) using better averaging and larger lattices has yielded an improved estimate of $T_f/\mathcal{J} = 1.12(2)$. The sizes that were used to produce Fig. 5.8 were necessarily quite small and good data with finer resolution simply show that there are still subtle finite size effects that cannot be discerned from the figure.) However, the most recent work (Hatano and Gubernatis, 1999) finds instead that $T_f/\mathcal{J} \approx 1.3$. Thus, the exact location of the critical temperature of a spin glass is still unknown today! From Fig. 5.8 it is evident that for $T < T_f$ well-equilibrated data exist only for rather small systems: the systems get very easily trapped in low-lying metastable states. In order to come as close to equilibrium as possible, one has to cool down the system very slowly. Similarly difficult, of course, is the search for the groundstates of the spin glass: again 'simulated annealing', i.e. equilibration at high temperatures combined with very slow cooling, turns out to be relatively efficient.

Fig. 5.8 Plot of g
against T for the
three-dimensional SK
$\pm J$ Ising model. The
lines are just guides to
the eye. (a) Plot of
cumulant intersections
for the mean-field spin
glass model; (b)
temperature
dependence of the
cumulant; (c) scaling
plot for the cumulant.
From Bhatt and
Young (1985).

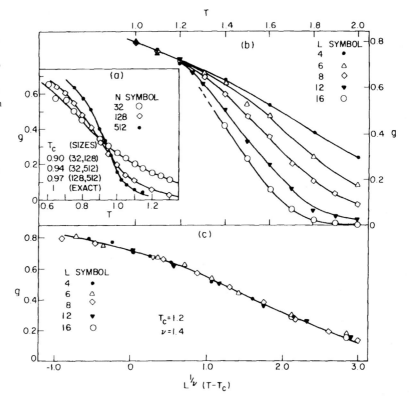

Fig. 5.8 Plot of g against T for the three-dimensional SK $\pm J$ Ising model. The lines are just guides to the eye. (a) Plot of cumulant intersections for the mean-field spin glass model; (b) temperature dependence of the cumulant; (c) scaling plot for the cumulant. From Bhatt and Young (1985).

Finding the ground state energy of a spin glass is like solving an optimization problem, where the Hamiltonian is treated as a functional of the spin configuration, and one wishes to minimize this functional. Similar optimization problems occur in economics: e.g. in the 'traveling salesman problem' a salesman has to visit n cities (with coordinates $\{x_k y_k\}$) successively in one journey and wishes to travel such that the total distance $d = \sum_{\ell=1}^{n-1} d_\ell$, $\{d_\ell = \sqrt{(x_k - x_k')^2 + (y_k - y_k')^2}\}$ becomes a minimum: clearly the salesman then saves time, mileage and gasoline costs, etc. A pictorial view of the 'traveling salesman problem' is shown for a small number of cities in Fig. 5.9. Now one can generalize this problem, treating this cost function like a Hamiltonian in statistical mechanics, and introduce 'temperature' into the problem, a term which originally was completely absent from the optimization literature. A Monte Carlo simulation is then used to modify the route in which the order of the visits of adjacent cities is reversed in order to produce a new trial state, and a Metropolis, or other, acceptance criterion is used. At high temperature the system is able to get out of 'local minima' and as the temperature is lowered it will hopefully settle to the bottom of the lowest minimum, i.e. the shortest route. This simulated annealing approach, introduced by

Kirkpatrick *et al.* (1983) to solve global optimization problems, has developed into a valuable alternative to other schemes for solving optimization problems. It is thus a good example of how basic science may have unexpected economic 'spin-offs'.

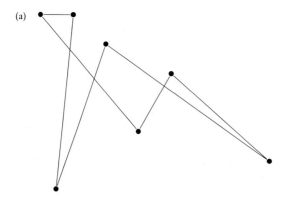

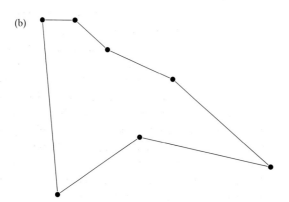

Fig. 5.9 Schematic view of the traveling salesman problem: (a) unoptimized route: (b) optimized route.

5.5 MODELS WITH MIXED DEGREES OF FREEDOM: Si/Ge ALLOYS, A CASE STUDY

There are many important models for which both discrete and continuous degrees of freedom must be incorporated. One example is an impure Heisenberg model for which Ising degrees of freedom specify whether or not a site is occupied by a magnetic ion and continuous variables describe the behavior of the magnetic spins at the sites which are occupied. A Monte Carlo study must then include possible changes in both variables. A more complex situation arises when all states of the discrete variable are interesting and the potential associated with the continuous variable is complicated. A simple example is Si/Ge alloys. These systems are examples of semiconductor alloys which play an extremely important role in technological development. For purposes of industrial processing we need to know just what the phase diagram looks like, and more realistic models than simple lattice alloy models are desirable. These systems may be modeled by an Ising degree of freedom, e.g. $S_i = +1$ if the site is occupied by Si and $S_i = -1$ if a Ge is present, and $S_i = 0$ corresponds to a vacancy at a site. The second, continuous variable describes the movement of the nodes from a perfect lattice structure to model the disorder due to the atomic displacements of a crystal that is compressible rather than rigid. Elastic interactions are included so both the local and global energies change as the system distorts. These systems are known to have strong covalent bonding so the interactions between atoms are also strongly directional in character. The empirical potentials which seem to describe the behavior of these systems effectively thus include both two-body and three-body terms. In order to limit the effort involved in calculating the energies of states, a cutoff was implemented beyond which the interaction was set to zero. This model was studied by Dünweg and Landau (1993) and Laradji *et al.* (1995) using a 'semi-grand canonical ensemble' in which the total number of atoms was fixed but the relative numbers of Si and Ge atoms could change. Monte Carlo 'moves' allowed an atom to be displaced slightly or to change its species, i.e. Si → Ge or Ge → Si. (The chemical potential μ represented the difference between the chemical potentials for the two different species; the chemical potential for vacancies was made so large that no vacancies appeared during the course of the simulation.) The simulation was carried out at constant pressure by allowing the volume to change and accepting or rejecting the new state with an effective Hamiltonian which included the translational entropy, i.e.

$$\Delta\mathcal{H}_{\text{eff}} = \Delta\mathcal{H} - Nk_{\text{B}}T\ln\frac{\Lambda'_x\Lambda'_y\Lambda'_z}{\Lambda_x\Lambda_y\Lambda_z}, \tag{5.35}$$

where Λ and Λ' represent the dimensions of the simulation box and of the trial box, respectively.

The data were analyzed using the methods which have been discussed for use in lattice models and showed, somewhat surprisingly, that the transition was mean field in nature! The analysis was not altogether trivial in that the

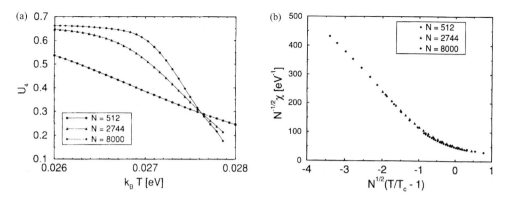

Fig. 5.10. Elastic Ising Si/Ge model data obtained using a Keating potential: (a) fourth order cumulant crossing; (b) finite size scaling of the 'susceptibility' using mean-field exponents. After Dünweg and Landau (1993).

critical point was located using a two-dimensional search in $(\mu - T)$ space (using histograms which will be described in Chapter 7). The behavior of the fourth order cumulant of the order parameter and the finite size scaling of the 'susceptibility' are shown in Fig. 5.10; both properties demonstrate clearly that the critical point is mean-field in nature. The first study, carried out with the Keating valence field potential yielded a somewhat surprising and unphysical result in that the lattice constant shrank continuously as the temperature was raised. When the calculations were repeated with the Stillinger–Weber potential, this effect disappeared. This showed the importance of not relying solely on fitting low temperature properties in designing phenomenological potentials for the description of real alloys.

5.6 SAMPLING THE FREE ENERGY AND ENTROPY

5.6.1 Thermodynamic integration

There are circumstances in which knowledge of the free energy itself, and not just its derivatives, is important. For example, at a strongly first order transition the bulk properties of a system will generally show pronounced hysteresis which makes a precise determination of the equilibrium location of the transition problematic. This problem can be largely avoided, however, by the determination and subsequent comparison of the free energies of different phases. The expressions given in Chapter 2 which provide a thermodynamic definition of the free energy F, can be used rather straightforwardly to actually provide numerical estimates for F. Since the internal energy U can be measured directly in a Monte Carlo simulation and the entropy can be obtained by integrating the specific heat C, i.e.

$$S(T) = \int_0^T \frac{C(T')}{T'} \, dT'.$$

(5.36)

Of course, Eqn. (5.33) only makes sense for Ising-type systems, for which $C(T \to 0) \to 0$, but not for 'classical' systems for which $C(T \to 0) \to \text{const.}$ and the entropy $S(T \to 0) \to -\infty$! In some cases the free energy in a low temperature state can be accurately estimated and used to determine the free energy at finite temperature. (For example, in an Ising model it will be given by the internal energy at $T = 0$.) Alternatively, the free energy may be estimated in the high temperature, disordered state by integrating the internal energy, i.e.

$$\frac{F(T)}{k_B T} = \frac{S(\infty, H)}{k_B} + \int_0^{1/k_B T} U d(1/k_B T).$$

(5.37)

The intersection of these two free energy branches (see Fig. 5.11a) determines the location of the transition. In some cases, however, the transition is encountered not by varying the temperature but rather by varying an applied field or chemical potential. In this case the appropriate thermodynamic integration becomes a two-step process as shown in Fig. 5.11b. Two different paths of constant field on opposite sides of the transition line are followed up to the desired temperature T and the free energies are computed. The temperature is then fixed and the field is then swept across the transition so that

$$F(H) = F(H_1, T) + \int_{H_1}^{H} M \, dH',$$

(5.38a)

(a)

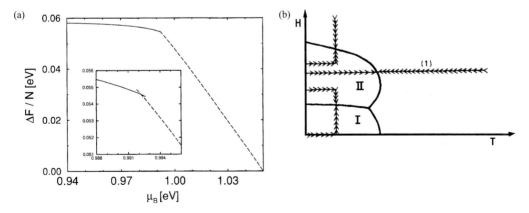

(b)

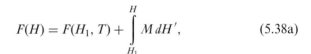

Fig. 5.11 (a) Comparison of free energies obtained with the chemical potential swept in opposite directions for the model of Si/Ge in the previous section. (b) Schematic view of paths for thermodynamic integration. In this figure there are three first order phase boundaries separating a high temperature disordered phase, and two low temperature ordered phases.

$$F(H) = F(H_2, T) + \int_{H_2}^{H} M \, dH', \qquad (5.38b)$$

and again the point of intersection locates the transition. The accuracy of this method is limited by the errors of the data points which are used for the integration and by residual finite size effects. (In addition to reducing the usual finite size effect, one must make the system large enough that there are no excursions to the other phase during a simulation run. The actual size that is needed depends on the magnitude of the discontinuities which occur at the transition.) Since the fluctuations are generally small near a first order transition, quite accurate data for large systems can be generated without too much difficulty, so the transition can be located quite accurately.

5.6.2 Groundstate free energy determination

For discrete spins the groundstate free energy is given simply by the internal energy. For systems with continuous variables, however, the groundstate entropy of classical systems is $-\infty$ and the determination of the entropy at low temperatures is non-trivial. One way to accomplish this is to divide the Hamiltonian into two parts, one for which the groundstate free energy can be calculated theoretically, and the second part is a perturbation which is slowly turned on. The free energy change can be determined by integration over the prefactor describing the magnitude of the perturbation. One specific application of this approach is the method used by Frenkel and Ladd (1984) in which an Einstein crystal (whose free energy may be calculated exactly) is taken as the unperturbed system with the interparticle interactions slowly turned on to produce a harmonic solid. Integration as a function of the added interaction produces the desired estimate for the free energy. Dünweg and Landau (1993) introduced an alternative method which relied on the Monte Carlo sampling of the ratio of the partition functions for the two different phases using a form of umbrella sampling. This worked quite well for the Keating potential but is not necessarily effective for all potentials.

5.6.3 Estimation of intensive variables: the chemical potential

In most of the methods that we have already discussed the intensive variable, e.g. magnetic field or chemical potential, was held fixed and the conjugate extensive variable, e.g. the magnetization or density, was measured. The inverse procedure, although more difficult, can also be carried out (Alexandrowicz, 1975; Meirovitch and Alexandrowicz, 1977). In the following we shall work in the language of a lattice gas model with nearest neighbor bond energy $-\varepsilon$, although the procedure for an Ising model would be completely equivalent. As previously seen in the discussion of the N-fold way, an occupied site would have five different possible 'local states' α depending on the number of nearest neighbor sites which were also occupied and would have energy E_α. We then define a set of five conjugate states by removing the 'central' atom. This means that $E_{\alpha'} = 0$. If the probabilities of occurrence of each state are $P(\alpha)$ and $P(\alpha')$, detailed balance requires that

$$\frac{P(\alpha)}{P(\alpha')} = \exp[(-E_\alpha + \mu)/k_B T], \qquad (5.39)$$

so that

$$\frac{\mu}{k_B T} = \ln\left(\frac{P(\alpha)}{P(\alpha')}\right) + \frac{E_\alpha}{k_B T}. \qquad (5.40)$$

By averaging over all five different local states rather good statistical precision can be obtained for the estimate of the chemical potential. As we shall see in Chapter 6 there are specialized 'particle insertion' techniques which can be used to estimate the chemical potential when the lattice restriction is removed.

5.6.4 Lee–Kosterlitz method

The correct identification of the order of a transition can become particularly tricky if the transition is actually weakly first order. Lee and Kosterlitz (1990) proposed a very simple scheme which can be remarkably effective, even for quite small systems. A long simulation run is made at some value of the extensive 'field', e.g. temperature, which is quite near to the phase transition and a histogram of the order parameter values is constructed. If there are two peaks in the distribution, the distribution is reweighted to a different field value (see Chapter 7 for a detailed description of reweighting) until the two peaks are the same height, and the difference between the maxima and the minimum between the two peaks is used to estimate the free energy barrier ΔF

$$\Delta F = \ln \frac{P_L(E_1)}{P_L(E_2)}, \qquad (5.41)$$

where $P_L(E_1)$ and $P_L(E_2)$ are the probabilities at the maximum and minimum values respectively. This procedure is repeated for different lattice sizes and if ΔF diverges with increasing size, the transition is first order in the thermo-dynamic limit. Otherwise, the transition is second order. This procedure was quite effective for small $q = 5$ Potts models even though a finite size scaling analysis for systems as large as $L = 240$ suggested that the transition was (incorrectly) second order.

5.6.5 Free energy from finite size dependence at T_c

A somewhat specialized but novel approach to the calculation of free energies at a critical point was proposed by Mon (1985). He considered the finite size variation of the free energy at the critical point for an L^d system with periodic boundary conditions for which it is expected that

$$f_{sing} \approx U_0 L^{-d}, \tag{5.42}$$

where U_0 is a scaling amplitude. The system is then decomposed into 2^d systems each of size $(L/2)^d$ and the ratio of the partition functions of the two systems is given by

$$\frac{Z_{L/2}}{Z_L} = \frac{\mathrm{Tr}\exp(-\beta H_{L/2})}{\mathrm{Tr}\exp(-\beta H_L)} = \langle\exp[-\beta(H_{L/2} - H_L)]\rangle, \tag{5.43}$$

where H_L represents the Hamiltonian for the original system and $H_{L/2}$ is the Hamiltonian for the divided system. From Eqn. (5.43) we can see that the free energy difference between the two lattices is

$$f_L - f_{L/2} = \frac{\ln\langle\exp[-\beta(H_{L/2} - H_L)]\rangle}{L^d} \tag{5.44}$$

and this relation, together with Eqn. (5.42) can be used to determine the singular part of the free energy. The estimation of the free energy difference in Eqn. (5.44) may not be easy to do directly but may be calculated using 'umbrella sampling', a method which will be described in the first part of Chapter 7.

5.7 MISCELLANEOUS TOPICS

5.7.1 Inhomogeneous systems: surfaces, interfaces, etc.

If a system contains surfaces or interfaces, its properties become position dependent. One particular strength of Monte Carlo simulation methods is that such effects can be studied in full detail and under perfectly well controlled conditions. For instance, let us stick to the example of the Ising ferromagnet that undergoes a phase transition at some critical temperature T_{cb} in the bulk, characterized by the power laws already discussed in Chapter 2, e.g. the bulk magnetization $m_b = \hat{B}(1 - T/T_{cb})^\beta$, the bulk susceptibility $\chi_b = \hat{\Gamma}_\pm|1 - T/T_{cb}|^{-\gamma}$, etc. We now may ask (Binder, 1983) how this behavior gets modified when we consider the local counterparts of these quantities right in the surface plane (m_1, $\chi_1 = (\partial m_1/\partial H)_T$) or in the nth layer away from the free surface (m_n, χ_n). Under which conditions does the surface order at a temperature T_{cs} higher than the bulk, i.e. ($M_1 \propto (1 - T/T_{cs})^{\beta_{2d}}$ with $T_{cs} > T_{cb}$ and β_{2d} is the two-dimensional Ising exponent)? If the surface layer orders at the same critical temperature as the bulk does, what are the associated exponents? ($m_1 \propto (1 - T/T_{cb})^{\beta_1}$, $\chi_1 \propto (1 - T/T_{cb})^{-\gamma_1}$). Actually, the surface involves many more exponents than the bulk does, since one can also consider the response to a local field H_1 ($\chi_{11} = (\partial m_1/\partial H_1)_T \propto (1 - T/T_c)^{-\gamma_{11}}$) and the critical behavior of surface excess quantities: the surface excess magnetization m_s is defined in terms of the profile m_n as $m_s = \sum_{n=1}^{\infty}(m_b - m_n)$, etc. In the simulation, all such questions can be addressed at once for well-defined models, control parameters (including local fields H_n in arbitrary layers indexed by n, suitable changes $\Delta \mathcal{J} = \mathcal{J}_s - \mathcal{J}$ of the exchange coupling in the surface plane, etc.) can be varied at will, etc. Moreover, one can choose an absolutely ideal, perfect surface (no adsorbed 'dirt', no surface roughness, no dislocations, no grain boundaries, no surface steps, . . .). In all these respects, simulations have a huge advantage over experiments, and hence the testing of corresponding theory has proceeded for the most part by simulation methods. As Nobel laureate Pauli had put it a long time ago, 'While God has created solids as perfectly ideal crystals, the devil is responsible for their imperfect surfaces': of course, the simulations can make contact with this complex reality as well, putting into the model more and more of these non-ideal effects (which can again be varied in a controlled manner to check their relevance).

Of course, surfaces and interfacial effects are not only of great interest near critical points, but in a much wider context. Just as in an Ising ferromagnet we may ask how the magnetization m_n varies as a function of the layer index n, in a fluid we may ask what is the profile of the local density $\rho(z)$ as function of the distance z from a solid wall (due to a container for instance), etc. Further, if we model flexible macromolecules as self-avoiding random walks (see Chapters 3, 4), we may consider the adsorption of flexible macromolecules at a hard wall in terms of a model where a monomer adjacent to a wall wins an energy ε, and this enthalpic gain may outweigh the entropic loss due to the reduced number of SAW configurations near a wall (Binder, 1983).

While many of the technical aspects of simulations of models addressing the effects of free surfaces or other boundaries are rather similar to simulations targeted to sample bulk properties, where surface effects are deliberately eliminated by the use of periodic boundary conditions, sometimes the demands for computational resources become exorbitant, since large (mesoscopic rather than of atomic scale) lengths occur. A typical example is the phenomenon of 'wetting', i.e. when a saturated gas below the critical temperature is exposed to a wall, in which a fluid layer may condense at the wall without accompanying condensation in the bulk. In the ideal case (and in the thermodynamic limit) the thickness of this 'wet' layer at the wall is infinite at all temperatures above the wetting temperature T_w. Of course, this is true only in the absence of gravity, and the chemical potential gas $\mu_{gas}(T)$ must always be held at its coexistence value $\mu_{coex}(T)$ for gas–liquid phase coexistence. For non-zero $\Delta\mu = \mu_{coex}(T) - \mu_{gas}(T)$, a fluid layer may also condense, but it is not infinitely thick, $\ell_{wetting} \propto (\Delta\mu)^{-p_{co}}$ wetting where p_{co} is some exponent that depends on the character of the forces between the wall and the particles in the gas (Dietrich, 1988). The approach to the wet state (for $T > T_w$) where $\ell_{wetting} \to \infty$ as $\Delta\mu \to 0$ is called 'complete wetting'. On the other hand, if we approach the wetting transition for $\Delta\mu = 0$ and let T approach T_w from below, we may distinguish two situations: $\ell_{wetting}(T \to T_w)$ may approach a finite value at T_w and then jump discontinuously to infinity ('first order wetting'); or $\ell_{wetting}$ may show a critical divergence, $\ell_{wetting} \propto (T_w - T)^{-p}$ where p is another exponent ('critical wetting'). To avoid confusion, we mention that for short range forces in $d = 3$ dimensions actually both exponents p_{co}, p are zero (which implies logarithmic divergences).

Simulation of such wetting phenomena is very difficult: not only must the linear dimension perpendicular to the wall be very large, much larger than $\ell_{wetting}$, but also the linear dimension of the system in the directions parallel to the wall must be huge, since a very large correlation length $\xi_\parallel \propto \Delta\mu^{-\nu_{co}}$ or $\xi_\parallel \propto (T_w - T)^{-\nu}$ appears (where ν_{co}, ν are exponents appropriate for 'complete wetting' or 'critical wetting', respectively). The occurrence of this large length can be understood qualitatively by recalling the interpretation of wetting phenomena as 'interface unbinding transitions' (Dietrich, 1988): as the fluid layer at the wall gets thicker, the gas–liquid interface gets more and more remote from the wall, and capillary wave excitations of larger and larger wavelength – up to $\xi_\parallel$ – become possible. This problem is very closely related to the finite size effects encountered in simulations set up to study interfaces between coexisting phases, already discussed in Section 4.2.3.6. Again the lesson is that a rather good qualitative understanding of the physics of a problem is already mandatory when one sets up the model parameters for a simulation of that problem!

Next we mention that even wetting phenomena can be studied with the simple Ising lattice model. We only have to remember the correspondence with the lattice gas interpretation: 'spin down' represents liquid, 'spin up' represents gas, and gas–liquid phase coexistence ($\Delta\mu = \Delta\mu_{coex}(T)$) in the Ising magnet then simply means that the bulk magnetic field H is zero. A wetting transition can be induced by applying a negative surface field $H_1 < 0$ – favoring hence liquid at the wall – at the surface of a ferromagnet with a positive spontaneous magnetization (i.e. gas) in the bulk.

We now make some more specific comments on the technical aspects of the simulation of such systems and also present a few typical examples.

If surface properties are of interest in a model for which the bulk values are well known, it may be preferable to sample layers near the surface more frequently than those far from the surface in order to reduce the statistical error in estimates for surface related properties. If the interior is sampled too infrequently, however, the bulk may not reach equilibrium, and this, in turn, will bias the surface behavior. If, instead, a slowly fluctuating interface is present, it may be preferable to sample layers in the interior, in the vicinity of the interface, more often than those near the surface. Both variants of preferential sampling have been used successfully; in a new problem it may be useful to first make some test runs before choosing the layer sampling probabilities. Note that there are a large number of interesting phenomena which may be seen in quite simple systems confined between two surfaces simply by varying the interaction in the surface layers and applying either surface or bulk fields, or both (see e.g. Landau (1996); Binder et al. (1988; 1989; 1992; 1996)). Perhaps the simplest model which shows such effects is the $L \times L \times D$ Ising film with Hamiltonian

$$\mathcal{H} = -\mathcal{J} \sum_{\langle i,j\rangle \in \text{bulk}} \sigma_i \sigma_j - \mathcal{J}_s \sum_{\langle i,j\rangle \in \text{surf}} \sigma_i \sigma_j - H \sum_i \sigma_i - H_1 \sum_{i \in \text{surf}} \sigma_i, \quad (5.45)$$

which in the limit of $D \to \infty$ becomes equivalently a semi-infinite system with $L \times L$ surface. Thus, capillary condensation in thin films and layering and critical wetting in thick films have both been studied in simple nearest neighbor Ising models between two confining surfaces using preferential sampling. For example, the data for the layer magnetization, shown in Fig. 5.12 demonstrate quite clearly the onset of wetting as the surface field H_1 is varied. Perhaps the most interesting consequence of these studies is the discovery that critical wetting is apparently mean-field-like in contradiction to theoretical predictions of non-universal, non-mean-field-like behavior. The discrepancy between the Monte Carlo result and the theoretical renor-malization group calculation, which used as the characteristic length the distance of the interface from the wall, was rather perplexing and helped spark new theoretical efforts. It currently appears likely that there is an additional characteristic length involved (the distance from the wall to the metastable state) which renormalized the 'bare' result and that the simula-tional result was indeed correct (Boulter and Parry, 1995). A clue to this possiblity is actually visible in Fig. 5.12 where a small 'kink' can be seen in the profile near the surface, even when the wetting interface has moved quite some distance into the bulk. The Monte Carlo data which yielded this unex-pected result were not simple to obtain or analyze because of the large fluctuations to which we alluded earlier. For example, data for the sucept-ibility of $L \times L \times D$ systems with $L = 50, D = 40$ using standard Metropolis sampling showed huge fluctuations (see Fig. 5.13). When a multispin coding technique was used to make much longer runs with $L = 128, D = 80$, Fig. 5.13 shows that the results were much improved. Note, however, that even though these data were taken far from the transition, only roughly a factor of two increase in linear dimension was possible with an improvement of roughly 10^2 in the implementation of the sampling algorithm.

Fig. 5.12 Profiles of the layer magnetization for a $128 \times 128 \times 160$ Ising film with $\mathcal{J}_s/\mathcal{J} = 1.33$, $\mathcal{J}/k_B T = 0.226$. The arrows show the values of the bulk magnetization in the spin-up and spin-down phases. From Binder et al. (1989).

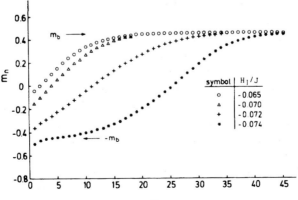

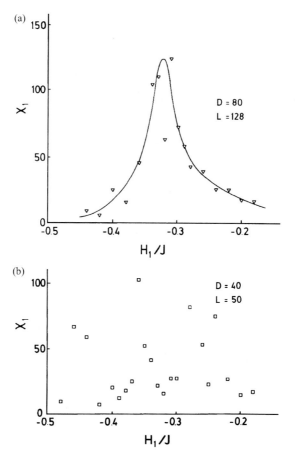

Fig. 5.13 Monte Carlo data for the surface layer suceptibility near the critical wetting transition of an Ising film for $\mathcal{J}_s = \mathcal{J}$, at $\mathcal{J}/k_B T = 0.23$: (a) Metropolis data for $L = 50$, $D = 40$; (b) multispin coding data for $L = 128$, $D = 80$. From Binder *et al.* (1989).

A novel effect which occurs in interacting statistical systems between two walls is the 'Casimir effect'; an overview of this unusual but increasingly popular effect was given by Krech (1994). This effect is the equivalent of the phenomenon in electromagnetism in which a force is produced between two conducting plates separated by a vacuum due to quantum fluctuations in the electromagnetic field in the vacuum. The free energy of a system $\mathcal{F}$ at temperature T and between plates of area A and separated by a distance D can be expressed as the sum of four terms,

$$\lim_{A \to \infty} \frac{\mathcal{F}(T, D)}{k_B T_{c,b} A} = D F_{\text{bulk}}(T) + F_{s,a}(T) + F_{s,b}(T) + \delta F_{a,b}(T, D), \quad (5.46)$$

where $F_{s,a}$ and $F_{s,b}$ are surface free energies, and $\delta F_{a,b}$ is the finite separation contribution which in d-dimensions at the bulk critical point has a contribution

$$\delta F_{a,b}^{\text{sing}}(t = 0, D) = \Delta_{a,b} D^{-(d-1)}, \quad (5.47)$$

where $\Delta_{a,b}$ is the critical Casimir amplitude. The determination of the Casimir amplitude by simulations is quite difficult since it represents only a very small correction to the bulk and surface free energies. This has been done quite successfully by Krech and Landau (1996) using a variation of a method proposed by Mon (1985). As an example we consider two $L \times D$ square lattice systems: in the x-direction there are periodic boundary conditions for both systems, the second system is split into two horizontal strips, each of width L and thickness $D/2$. The top strip has the same boundary conditions as the original lattice and the second one has periodic boundaries. An interaction λ between the strips is used to interpolate between the two systems using umbrella sampling and the sums of different Casimir amplitudes are extracted from the difference. Using this scheme, however, there are finite size effects due to both L and D so an additional extrapolation is needed.

The behavior of the interface itself may be of interest. One well studied problem is that of interface roughening as the temperature is raised for a system which has a 'smooth' interface at low temperatures. (See, e.g. Ising model simulations by Mon et al. (1988, 1990).) At first glance it would seem that the simplest way to impose an interface would be by fixing the top and bottom walls of the system to point in opposite directions. As the interface roughness grows, however, it is possible that there will be excursions of the interface which will hit one of the walls; the 'confinement' of the interface may thus modify its behavior. Instead an antiperiodic boundary may be imposed so that the interface may wander in an unrestricted manner. A periodic boundary may then be used in the directions parallel to the interface (see Fig. 4.11). The interfacial fluctuations may be quite slow and correlation times become quite long for large systems. An 'interface flipping' method has been developed by Hasenbusch and Meyer (1991) for the treatment of interfaces in solid-on-solid models. The essential ingredient in this method is that the entire interface is simply reflected about the mean interface position. This approach has been shown to greatly reduce the correlation time for fluctuations involving the interface near the roughening transition. For the three-dimensional Ising model they found an effective dynamic critical exponent of about 0.4 whereas Metropolis yields $z \sim 2$. Swendsen–Wang updating is even worse than local updating.

5.7.2 Other Monte Carlo schemes

5.7.2.1 Damage spreading

An example of a method which uses existing simulation techniques in a novel way is that of 'damage spreading' (Herrmann, 1990). (This phenomenon was first observed in cellular automata.) Two initial states of the system are prepared in such a way that they differ only slightly. Both systems are then simulated with the same algorithm and the same random number sequence and the difference between the two systems, or the 'damage' is monitored to see if it disappears, stays about the same, or spreads; the spreading of the damage is an indication of the onset of a phase transition. A useful quantitative metric is given by the 'Hamming distance'

$$\Delta(t) = \frac{1}{2N} \sum_i |\sigma_i(t) - \rho_i(t)|, \tag{5.48}$$

where $\{\sigma_i(t)\}$ and $\{\rho_i(t)\}$ are the two 'parallel' time-dependent configurations and N is the number of sites. The dynamical behavior will then be determined 'chaotic' if $\Delta(t)$ goes to a finite value as $t \to \infty$ for $\Delta(0) \to 0$. Studies on the Ising model show that the Hamming distance goes nicely to zero at the critical point. This is an example of a process which cannot be studied theoretically but is quite well suited for Monte Carlo simulation and delivers information in a quite unusual way. One interesting consequence of this process is that it shows that in multilattice coding it can be dangerous to adopt the time saving practice of using the same random number for every lattice if all the lattices are at the same temperature; except exactly at the critical point, all the lattices will eventually reach the same state.

5.7.2.2 Gaussian ensemble method

Challa and Hetheringon (1988) introduced a 'Gaussian ensemble' method which interpolates between the canonical and microcanonical ensemble. A system of N spins is coupled to a bath of N' spins which has a particular functional form for the entropy. (For $N' = 0$ the microcanonical is obtained, for $N' = \infty$ the canonical ensemble results.) The total, composite system is then simulated with the result that 'van der Waals loops' can be traced out clearly in (E,T) space and the small system acts in many ways as a probe. The relative probability of two different states ν and μ is given by

$$\frac{P_\nu}{P_\mu} = \frac{\exp[-a(E_\nu - E_t)^2]}{\exp[-a(E_\mu - E_t)^2]}, \tag{5.49}$$

where E_t is the total energy of the system plus bath and $a \propto 1/N'$. This method may offer certain advantages for the study of first order transitions, but more careful finite size analyses, considering both the size of the system as well as that of the heat bath, still need to be performed. We mention this approach here because it provides an example of how the theoretical concepts of reservoirs and walls, etc. can be used to develop a simple, new simulation technique with properties that differ from their more obvious predecessors.

5.7.3 Finite size effects: a review and summary

We have already seen a number of different cases where the finite system size affects the nature of the results. This may come about because of a discretization of the excitation spectrum in a classical system or due to a limitation of the correlation length near a phase transition in any system. In the latter case a cursory inspection of the data may be incapable of even determining the order of the transition, but we have seen that finite size scaling provides a theoretically well grounded mechanism for the extraction of system behavior in the thermodynamic limit. We have now observed multiple finite size scaling forms and have seen that they are clearly effective. The general feature of all of them is that if appropriate scaling variables are chosen, both the location of the transition temperature as well as a description of the behavior in the infinite system can be accurately extracted. Thus, for example, near a second order phase transition that is reached by changing the temperature, the appropriate scaling variable is $\varepsilon L^{1/\nu}$ as long as the lattice dimension is below the upper critical dimension. If instead the transition is approached by varying a field h that is conjugate to the order parameter, the scaling variable is $hL^{\beta\delta/\nu}$. The complication in all of this is that as the size of the system increases, statistical errors become a problem because of correlations. Thus, the two effects of finite system size and finite sampling time become intertwined. For a temperature driven first order transition the relevant scaling variable becomes εL^d where d is the spatial dimension. In all cases, however, scaling is valid only in some asymptotic size regime which may vary from model to model. There have also been attempts to recast finite size scaling in a form which will enable the extraction of thermodynamic information at much longer size scales. Kim (1993) proposed using the ratio of the finite lattice correlation length and the lattice size as a new scaling variable and Kim et al. (1996) showed that, close enough to the phase transition of the Ising model, the behavior for lattices which were much larger than those which could be measured was accurately predicted.

In summary, then, the key to a successful finite size scaling analysis is the careful examination of the quality of the scaling with particular care given to the identification of systematic deviations (however small) from scaling. This means that the statistical accuracy of the data is important and some compromise must be made between large lattices and high statistics.

5.7.4 More about error estimation

In Chapter 2 we introduced simple concepts of the estimation of errors in Monte Carlo data. A more sophisticated method of error estimation, the 'jackknife' was introduced by Quenouille and Tukey (see e.g. Miller, 1974) to reduce bias in an estimator and to provide a measure of the variance of the resulting estimator by reusing the individual values in the sample. This method, which has its origin in the theory of statistics, is of quite general applicability. Let $A_1, \ldots, A_n$ represent a sample of n independent and identically distributed random variables and $\langle \theta \rangle$ is the estimator calculated using all members of the sample. The data are divided into g groups of size h each, i.e. $n = gh$. Deleting the ith group of h values, one then calculates the estimator $\langle \theta_{-i} \rangle$ based on the remaining groups. The estimator

$$\bar{\theta} = \frac{1}{g}\sum_{i=1}^{g} \tilde{\theta}_i, \qquad (5.50)$$

where

$$\tilde{\theta}_i = g\langle\theta\rangle - (g-1)\langle\theta_{-i}\rangle. \tag{5.51}$$

This estimate eliminates the order n^{-1} bias from the estimator. Error estimates can also be extracted from a jackknife analysis.

5.7.5 Random number generators revisited

In Chapter 2 we briefly touched on the entire matter of random number generation and testing for quality. We now wish to return to this topic and cite a specific example where the deficiencies of several generators could only be clearly seen by careful inspection of the results of a Monte Carlo simulation which was carried out using the generator in question. The Wolff cluster flipping algorithm was used to study 16×16 Ising square lattices using the generators defined in Chapter 2. Most of the simulations were performed exactly at $T = T_c$, and between 5 and 10 runs of 10^7 updates were performed. (Note that for the Swendsen–Wang and Metropolis algorithms, one update means one complete update of the lattice (MCS); in the Wolff algorithm, one update is less than one MCS and depends on the temperature. For simulations at T_c, a Wolff update is ~ 0.55 MCS.) Surprisingly, the use of the 'high quality' generators together with the Wolff algorithm produces *systematically incorrect* results. Simulations using R250 produce energies which are systematically too *low* and specific heats which are too *high* (see Table 5.1). Each of the ten runs was made at the infinite lattice critical temperature and calculated averages over 10^6 MCS; the deviation from the exact value of the energy was over 40σ (standard deviations)! Runs made using the SWC generator gave better results, but even these data showed noticeable systematic errors which had the opposite sign from those produced using R250! In contrast, data obtained using the simple 32 bit congruential generator CONG produced answers which were correct to within the error bars. Even use of the mixed generator SWCW did not yield results which were free of bias, although the systematic errors were much smaller (2σ for the energy and 4σ for the specific heat). Use of another shift-register random number generator, R1279, resulted in data which were in substantially better agreement with exact values than were the R250 values. These data may be contrasted to those which were obtained using the identical random number generators in conjunction with the single spin-flip Metropolis method and the multicluster flipping approach of Swendsen and Wang (1987). For all combinations of simulation methods and random number generators, the energy and specific heat values (shown in Table 5.2) are correct to within a few σ of the respective simulations; except for the CONG generator with Metropolis and R250 with Swendsen–Wang, the answers agree to within 1σ.

The problems which were encountered with the Wolff method are, in principle, a concern with other algorithms. Although Metropolis simulations are not as sensitive to these correlations, as resolution improves some very small bias may appear. In fact, some time after the errors with the Wolff

Table 5.1 *Values of the internal energy for 10 independent runs with the Wolff algorithm for an $L = 16$ Ising square lattice at K_c. The last number in each column, labeled 'dev', gives the difference between the simulation value and the exact value, measured in terms of the standard deviation σ of the simulation.*

	CONG	R250	R1279	SWC	SWCW
	1.453 089	1.455 096	1.453 237	1.452 321	1.453 058
	1.453 107	1.454 697	1.452 947	1.452 321	1.453 132
	1.452 866	1.455 126	1.453 036	1.452 097	1.453 330
	1.453 056	1.455 011	1.452 910	1.452 544	1.453 219
	1.453 035	1.454 866	1.453 040	1.452 366	1.452 828
	1.453 198	1.455 054	1.453 065	1.452 388	1.453 273
	1.453 032	1.454 989	1.453 129	1.452 444	1.453 128
	1.453 169	1.454 988	1.453 091	1.452 321	1.453 083
	1.452 970	1.455 178	1.453 146	1.452 306	1.453 216
	1.453 033	1.455 162	1.452 961	1.452 093	1.453 266
$-\langle E \rangle$	1.453 055	1.455 017	1.453 056	1.452 320	1.453 153
error	0.000 030	0.000 046	0.000 032	0.000 044	0.000 046
dev.	$-0.31\,\sigma$	$42.09\,\sigma$	$-0.27\,\sigma$	$-16.95\,\sigma$	$1.94\,\sigma$

algorithm were first noticed, a separate simulation of the Blume–Capel model (spin-1 Ising model with single ion anisotropy) near the tricritical point revealed asymmetries in the resultant distribution of states between $+1$ and -1 which were clearly with the Metropolis method traced to problems with the random number generator (Schmid and Wilding, 1995). Hidden errors obviously pose a subtle, potential danger for many simulations such as percolation or random walks of various kinds which generate geometric structures using similar 'growth algorithms' as the Wolff method.

The problems with the widely used shift register generator have been attributed to triplet correlations (Heuer *et al.*, 1997; Compagner, 1991). This problem can be simply removed by XORing together two shift register generators with different pairs of lags without too great a loss in speed. The 'universal' properties have been analyzed by Shchur *et al.* (1998) and we refer the reader to this paper and to Heuer *et al.* (1997) for a deeper description of the problems and tests. We do wish to comment that it is nonetheless unclear just how and why these correlations affect specific algorithms in the manner that they do.

To summarize, extensive Monte Carlo simulations on an Ising model for which the exact answers are known have shown that ostensibly high quality random number generators may lead to subtle, but *dramatic*, systematic errors for some algorithms, but not others. Since there is no reason to believe that this model has any special idiosyncrasies, this result should be viewed as another stern warning about the need to very carefully test the implementation of new algorithms. In particular, each specific algorithm must be tested

Table 5.2 *Values of the internal energy (top) and specific heat (bottom) for an*
L = 16 Ising square lattice at K_c. Data were obtained using different random
number generators together with Metropolis and Swendsen–Wang algorithms. The
values labeled 'dev.' show the difference between the simulation results and the
exact values in terms of standard deviations σ of the simulations.

	Metropolis CONG	SW CONG	Metropolis R250	SW R250	Metropolis SWC	SW SWC
$-\langle E \rangle$	1.452 783	1.453 019	1.453 150	1.452 988	1.453 051	1.453 236
error	0.000 021	0.000 053	0.000 053	0.000 056	0.000 080	0.000 041
dev.	$-13.25\ \sigma$	$-0.86\ \sigma$	$1.62\ \sigma$	$-1.36\ \sigma$	$-0.17\ \sigma$	$4.16\ \sigma$
$-\langle C \rangle$	1.497 925	1.498 816	1.498 742	1.496 603	1.498 794	1.499 860
error	0.000 179	0.000 338	0.000 511	0.000 326	0.000 430	0.000 433
dev.	$-4.40\ \sigma$	$0.31\ \sigma$	$0.06\ \sigma$	$-6.47\ \sigma$	$0.19\ \sigma$	$2.65\ \sigma$

together with the random number generator being used *regardless* of the tests
which the generator has previously passed!

5.8 SUMMARY AND PERSPECTIVE

We have now seen a quite broad array of different simulational algorithms
which may be applied to different systems and situations. Many new
approaches have been found to circumvent difficulties with existing methods,
and together with the rapid increase in computer speed the overall increase in

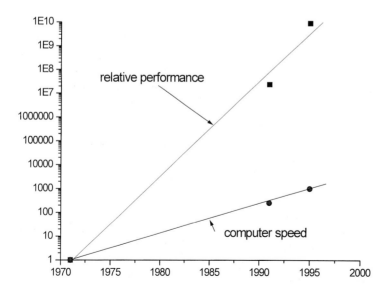

Fig. 5.14 Approximate
variation of Ising
model simulation
performance with
time: (upper curve)
total relative
performance; (lower
curve) relative
improvement in
computer speed.

our capabilities has been enormous. In fact, a brief overview of progress made for the Ising model over the last 25 years, shown in Fig. 5.14, indicates that the improvement due to algorithmic improvements far exceeds that due to raw computer speed alone. Of course, it is not only the improvement in speed which matters but also the net cost. Over the last decade alone the cost of purchasing a machine divided by the speed of the Monte Carlo algorithm has decreased by a factor of 10^4! Ultimately, however, the choice of method depends on the problem being considered, the type of computer which is available, and the judgement of the researcher.

REFERENCES

Alexandrowicz, Z. (1975), J. Stat. Phys. 13, 231.

Berg, B. A. and Neuhaus, T. (1991), Phys. Lett. B 267, 241.

Berg, B. A. and Neuhaus, T. (1992), Phys. Rev. Lett. 68, 9.

Bhatt, R. N. and Young, A. P. (1985) Phys. Rev. Lett. 54, 924.

Binder, K. (1977), Z. Phys. B 26, 339.

Binder, K. (1983), in *Phase Transitions and Critical Phenomena*, Vol.VIII, eds. C. Domb and J. L. Lebowitz (Academic Press, London) p.1.

Binder, K. and Landau, D. P. (1988), Phys. Rev. B 37, 1745.

Binder, K. and Landau, D. P. (1992), J. Chem. Phys. 96, 1444.

Binder, K. and Schröder, K. (1976), Phys. Rev. B 14, 2142.

Binder, K. and Young, A. P. (1986), Rev. Mod. Phys. 58, 801.

Binder, K., Landau, D. P., and Wansleben, S. (1989), Phys. Rev. B 40, 6971.

Binder, K., Evans, R., Landau, D. P., and Ferrenberg, A. M. (1996), Phys. Rev. E 53, 5023.

Bortz, A. B., Kalos, M. H., and Lebowitz, J. L. (1975), J. Comput. Phys. 17, 10.

Boulter, C. J. and Parry, A. O. (1995), Phys. Rev. Lett. 74, 3403.

Brown, F. R. and Woch, T. J. (1987), Phys. Rev. Lett. 58, 2394.

Cardy, J. L. and Jacobsen, J. L. (1997), Phys. Rev. Lett. 79, 4063.

Challa, M. S. S. and Hetherington, J. H. (1988), Phys. Rev. Lett. 60, 77.

Chen, K., Ferrenberg, A. M., and Landau, D. P. (1993), Phys. Rev. B 48, 3249.

Chen, S., Ferrenberg, A. M., and Landau, D. P. (1995), Phys. Rev. E 52, 1377.

Compagner, A. (1991), J. Stat. Phys. 63, 883.

Creutz, M. (1980), Phys. Rev. D 21, 2308.

Creutz, M. (1983), Phys. Rev. Lett. 50, 1411.

Creutz, M. (1987), Phys. Rev. D 36, 515.

De Meo, M., D'Onorio, M., Heermann, D., and Binder, K. (1990), J. Stat. Phys. 60, 585.

Dietrich, S. (1988), in *Phase Transitions and Critical Phenomena*, Vol. XII, eds. C. Domb and J. L. Lebowitz (Academic Press, London) p.1.

Dünweg, B. and Landau, D. P. (1993), Phys. Rev. B 48, 14182.

Edwards, S. F. and Anderson, P. W. (1975), J. Phys. F 5, 965.

Evertz, H. G. and Landau, D. P. (1996), Phys. Rev. B 54, 12,302.

Fortuin, C. M. and Kasteleyn, P. W. (1972), Physica 57, 536.

Frenkel, D. and Ladd, A. J. C. (1984), J. Chem. Phys. 81, 3188.

Goodman, J. and Sokal, A. (1986), Phys. Rev. Lett. 56, 1015.

Griffiths, R. B. (1969), Phys. Rev. Lett. **23**, 17.

Hasenbusch, M. and Meyer, S. (1990), Phys. Lett. B **241**, 238.

Hasenbusch, M. and Meyer, S. (1991), Phys. Rev. Lett. **66**, 530.

Hasenbusch, M. (1995), Nucl. Phys. B **42**, 764.

Hatano, N. and Gubernatis, J. E. (1999) AIP Conf. Proc. **469**, 565

Heermann, D. W. and Burkitt, A. N. (1990), in *Computer Simulation Studies in Condensed Matter Physics II*, eds. D. P. Landau, K. K. Mon, and H.-B. Schüttler, (Springer Verlag, Heidelberg).

Herrmann, H. J. (1990), in *Computer Simulation Studies in Condensed Matter Physics II*, eds. D. P. Landau, K. K. Mon, and H.-B. Schüttler (Springer Verlag, Heidelberg).

Heuer, A., Dünweg, B., and Ferrenberg, A. M. (1997), Comput. Phys. Commun. **103**, 1.

Holm, C. and Janke, W. (1993), Phys. Lett. A **173**, 8.

Hui, K. and Berker, A. N. (1989), Phys. Rev. Lett. **62**, 2507.

Imry, Y. and Ma, S. (1975), Phys. Rev. Lett. **35**, 1399.

Kandel, D., Domany, E., Ron, D., Brandt, A., and Loh, Jr., E. (1988), Phys. Rev. Lett. **60**, 1591.

Kandel, D., Domany, E., and Brandt, A. (1989), Phys. Rev. B **40**, 330.

Kasteleyn, P. W. and Fortuin, C. M. (1969), J. Phys. Soc. Japan Suppl. **26s**, 11.

Kawashima, N. and Young, A. P. (1996), Phys. Rev. B **53**, R484.

Kim, J.-K. (1993), Phys. Rev. Lett. **70**, 1735.

Kim, J.-K., de Souza, A. J. F., and Landau, D. P. (1996), Phys. Rev. E **54**, 2291.

Kirkpatrick, S., Gelatt, Jr., S. C., and Vecchi, M. P. (1983), Science **220**, 671.

Kolesik, M., Novotny, M. A., and Rikvold, P. A. (1998), Phys. Rev. Lett. **80**, 3384.

Krech, M. (1994), *The Casimir Effect in Critical Systems* (World Scientific, Singapore).

Krech, M. and Landau, D. P. (1996), Phys. Rev. E **53**, 4414.

Landau, D. P. (1992), in *The Monte Carlo Method in Condensed Matter Physics*, ed. K. Binder (Springer, Berlin).

Landau, D. P. (1994), Physica A **205**, 41.

Landau, D. P. (1996), in *Monte Carlo and Molecular Dynamics of Condensed Matter Systems*, eds. K. Binder and G. Ciccotti (Società Italiana de Fisica, Bologna).

Landau, D. P. and Binder, K. (1981), Phys. Rev. B **24**, 1391.

Landau, D. P. and Krech, M. (1999), J. Phys. Cond. Mat. **11**, 179.

Laradji, M., Landau, D. P., and Dünweg, B. (1995), Phys. Rev. B **51**, 4894.

Lee, J. and Kosterlitz, J. M. (1990), Phys. Rev. Lett. **65**, 137.

Marsaglia, G. (1972), Ann. Math. Stat. **43**, 645.

Meirovitch, H. and Alexandrowicz, Z. (1977), Mol. Phys. **34**, 1027.

Miller, R. G. (1974), Biometrika **61**, 1.

Mon, K. K. (1985), Phys. Rev. Lett. **54**, 2671.

Mon, K. K., Wansleben, S., Landau, D. P., and Binder, K. (1988), Phys. Rev. Lett. **60**, 708.

Mon, K. K., Landau, D. P., and Stauffer, D. (1990), Phys. Rev. B **42**, 545.

Novotny, M. A. (1995a), Phys. Rev. Lett. **74**, 1.

Novotny, M. A. (1995b), Computers in Physics **9**, 46.

Ogielski, A. T. (1985), Phys. Rev. B **32**, 7384.

Schmid, F. and Wilding, N. B. (1995), Int. J. Mod. Phys. C **6**, 781.

Selke, W., Shchur, L. N., and Talapov, A. L. (1994), in *Annual Reviews of Computer Science I*, ed. D. Stauffer (World Scientific, Singapore) p.17.

Shchur, L. N. and Butera, P. (1998), Int. J. Mod. Phys. C **9**, 607.

Sweeny, M. (1983), Phys. Rev. B **27**, 4445.

Swendsen, R. H. and Wang, J.-S. (1987), Phys. Rev. Lett. **58**, 86.

Wang, J.-S., Swendsen, R. H., and Kotecký, R. (1989), Phys. Rev. Lett. **63**, 109.

Wang, J.-S. (1989), Physica A **161**, 249

Wang, J.-S. (1990), Physica A **164**, 240.

Wansleben, S. (1987), Comput. Phys. Commun. **43**, 315.

Wiseman, S. and Domany, E. (1995), Phys. Rev. E **52**, 3469.

Wiseman, S. and Domany, E. (1998), Phys. Rev. Lett. **81**, 22.

Wolff, U. (1988), Nucl. Phys. B **300**, 501.

Wolff, U. (1989a), Phys. Rev. Lett. **62**, 361.

Wolff, U. (1989b), Nucl. Phys. B **322**, 759.

Wolff, U. (1990), Nucl. Phys. B **334**, 581.

Young, A. P. (1996), in *Monte Carlo and Molecular Dynamics of Condensed Matter Systems*, eds. K. Binder and G. Ciccotti (Società Italiana di Fisica, Bologna).

Young, A. P. (1998), (ed.) *Spin Glasses and Random Fields* (World Scientific, Singapore).

Zorn, R., Herrmann, H. J., and Rebbi, C. (1981), Comput. Phys. Commun. **23**, 337.

6   Off-lattice models

6.1 FLUIDS

6.1.1 *NVT* ensemble and the virial theorem

The examination of the equation of state of a two-dimensional model fluid (the hard disk system) was the very first application of the importance sampling Monte Carlo method in statistical mechanics (Metropolis *et al.*, 1953), and since then the study of both atomic and molecular fluids by Monte Carlo simulation has been a very active area of research. Remember that statistical mechanics can deal well analytically with very dilute fluids (ideal gases!), and it can also deal well with crystalline solids (making use of the harmonic approximation and perfect crystal lattice periodicity and symmetry), but the treatment of strongly correlated dense fluids (and their solid counterparts, amorphous glasses) is much more difficult. Even the description of short range order in fluids in a thermodynamic state far away from any phase transition is a non-trivial matter (unlike the lattice models discussed in the last chapter, where far away from phase transitions the molecular field approximation or a variant thereof is usually both good enough and easily worked out, and the real interest is generally in phase transition problems).

We are concerned here with classical mechanics only (the extension to the quantum case will be treated in Chapter 8) and then momenta of the particles cancel out from the statistical averages of any observables A, which are given as

$$\langle A \rangle_{N,V,T} = \frac{1}{Z} \int d\mathbf{X} A(\mathbf{X}) e^{-U(\mathbf{X})/k_B T}. \qquad (6.1)$$

Here we have specialized the treatment to a case where there are N point particles in a box of volume V in thermal equilibrium at a given temperature T: This situation is called the NVT-ensemble of statistical mechanics. The phase space $\{\mathbf{X}\}$ is spanned by all the coordinates $\mathbf{r}_i$ of the N point particles, i.e. $\{\mathbf{X}\} = \{\mathbf{r}_1, \mathbf{r}_2, \ldots, \mathbf{r}_N\}$ and is $3N$-dimensional. Each point in that space contributes to the average Eqn. (6.1) with the Boltzmann weight,

$$P(\mathbf{X}) = e^{-U(\mathbf{X})/k_B T}/Z, \qquad (6.2)$$

which is the continuum analog of the weight that we have encountered for the discrete lattice models (Eqn. (4.5)). Here $U(\mathbf{X})$ is not the total energy, of course, but only the total potential energy (the kinetic energy has cancelled out from the average). Often it is assumed that $U(\mathbf{X})$ is simply a sum of pair-wise interactions $U(\mathbf{r}_i - \mathbf{r}_j)$ between point particles at positions $\mathbf{r}_i$, $\mathbf{r}_j$,

$$U(\mathbf{X}) = \sum_{i<j} u(\mathbf{r}_i - \mathbf{r}_j), \tag{6.3}$$

but sometimes three-body and four-body interactions, etc., are also included. A standard choice for a pair-wise potential is the Lennard-Jones interaction

$$U_{LJ}(r) = 4\varepsilon\big[(\sigma/r)^{12} - (\sigma/r)^6\big], \tag{6.4}$$

ε being the strength and σ the range of this potential.

Problem 6.1 Determine the location and depth of the minimum of the Lennard–Jones potential. At which distance has this potential decayed to about $1/1000$ of its depth in the minimum?

Many other potentials have also been used in the literature; examples include the use of hard-core interactions to represent the repulsion at short distances,

$$u(r) = \infty, \quad r < r_0, \quad u(r) = 0, \quad r > r_0, \tag{6.5}$$

additional soft-sphere attractions,

$$u(r) = \infty, \quad r < r_0, \quad u(r) = -\varepsilon, \quad r_0 \le r < r_1, \quad u(r) = 0, \quad r > r_1 \tag{6.6}$$

and inverse power law potentials,

$$u(r) = (\sigma/r)^n, \quad n = \text{integer}, \tag{6.7}$$

etc.

Just as in the Monte Carlo algorithm for a lattice classical spin model, where a spin $(\mathbf{S}_i)$ was randomly selected and a new spin orientation was proposed as the basic Monte Carlo step, we now select a particle i at random and consider a random displacement δ from its old position $\mathbf{r}_i' = \mathbf{r}_i + \delta$ to a new position. This displacement vector δ is chosen randomly and uniformly from some volume region, ΔV, whose size is fixed such that the acceptance probability for the proposed move is on average neither close to unity nor close to zero. As in the case of the lattice model, the acceptance probability for the move, $W(\mathbf{r}_i \to \mathbf{r}_i')$, depends on the energy change $\Delta U = U(\mathbf{r}_i') - U(\mathbf{r}_i)$, given by the Boltzmann factor

$$W(\mathbf{r}_i - \mathbf{r}_i') = \min\{1, \exp(-\Delta U/k_B T)\}. \tag{6.8}$$

The implementation of the algorithm is thus quite analogous to the lattice case and can be summarized by the following steps:

> ### 'Off-lattice' Metropolis Monte Carlo method
>
> **(1)**   Choose an initial state (to avoid difficulties when particles are very close to each other and U thus very large, one frequently distributes particles on the sites of a regular face-centered cubic lattice).
> **(2)**   Consider a particle with a randomly chosen label i and calculate a trial position $\mathbf{r}'_i = \mathbf{r}_i + \delta$.
> **(3)**   Calculate the energy change ΔU which results from this displacement.
> **(4)**   If $\Delta U < 0$ the move is accepted; go to (2).
> **(5)**   If $\Delta U > 0$, a random number η is chosen such that $0 < \eta < 1$.
> **(6)**   If $\eta < \exp(-\Delta U / k_B T)$, accept the move and in any case go then to (2). Note that if such a trial move is rejected, the old configuration is again counted in the averaging.

From this algorithm it is straightforward to calculate quantities like the average potential energy $\langle U \rangle_{NVT}$, or structural information like the radial pair distribution function $g(r)$, but in order to obtain the equation of state, one would also like to know the pressure p. Since this is an intensive variable, it is not so straightforward to obtain it from Monte Carlo sampling as it would be for any density of extensive variable. Nevertheless there is again a recipe from statistical mechanics that helps us, namely the virial theorem

$$p = \rho k_B T + \frac{1}{dV} \left\langle \sum_{i<j} \mathbf{f}(\mathbf{r}_i - \mathbf{r}_j) \cdot (\mathbf{r}_i - \mathbf{r}_j) \right\rangle, \tag{6.9}$$

where $\rho \equiv N/V$ is the particle density, $\mathbf{f}(\mathbf{r}_i - \mathbf{r}_j)$ is the force between particles i and j, and d is the spatial dimension. Since for the continuous, pairwise interactions considered above, such as those in Eqns. (6.4) and (6.7), the forces are easily related to derivatives du/dr of these potentials and one can re-express the virial theorem in terms of the pair distribution function. In $d = 3$ dimensions this yields

$$p = \rho k_B T - \frac{2}{3}\pi \rho^2 \int_0^\infty dr\, r^3 \frac{du(r)}{dr} g(r). \tag{6.10}$$

Of course, these expressions for the pressure do not work for potentials that are discontinuous, such as those in Eqns. (6.5) and (6.6), and other techniques then have to be used instead. Finally, we note that the internal energy and the compressibility can also be conveniently expressed in terms of the pair distribution function which in $d = 3$ dimensions is

$$\langle U \rangle / N = 2\pi\rho \int_0^\infty dr\, r^2 u(r) g(r), \tag{6.11a}$$

$$\kappa/\kappa_{\mathrm{id}} = 1 + 4\pi\rho \int_0^\infty dr\, r^2 [g(r) - 1], \qquad (6.11\mathrm{b})$$

where κ_{id} is the ideal gas compressibility.

Problem 6.2 Write a program that approximates $g(r)$ via a histogram, binning together particles that fall within a distance interval $[r, r + \Delta r]$ from each other.

Problem 6.3 Generalize Eqns. (6.10)–(6.11) to dimensions $d = 2$ and $d = 4$.

6.1.2 NpT ensemble

The isobaric–isothermal ensemble is very often used in Monte Carlo simulations of fluids and solids, in particular when one wishes to address problems such as the fluid–solid transition or transitions among different solid phases. At such first order transitions, first derivatives (such as internal energy U, volume V) of the appropriate thermodynamic potential exhibit a jump (e.g. ΔU, ΔV). Using such an extensive variable (like the volume V) as a control parameter of a simulation, however, causes particular problems if the chosen value of V falls in the 'forbidden region' of this jump. It means that in thermal equilibrium the system should separate into two coexisting phases (e.g. if we cool down a box containing water molecules from high temperature to room temperature at any intermediate density N/V between that of water vapor and that of water at room temperature). This separation can be observed in the framework of NVT simulations in simple cases, e.g. for a two-dimensional Lennard-Jones fluid this is seen in the snapshots (Rovere *et al.*, 1990) in Fig. 6.1, but reaching equilibrium in such a computer simulation of phase separation is rather cumbersome. Also, averaging any observables in such a two-phase coexistence regime is a tricky business – obviously in Fig. 6.1 it would be hard to disentangle which features are due to the gas phase, which are due to the liquid phase, and which are attributed to the interface.

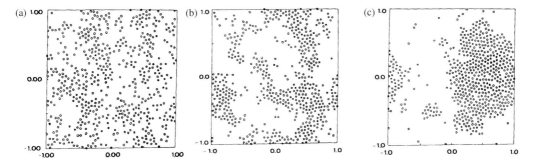

Fig. 6.1 Snapshots of 576 particles at a density $\rho^* = 0.3$ for $T^* =$ (a) 0.7, (b) 0.5, and (c) 0.45. Here ρ^*, T^* are density and temperature in reduced units, i.e. the Lennard-Jones parameters σ and ε/k_B are used as units of length and temperature, respectively. From Rovere *et al.* (1990).

(By the way, interfaces are slowly fluctuating objects and are hard to characterize quantitatively, see Section 4.2.3.6.) Sometimes phase separation is
even missed, either because the system is too small, or because of hysteresis.
As a result, for a study of phase transitions in off-lattice systems it is often
preferable to use the NpT ensemble (or the grand canonical μVT ensemble
where the chemical potential μ rather than the pressure p is used as a second
intensive thermodynamic variable to characterize the static system). For systems with continuous potentials the first use of the NpT ensemble dates back
to 1972 (McDonald, 1972). We follow Frenkel and Smit (1996) in deriving it
from statistical mechanics. To begin with we consider the partition function
$Z(N, V, T)$ in the canonical (NVT) ensemble for a box $V = L^3$ in three
dimensions,

$$
Z(N, V, T) = \frac{1}{\Lambda^{3N} N!} \int_0^L \cdots \int_0^L d\mathbf{r}_1 \ldots d\mathbf{r}_N \exp[-U(\mathbf{r}_1, \ldots, \mathbf{r}_N)/k_B T],
$$

(6.12)

where the prefactors ensure the proper normalization of entropy via the
quasi-classical limit of quantum mechanics (Λ is the thermal de Broglie
wavelength of the atoms and the factor $1/N!$ accounts for the indistinguish-
ability of the particles).

In the NpT ensemble the volume V, and hence the linear dimension L, is
not fixed but is a fluctuating quantity. It is convenient to define scaled
coordinates $\mathbf{s}_i$ by

$$
\mathbf{r}_i = L\mathbf{s}_i, \qquad \text{for } i = 1, 2, \ldots, N, \tag{6.13}
$$

and treat the $\{\mathbf{s}_i\}$ and the linear dimension L as separate variables $\{\mathbf{s}_i, L\}$. The
(Helmholtz) free energy $F(N, V, T)$ thus is written as

$$
\begin{aligned}
F(N, V, T) &= -k_B T \ln Z(N, V, T) \\
&= -k_B T \ln \left\{ \frac{[V/\Lambda^3]^N}{N!} \right\} \\
&\quad - k_B T \ln \int_0^1 \cdots \int_0^1 d\mathbf{s}_N \exp\left[-\frac{U(\mathbf{s}_1, \ldots, \mathbf{s}_N, L)}{k_B T} \right] \\
&= F_{ig}(N, V, T) + \Delta F(N, V, T),
\end{aligned}
$$

(6.14)

where the first term has been identified as the well-known expression for the
free energy of the ideal gas, $F_{ig}(N, V, T)$, and $\Delta F(N, V, T)$ is the non-
trivial part involving all the interactions among the particles. Of course, U
depends originally on the actual coordinates $\mathbf{r}_1, \ldots, \mathbf{r}_N$, and when we write U
in terms of the $\{\mathbf{s}_i\}$ we must allow for L as an additional variable.

Now we consider the situation in which the system under consideration is
actually a subsystem of a much larger ideal gas system of volume V_0, with
$V_0 \gg V$, which acts as a heat bath (exchange of energy but not of particles is
possible), and from which it is separated by a piston which is free to move.

Denoting the total number of atoms as M, we find that there are hence $(M - N) \gg N$ atoms in the reservoir. The partition function of the total system is simply the product of the partition function of these two subsystems,

$$Z(N, M - N, V, V_0 - V, T)$$

$$= \frac{V^N (V_0 - V)^{M-N}}{N!(M - N)!} \Lambda^{-3M} \int_0^1 \cdots \int_0^1 d\mathbf{s}_1' \dots d\mathbf{s}_{M-N}' \int_0^1 \cdots \int_0^1 d\mathbf{s}_1 \dots d\mathbf{s}_N \qquad (6.15)$$

$$\exp\left[-\frac{U(\mathbf{s}_1, \dots, \mathbf{s}_N, L)}{k_B T}\right].$$

Note that the integral over the $3(M - N)$ scaled coordinates $\mathbf{s}_1', \dots, \mathbf{s}_{M-N}'$ of the ideal gas particles simply yields unity. The probability density $P(V)$ that the N-particle subsystem has the volume V then is

$$P(V) =$$

$$\frac{V^N (V_0 - V)^{M-N} \int_0^1 \cdots \int_0^1 d\mathbf{s}_1 \dots d\mathbf{s}_N \exp[-U(\mathbf{s}_1, \dots, \mathbf{s}_N, L)/k_B T]}{\int_0^{V_0} dV' V'^N (V_0 - V')^{M-N} \int_0^1 \cdots \int_0^1 d\mathbf{s}_1 \dots d\mathbf{s}_N \exp[-U(\mathbf{s}_1, \dots, \mathbf{s}_N, L')/k_B T]}.$$

$$(6.16)$$

Let us now exploit the fact that we consider the limit $V_0 \to \infty$, $M \to \infty$ but with $(M - N)/V_0 = \rho$ held fixed. In that limit, a minor volume change of the small system does not alter the pressure p of the large system. In order to introduce the pressure p in Eqns. (6.15) and (6.16), in the limit $V/V_0 \to 0$ we can write

$$(V_0 - V)^{M-N} = V_0^{M-N}[1 - (V/V_0)]^{M-N}$$
$$\approx V_0^{M-N} \exp[-(M - N)V/V_0] = V_0^{M-N} \exp[-\rho V] \qquad (6.17)$$

and simply use the ideal gas law $\rho = p/k_B T$ to replace the exponential factor in Eqn. (6.17) by $\exp(-pV/k_B T)$. Integrating the partition function over the volume V and splitting off the partition function of the reservoir, $V_0^{M-N}/[(M - N)! \Lambda^{3(M-N)}]$, we obtain the partition function $Y(N, p, T)$ in the NpT ensemble

$$Y(N, p, T) \equiv \frac{p/k_B T}{\Lambda^{3N} N!} \int dV\, V^N \exp(-pV/k_B T) \int_0^1 \cdots \int_0^1 d\mathbf{s}_1 \dots d\mathbf{s}_N \qquad (6.18)$$

$$\exp[-U(\mathbf{s}_1, \dots, \mathbf{s}_N, L)/k_B T].$$

The probability density $P(V)$ then becomes

$$P(V) =$$

$$\frac{V^N \exp(-pV/k_\mathrm{B}T) \int_0^1 \cdots \int_0^1 d\mathbf{s}_1 \ldots d\mathbf{s}_N \exp[-U(\mathbf{s}_1, \ldots, \mathbf{s}_N, L)/k_\mathrm{B}T]}{\int_0^{V_0} dV' \, V'^N \exp(-pV'/k_\mathrm{B}T) \int_0^1 \cdots \int_0^1 d\mathbf{s}_1 \ldots d\mathbf{s}_N \exp[-U(\mathbf{s}_1, \ldots, \mathbf{s}_N, L)/k_\mathrm{B}T]}.$$

$$(6.19)$$

The partition function $Y(N, p, T)$ yields the Gibbs free energy as usual, $G(N, p, T) = -k_\mathrm{B}T \ln Y(N, p, T)$. Equation (6.19) is now the starting point for the NpT Monte Carlo method. We note that the probability density of finding the subsystem in a specific configuration of the N atoms (as specified by $\mathbf{s}_1, \ldots, \mathbf{s}_N$) and a volume V is

$$P(\mathbf{s}_1, \ldots, \mathbf{s}_N, V) \propto V^N \exp(-pV/k_\mathrm{B}T) \exp[-U(\mathbf{s}_1, \ldots, \mathbf{s}_N, L)/k_\mathrm{B}T]$$
$$= \exp\{-[U(\mathbf{s}_1, \ldots, \mathbf{s}_N, L) + pV - Nk_\mathrm{B}T \ln V]/k_\mathrm{B}T\}.$$

$$(6.20)$$

Equation (6.20) looks like the Boltzmann factor for traditional Monte Carlo sampling if the square bracket is interpreted as a generalized 'Hamiltonian', involving an extra variable, $V = L^3$. Thus, trial moves which change V have to be carried out, and these must satisfy the same rules as trial moves in the particle positions $\{\mathbf{s}_i\}$. For example, consider attempted changes from V to $V' = V + \Delta V$, where ΔV is a random number uniformly distributed in the interval $[-\Delta V_{max}, +\Delta V_{max}]$ so that $V' = L'^3$. In the Metropolis scheme, the acceptance probability of such a volume changing move is hence

$$W(V \to V') = \min\left\{1, \exp\left(-\frac{1}{k_\mathrm{B}T}[U(\mathbf{s}_1, \ldots, \mathbf{s}_N; L')\right.\right.$$
$$\left.\left. - U(\mathbf{s}_1, \ldots, \mathbf{s}_N; L) + p(V' - V) - k_\mathrm{B}TN \ln(V'/V)]\right)\right\}.$$

$$(6.21)$$

The frequency with which 'volume moves' should be tried in place of the standard particle displacements $\mathbf{r}_i \to \mathbf{r}_i'$ depends on the efficiency with which phase space is then sampled by the algorithm. In general, a volume trial move could mean that all interatomic interactions are recomputed, which would need a cpu time comparable to N trial moves on the atomic positions. Fortunately, for potentials which can be written as a sum over terms U_n that are simple inverse powers of interatomic distances there is a scaling property that makes the volume changing trial move much 'cheaper'. We can see this by writing

$$U_n = \sum_{i<j} \varepsilon(\sigma/|\mathbf{r}_i - \mathbf{r}_j|)^n = L^{-n} \sum_{i<j} \varepsilon(\sigma/|\mathbf{s}_i - \mathbf{s}_j|)^n, \qquad (6.22)$$

from which we can infer that $U_n(L') = (L/L')^n U_n(L)$. Note, however, that Eqn. (6.22) is only true for an *untruncated* potential (cf. Section 6.2).

In order to check the equilibration of the system (and the validity of the implementation of the algorithm!) it is also advisable to calculate the pressure p from the virial theorem (see Eqns. (6.9, 10)) in such an NpT ensemble, since one can prove that the virial pressure and the externally applied pressure (that appears in the probability, Eqn. (6.21)) must agree. Finally, we mention that in solids (which are intrinsically anisotropic!) a generalization of this algorithm applies where one does not consider isotropic volume changes but anisotropic ones. For an orthorhombic crystal it is thus necessary to have a box with three different linear dimensions L_x, L_y, L_z, and in the NpT ensemble these three linear dimensions may change separately. We shall return to an example for this case in Section 6.6.

6.1.3 Grand canonical ensemble

The grand canonical ensemble μVT uses the volume V and the chemical potential μ as independent thermodynamic variables along with the temperature T. While in the NpT ensemble the particle number N was fixed and the volume could fluctuate, here it is exactly the other way around. Of course, in the thermodynamic limit ($N \to \infty$ or $V \to \infty$, respectively) fluctuations are negligible, and the different ensembles of statistical mechanics yield equivalent results. However, in computer simulations one often wishes to choose N and/or V as small as possible, in order to save cpu time. Then the optimal choice of statistical ensembles is a non-trivial question, the answer to which depends both on the type of physical system being studied and the type of properties to be calculated. As an example, consider the study of adsorption of small gas molecules in the pores of a zeolite crystal (see e.g. Catlow, 1992; Smit, 1995). Then the adsorbate in an experiment is in fact in contact with a gas reservoir with which it can exchange particles, and this is exactly the type of equilibrium described by the μVT ensemble. Choosing this ensemble to simulate an 'adsorption isotherm' (describing the amount of adsorbed gas as a function of the gas pressure in the reservoir) has the advantage that the simulation closely parallels the experiment. It may also be advantageous to choose the μVT ensemble for other cases, e.g. for a study of the liquid/gas transition and critical point of a bulk fluid (Wilding, 1997). Experimental studies of this problem typically are done in the NVT or NpT ensembles, respectively. Simulations of fluid criticality have been attempted as well, both in the NVT ensemble (Rovere *et al.*, 1990) and the NpT ensemble (Wilding and Binder, 1996), but these approaches are clearly less efficient than the simulations in the μVT ensemble (Wilding, 1997).

The grand canonical partition function is written

$$
Y(\mu, V, T) = \sum_{N=0}^{\infty} \frac{1}{N!} (V/\Lambda^3)^N \exp(\mu N/k_B T) \int d\mathbf{s}_1, \ldots, \int d\mathbf{s}_N
$$
$$
\exp[-U(\mathbf{s}_1, \ldots, \mathbf{s}_N)/k_B T], \tag{6.23}
$$

where the $\mathbf{s}_i$ are the scaled coordinates of the particles, Eqn. (6.14). Note that we again consider only a cubic box in $d = 3$ dimensions here, $V = L^3$. Then the corresponding probability density is

$$
\mathcal{N}_{\mu V T}(\mathbf{s}_1, \ldots, \mathbf{s}_N, N) \propto \frac{1}{N!} \left(\frac{V}{\Lambda^3}\right)^N \exp\{-[U(\mathbf{s}_1, \ldots, \mathbf{s}_N) - \mu N]/k_B T\}. \tag{6.24}
$$

This probability density can be sampled by a Metropolis Monte Carlo method (see Chapter 4). In addition to trial moves that displace particles (the acceptance probability for such moves is still given by Eqn. (6.8)) trial moves for the insertion or removal of particles from the reservoir are also introduced. The insertion of a particle at a randomly selected position $\mathbf{s}_{N+1}$ is accepted with the probability (Norman and Filinov, 1969)

$$
W(N \to N + 1) = \min\left\{1, \frac{V}{\Lambda^3(N+1)} \exp\{-[U(\mathbf{s}_1, \ldots, \mathbf{s}_{N+1})\right.
$$
$$
\left. -U(\mathbf{s}_1, \ldots, \mathbf{s}_N) - \mu]/k_B T\}\right\}, \tag{6.25}
$$

while the removal of a randomly chosen particle is accepted with the probability

$$
W(N \to N - 1) = \min\left\{1, \frac{\Lambda^3 N}{V} \exp\{-[U(\mathbf{s}_1, \ldots, \mathbf{s}_N) - U(\mathbf{s}_1, \ldots, \mathbf{s}_{N-1})\right.
$$
$$
\left. + \mu]/k_B T\}\right\}. \tag{6.26}
$$

Since the particles are indistinguishable, their labeling is arbitrary, and hence in Eqn. (6.26) we have given the particle that was removed the index N. Obviously, two successive (successful) events in which a particle is removed at a site $\mathbf{s}_N$ and inserted at a site $\mathbf{s}_N'$ have the same effect as a (random) move from $\mathbf{s}_N$ to $\mathbf{s}_N'$ according to Eqn. (6.8). Therefore, these displacement moves are not actually necessary, and one can set up a simulation program that includes random insertions and removals exclusively. For densities which are not too large (but including the critical density of a fluid (Wilding, 1997)), such an algorithm is in effect very efficient, much more so than the simple random displacement algorithm of Eqn. (6.8). This is true because the effective displacements generated are of the order of the linear dimension of the box while the displacements generated by the algorithm of Eqn. (6.8) are of the order δ, a length typically chosen of the same order as the range σ of the inter-particle potential. On the other hand, the efficiency of this straight-forward implementation of the grand canonical Monte Carlo algorithm deteriorates very quickly when the density increases – for dense fluids near their fluid–solid transition successful attempts of a particle insertion are extremely rare, and thus the method becomes impractical, at least in this straightforward form.

A particular advantage of grand canonical simulations of gas–fluid criticality is that the analysis in terms of finite size scaling is most natural in this ensemble (Wilding, 1997; see also Section 4.3.5). As has already been discussed in Section 4.3.5, for an accurate analysis of this situation one needs to properly disentangle density fluctuations and energy density fluctuations in terms of the appropriate 'scaling fields'. In this way, critical phenomena in fluids can be studied with an accuracy which is nearly competitive to that in corresponding studies of lattice systems (Chapter 4).

Extensions to binary (A, B) or multicomponent mixtures can also be straightforwardly considered. For the grand canonical simulation of a binary mixture, two chemical potentials μ_A, μ_B are needed, of course, and the term μN in Eqn. (6.24) is generalized to $\mu_A N_A + \mu_B N_B$. Then the moves in Eqns. (6.25) and (6.26) must distinguish between the insertion or removal of an A particle or a B particle. An important extension of the fully grand canonical simulation of mixtures is the so-called *semi-grand canonical simulation* technique, where the total particle number $N_{tot} = N_A + N_B$ is held fixed and only the chemical potential difference $\Delta\mu = \mu_A - \mu_B$ is an independent variable, since then $\mu_A N_A + \mu_B N_B = \Delta\mu N_A + \mu_B N_{tot}$ and the second term $\mu_B N_{tot}$ then cancels out from the transition probability. Thus, the moves consist of the removal of a B particle and insertion of an A particle at the same position, or vice versa. Alternatively, one can consider this move as an 'identity switch': an A particle transforms into B or a B into A. The obvious advantage of this algorithm is that it still can be efficient for very dense systems, where the standard grand canonical algorithm is bound to fail. Thus the semi-grand canonical method can be generalized from simple monatomic mixtures to such complex systems as symmetrical mixtures of flexible polymers (Sariban and Binder, 1987). An entire polymer chain then undergoes such an 'identity switch', keeping its configuration constant. In addition, other moves are needed to sample the possible configurations, and these will be described in Section 6.6 below.

Problem 6.4 Demonstrate that the algorithm defined by Eqns. (6.25) and (6.26) satisfies the detailed balance principle with the semi-grand canonical probability distribution, Eqn. (6.24).

Problem 6.5 Write down the transition probabilities and the grand canonical probability distribution for a Monte Carlo algorithm that samples the lattice gas model, Eqn. (2.49), at given volume of the lattice $V = L^3$, temperature T and chemical potential μ. Discuss the differences between the result and Eqns. (6.24)–(6.26).

6.1.4 Subsystems: a case study

In dense off-lattice systems particle insertions often are very hard to perform, and simulations in the grand canonical ensemble are impractical. Nevertheless, equivalent information often is easily deduced from a study of subsystems of a larger system that is simulated in the standard canonical NVT ensemble (Rovere *et al.*, 1990; Weber *et al.*, 1995). A study of subsystems is attractive because from a single simulation one can obtain information about both finite size behavior and response functions that is not accessible otherwise. In order to explain how this is done, we best proceed by way of an example, and for this purpose we choose the solid–liquid transition of hard disks in $d = 2$ dimensions. Actually this model system has been under study since the very first application of the importance sampling Monte Carlo method (Metropolis *et al.*, 1953), and many classic papers have appeared since then (e.g. Alder and Wainwright, 1962; Zollweg and Chester, 1992).

The total system of size $S \times S$ is divided into $L \times L$ subsystems of linear dimension $L = S/M_b$ with $M_b = 1, 2, 3, 4, \ldots$ up to a value at which the subsystem size becomes too small for a meaningful analysis. The boundaries of these subsystems have no physical effect whatsoever; they only serve to allow a counting of which particle belongs to which subsystems, so information on subsystem properties for all subsystem sizes is deduced simultaneously from the same simulation run. (Actually, one can also choose non-integer M_b to allow a continuous variation of L, choose subsystems of spherical rather than quadratic shape, if desired, etc.). Such subsystem properties are, first of all, the density ρ, and in the present example another quantity of interest is the bond orientational order parameter ψ defined as

$$\psi = \left| \sum_i \sum_j \exp(6i\phi_{ij}) \right| / N_{\text{bond}}, \tag{6.27}$$

where the sum over i runs over all particles in the subsystem and the sum over j runs over all neighbors of i (defined by the criterion that the distance is less than 1.3 times the close packing distance). ϕ_{ij} is the angle between the 'bond' connecting neighbors i and j and an arbitrary but fixed reference axis, and N_{bond} is the number of bonds included in the sums in Eqn. (6.27).

A study of the probability distribution $P_L(\psi, \rho)$ is illuminating (see Fig. 6.2) as it allows the estimation of various response functions. While we expect that ρ and ψ fluctuate independently of each other in the disordered phase, this is not so in the ordered phase where an increase of ρ also enhances ψ, and a cross-correlation $\langle \Delta\psi\Delta\rho \rangle$ is thus non-vanishing. For linear dimensions L much larger than the (largest) correlation length ξ we can assume a Gaussian probability distribution (Landau and Lifshitz, 1980; Weber *et al.*, 1995)

$$P_L(\psi, \rho) \propto \exp\left\{ -\frac{L^d}{2} \left[\frac{(\Delta\psi)^2}{\chi_{L,\rho}} - \frac{\Delta\psi\Delta\rho}{\gamma_L} + \frac{(\Delta\rho)^2}{\kappa_{L,\psi}} \right] \right\}, \tag{6.28}$$

Fig. 6.2 Contour plot of the joint probability distribution $P(\psi, \rho)$ of the bond–orientational order parameter ψ and the subsystem density ρ for subsystems with $M_b = 6$, at a system density (in units of the close packing density) of (a) $\rho = 0.78$ (fluid phase) and (b) $\rho = 0.95$ (solid phase). The total number of particles is $N = 2916$, and averages were taken over 600 000 MCS/particle. From the outermost to the innermost contour the probability increases as $i\Delta p$, $i = 1, 2, 3, 4, 5$, with (a) $\Delta p = 0.000\,965$ and (b) $\Delta p = 0.000\,216$. Note that in the disordered phase the peak of $P(\psi, \rho)$ occurs at a non-zero value of ψ, because ψ is the absolute value of a two-component order parameter. From Weber *et al.* (1995).

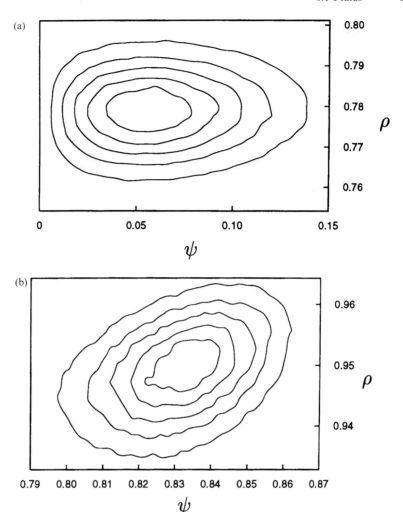

with the fluctuations $\Delta\psi \equiv \psi - \langle\psi\rangle_L$ and $\Delta\rho \equiv \rho - \langle\rho\rangle_L$. The bond orientational 'susceptibility' measured in a system of linear dimension L at constant density ρ is denoted by $\chi_{L,\rho}$, and γ_L^{-1} is the coupling parameter measured on the same length scale L, while $\kappa_{L,\psi}$ denotes the compressibility measured on length scale L at a constant value $\langle\psi\rangle$ of the order parameter. Note that factors $1/k_BT$ have been absorbed in these definitions throughout.

From Eqn. (6.28) we can derive an expression for the differences between the 'susceptibilities' at constant density $\chi_{L,\rho}$ and constant chemical potential $\chi_{L,\mu}$. Note that a subsystem with $L \ll S$ can freely exchange particles with a much larger 'reservoir' (remember that the walls of the subsystems are only virtual boundaries, of course), and hence is at constant chemical potential even if the total system is held at constant density ρ. Thus (for $L \to \infty$ the index L can be omitted)

$$\chi_\mu - \chi_\rho = L^d \langle\Delta\psi\Delta\rho\rangle^2 / \langle(\Delta\rho)^2\rangle. \tag{6.29}$$

In fact, a distinction between χ_μ and χ_ρ is expected only in the ordered phase, since

$$\langle \Delta\psi\Delta\rho\rangle_L = (\partial\langle\psi\rangle_L/\partial\mu)/(L^d/k_B T), \tag{6.30}$$

and $\langle\psi\rangle_{L\to\infty} \equiv 0$ in the disordered phase. As expected, the distribution in Fig. 6.2 has contours with the long axis parallel to the abscissa (no ψ–ρ coupling) in the disordered phase, while in the ordered phase the long axis forms a non-trivial angle with the abscissa, due to the presence of a coupling term in Eqn. (6.28). From Fig. 6.2 both χ_μ, $\langle\Delta\psi\Delta\rho\rangle$, and $\langle(\Delta\rho)^2\rangle$ can be measured, and one finds susceptibilities χ_μ, χ_ρ in both ensembles (from Eqn. (6.29)) and the isothermal compressibility

$$\kappa = L^d \rho^{-2}\langle(\Delta\rho)^2\rangle_L \tag{6.31}$$

from a *single* simulation run!

However, it is important to realize that the subsystem fluctuations 'cut off' correlations across the subsystem boundaries, and hence one has to carry out an extrapolation according to (Rovere *et al.*, 1990; Weber *et al.*, 1995)

$$\chi_{L,\mu} = \chi_\mu(1 - \text{const.}\xi/L), \qquad L \gg \xi, \tag{6.32}$$

where the constant is of order unity. Actually, both the compressibility κ (Fig. 6.3) and the susceptibility χ_μ (Fig. 6.4a) have to be found by an extrapolation of the form given by Eqn. (6.32), see Fig. 6.4b, and hence are denoted as κ_∞, χ_∞, in these figures. Figure 6.4b shows that the extrapolation suggested by Eqn. (6.32) does indeed work, but one must discard data for small L^{-1} which bend systematically down to smaller values. This effect is due to crossover from the grand canonical ensemble (small sub-boxes, $M_b \gg 1$) to the canonical ensemble (realized by $M_b = 1$, of course). Indeed, Eqn. (6.29) shows that $\chi_\mu > \chi_\rho$ in the ordered phase.

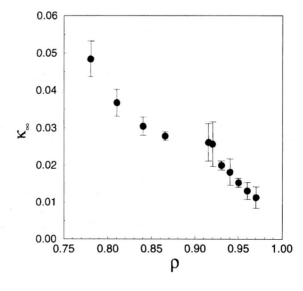

Fig. 6.3 Compressibility κ_∞ of the hard disk model as a function of density ρ, obtained by extrapolation from circular and rectangular subsystems in the solid and fluid phases, respectively. Total number of particles is $N = 576$. From Weber *et al.* (1995).

(a)

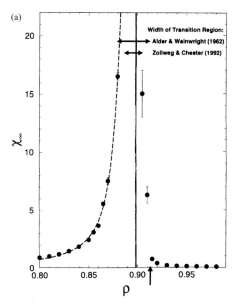

(b)

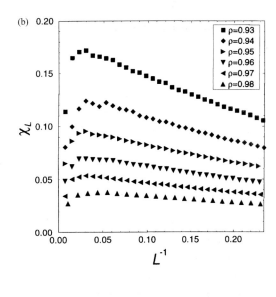

Fig. 6.4 (a) Extrapolated 'susceptibility' χ_∞ of the hard disk system versus density. The data in the fluid are fitted to $\chi_\infty \propto \exp\{b'(\rho_f - \rho)^{-1/2}\}$ where b' is a constant and $\rho_f = 0.913$ is marked with an arrow. The vertical solid line marks the estimated transition density $\rho_{\text{cross}} = 0.8985 \pm 0.0005$ obtained from cumulant intersections (Fig. 6.5). Previous estimates for the width of the two-phase region are indicated by horizontal arrows. Error bars are only shown when they exceed the size of the symbols. (b) Susceptibility χ_L as a function of the inverse linear subsystem size L^{-1} in the solid phase away from the transition, for $N = 16\,384$ particles. From Weber et al. (1995).

The major reason for the great interest in the solid–liquid transition of hard disks is a longstanding controversy about whether the Nelson–Halperin (1979) theory works for this model. According to this theory, melting in two dimensions is not a conventional first order transition (as it is in the three-dimensional case) but rather occurs via a sequence of two continuous transitions: by increasing the density one leaves the fluid phase through a divergence of the susceptibility χ_∞,

$$\chi_\infty \propto \exp\{b'(\rho_f - \rho)^{-1/2}\}, \tag{6.33}$$

where b' is a constant and at ρ_f a transition occurs to a rather unconventional phase, the hexatic phase. In this phase, for $\rho_f < \rho < \rho_f'$, the order parameter $\langle \psi \rangle$ is still zero in the thermodynamic limit $L \to \infty$, but correlation functions of this order parameter decay algebraically, i.e. the correlation length ξ (cf. Eqn. (6.32)) is infinite. Only for $\rho > \rho_f'$ would one have $\langle \psi \rangle > 0$, i.e. a conventional solid.

As Fig. 6.4a shows, Eqn. (6.33) provides a very good fit to the simulation data, but the 'critical' density ρ_f is larger than the density ρ_{cross}, which results from cumulant intersections (Fig. 6.5). As in the case of the Ising model, see Chapter 4, the cumulant of the bond orientational order parameter has been defined as (cf. Eqn. (4.12))

$$U_L = 1 - \langle \psi^4 \rangle_L / (3 \langle \psi^2 \rangle_L^2). \tag{6.34}$$

Figure 6.5 shows that the intersection occurs in the region $0.898 \leq \rho$ ≤ 0.899, and this estimate clearly is significantly smaller than $\rho_f \approx 0.913$ extracted from the fit to Eqn. (6.33), cf. Fig. 6.4a. Thus the implication is that at the (first order) transition χ_∞ is still finite, ρ_f only has the meaning of a 'spinodal point' (limit of metastability of the fluid phase). Of course, noting that ρ is the density of an extensive thermodynamic variable, we emphasize that in principle there should be a jump in density from ρ_l (where one leaves the fluid phase) to ρ_s (where one enters the solid phase). In the 'forbidden' region of densities in between ρ_l and ρ_s one finds two-phase coexistence (which for large enough L must show up in a double peak distribution for $\rho_L(\psi, \rho)$, rather than the single peaks seen in Fig. 6.2). Unfortunately, even with 16 384 particles no evidence for this ultimate signature of first order melting in two dimensions is found. The large values found for χ_∞ near the transition at ρ_{cross} in Fig. 6.4a imply that the system is indeed rather close to a continuous melting transition, and previous estimates for the width of the two-phase coexistence region (included in Fig. 6.4a) clearly are too large. This fact that the system is so close to continuous melting also explains why one cannot see a jump singularity of κ_∞ at the transition (Fig. 6.3 rather suggests only a discontinuity of the slope). However, the conclusions are called into question by a recent finite size scaling analysis for very large systems (Jaster, 1998) which studied χ_∞ much closer to the transition than in the data in Fig. 6.4a and which concluded that there is a continuous transition at $\rho_c \approx 0.900$ compatible with Fig. 6.5.

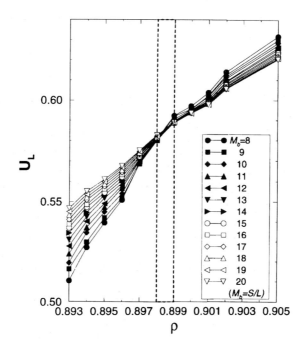

Fig. 6.5 Order parameter cumulants for the bond orientational order parameter, plotted as a function of the total density ρ for various subsystem sizes $L = S/M_b$. The vertical dashed lines mark the range within which the cumulant intersection occurs, i.e. they indicate the error in the estimated transition density of $\rho_{cross} = 0.8985$ $\pm\ 0.0005$. From Weber et al. (1995).

Originally, the subsystem analysis for off-lattice systems was used to study the gas–liquid transition (Rovere *et al.*, 1990), but it now is evident that for this problem the grand canonical simulation method is more efficient (Wilding, 1997). For very dense systems, however, the subsystem analysis clearly has its merits. Another recent, useful application concerns the analysis of capillary-wave type fluctuations of interfaces between coexisting phases in polymer mixtures (Werner *et al.*, 1997). Thus we suggest that the reader keep this technique in mind as an alternative to the more traditional approaches.

6.1.5 Gibbs ensemble

For a study of many fluids or fluid mixtures one is not primarily interested in a precise knowledge of critical properties, but rather in an overall description of phase diagrams, involving the description of phase coexistence between liquid and gas, or between an A-rich phase and a B-rich phase in a binary mixture (AB), respectively. The so-called 'Gibbs ensemble' method, pioneered by Panagiotopoulos (1987, 1995), is an efficient (and computationally 'cheap') approach to achieve that goal, and hence is of widespread use for a large variety of systems.

The basic idea of this method is very intuitive. Consider a macroscopic system where gas and fluid phases coexist in thermal equilibrium. The Gibbs ensemble attempts to simulate two microscopic regions within the bulk phase, away from an interface (Fig. 6.6). The thermodynamic requirements for phase coexistence are that each region should be in internal equilibrium and that temperature, pressure and the chemical potential are the same in both regions. The system temperature in Monte Carlo simulations is specified in advance. The remaining conditions are satisfied by three types of Monte Carlo moves: displacements of particles within each region (to ensure internal equilibrium), exchange of volume between the two regions (to ensure equality of pressures) and particle exchanges (to ensure equality of the chemical potentials).

From this discussion and from Fig. 6.6 it is evident that the Gibbs ensemble somehow interpolates between the NVT, NpT and μVT ensembles discussed above; and it is applicable only when grand canonical simulations (or semi-grand canonical ones, for the simulation of phase equilibrium in a mixture) are also feasible, since the transfer of particles from one box to the other one is an indispensable step of the procedure in order to maintain the equality of the chemical potentials of the two boxes. Therefore, its application is straightforward for fluid–fluid phase equilibria only and not for phase equilibria involving solid phases (or for complex fluids, such as very asymmetric polymer mixtures).

For a formal derivation of the acceptance rules of the moves shown in Fig. 6.6, one proceeds similarly as in the derivation of rules for the NpT and μVT ensembles. The total particle number $N = N_{\mathrm{I}} + N_{\mathrm{II}}$ and the total volume V

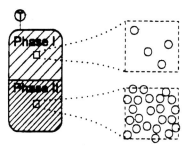

Fig. 6.6 Schematic
diagram of the Gibbs
ensemble technique. A
two-dimensional
system is shown for
simplicity. Broken
lines indicate
boundaries where
periodic boundary
conditions are applied.
From Panagiotopoulos
(1995).

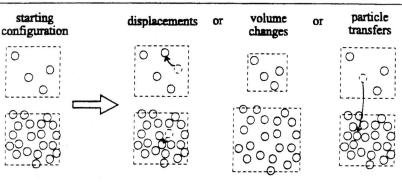

$= V_{\mathrm{I}} + V_{\mathrm{II}}$ of the two boxes are kept constant, and hence we apply the canonic partition function, cf. Eqns. (6.12) and (6.15)

$$Z_{NVT} = \frac{1}{\Lambda^{3N} N!} \sum_{N_{\mathrm{I}}=0}^{N} \binom{N}{N_{\mathrm{I}}} \int_{0}^{V} dV_{\mathrm{I}} \, V_{\mathrm{I}}^{N_{\mathrm{I}}} (V - V_{\mathrm{I}})^{(N-N_{\mathrm{I}})} \int ds_1 \ldots ds_{N_{\mathrm{I}}} e^{-U_{\mathrm{I}}/k_{\mathrm{B}}T}$$

$$\int ds_{N_{\mathrm{I}}+1} \ldots \int ds_N e^{-U_{\mathrm{II}}/k_{\mathrm{B}}T}.$$

$$(6.35)$$

U_{I} is the total intermolecular interaction potential of the N_{I} particles in V_{I}, and U_{II} the corresponding quantitiy in V_{II}. The probability density corresponding to Eqn. (6.35) is

$$P(N_{\mathrm{I}}, \mathbf{V}_{\mathrm{I}}; N, V, T) \propto \frac{N!}{N_{\mathrm{I}}(N - N_{\mathrm{I}})!} \exp\left\{ N_{\mathrm{I}} \ln V_{\mathrm{I}} + (N - N_{\mathrm{I}}) \ln (V - V_{\mathrm{I}}) \right.$$
$$\left. - \frac{U_{\mathrm{I}}}{k_{\mathrm{B}}T} - \frac{U_{\mathrm{II}}}{k_{\mathrm{B}}T} \right\}.$$

$$(6.36)$$

From Eqn. (6.36), one obtains the transition probability for the various types of moves as in Sections 6.1.1–3. For a displacement step in one of the regions, the situation is exactly the same as in a standard NVT simulation. For a volume exchange step, we have (cf. Eqn. (6.21))

$$W(V_{\mathrm{I}} \rightarrow V_{\mathrm{I}} + \Delta V, V_{\mathrm{II}} \rightarrow V_{\mathrm{II}} - \Delta V) = \min\left\{1, \exp\left[-\frac{\Delta U_{\mathrm{I}} + \Delta U_{\mathrm{II}}}{k_{\mathrm{B}}T}\right.\right.$$
$$\left. + N_{\mathrm{I}} \ln \frac{V_{\mathrm{I}} + \Delta V}{V_{\mathrm{I}}} \right.$$
$$\left.\left. + (N - N_{\mathrm{I}}) \ln \frac{(V - V_{\mathrm{I}} - \Delta V)}{V - V_{\mathrm{I}}}\right]\right\}.$$

(6.37)

The transition probability for particle exchanges (written here for a transfer from region II to region I) is

$$W(N_{\mathrm{I}} \rightarrow N_{\mathrm{I}} + 1, N_{\mathrm{II}} \rightarrow N_{\mathrm{II}} - 1) = \min\left\{1, \frac{(N - N_{\mathrm{I}})V_{\mathrm{I}}}{(N_{\mathrm{I}} + 1)(V - V_{\mathrm{I}})}\right.$$
$$\left. \exp\left[-\frac{\Delta V_{\mathrm{I}} + \Delta V_{\mathrm{II}}}{k_{\mathrm{B}}T}\right]\right\}.$$

(6.38)

Note that beforehand neither the vapor pressure at which phase coexistence occurs nor the associated chemical potential need to be known starting from suitable initial conditions (e.g. one box with density smaller than the gas density at phase coexistence, the other box with a density higher than the corresponding liquid density). The system will automatically develop towards phase coexistence, but of course, the total density N/V must be chosen such that the state point would fall inside of the two-phase coexistence region in the thermodynamic limit.

One practical difficulty is that in a long simulation run it can happen (and will inevitably happen close to criticality) that the box labeled by I will sometimes contain the gas phase and sometimes the liquid phase, and so one would not obtain any meaningful results (refering to properties of a pure phase) by simply taking running averages for the two boxes separately. Hence a safer way to analyze the results is to record the density distribution function: as long as it shows two clearly separated peaks, there is no difficulty in ascribing to them the properties of the two coexisting phases. Unlike canonical simulations of phase coexistence (Rovere et al., 1990), equilibrium is established very quickly and the data are not affected so much by interfacial contributions. Near the critical point, however, the accuracy of the method deteriorates, finite size effects are less straightforward to analyze, since both volumes and particle numbers of the individual boxes fluctuate. Given the current status of our knowledge, the grand canonical method in conjunction with finite size scaling yields clearly superior results (Wilding, 1997). Nevertheless, the Gibbs ensemble method has a suitable place in our 'bag of tricks'; due to its relative simplicity of implementation and modest cpu requirements it has been applied in numerous studies of simple fluids as well as of ionic, associating and reacting fluids and even for simple models of homopolymers (combining the technique with 'configurational bias' Monte

Carlo methods, see e.g. Mooij *et al.* (1992)). We do not give further details here, but draw the reader's attention to the recent extensive reviews presented by Panagiotopoulos (1995) and Frenkel and Smit (1996).

Problem 6.6 Generalize Eqn. (6.38) to a multicomponent system (where at phase coexistence the chemical potentials of all components should be equal).

6.1.6 Widom particle insertion method and variants

The test particle insertion method (Widom, 1963) is a technique which can be used to sample the chemical potential in a fluid. Remember that the chemical potential is defined by

$$\mu = (\partial F/\partial N)_{VT} = (\partial G/\partial N)_{pT}. \tag{6.39}$$

Consider first the case of the NVT ensemble where $F = -k_B T \ln Z(N, V, T)$ and the partition function $Z(N, V, T)$ is given by Eqn. (6.12). For $N \gg 1$ we can replace the partial derivative with respect to N by a difference, $\mu = k_B T \ln\{Z/(N+1, V, T)/Z(N, V, T)\}$. Again using scaled coordinates $\mathbf{s}_i$ (Eqn. (6.13)) and Eqn. (6.14) to split off the contribution of the ideal gas, $\mu_{id}(\rho) = -k_B T \ln\{V/[\Lambda^d(N+1)]\}$ with $\rho = N/V$, we find

$$\mu = \mu_{id}(\rho) + \mu_{ex} \tag{6.40}$$

where

$$\mu_{ex} = -k_B T \left\{ \int_0^1 d\mathbf{s}_1 \ldots \int_0^1 d\mathbf{s}_{N+1} \exp\left[-\frac{U(\mathbf{s}_1, \ldots, \mathbf{s}_{N+1}, L)}{k_B T}\right] \bigg/ \int_0^1 d\mathbf{s}_1 \ldots \int_0^1 d\mathbf{s}_N \right.$$
$$\left. \exp\left[-\frac{U(\mathbf{s}_1, \ldots, \mathbf{s}_N, L)}{k_B T}\right] \right\}.$$

We now separate the potential energy U of the $(N+1)$-particle system into the energy of the N-particle system and the interaction energy ΔU of the $(N+1)$th particle with the rest of the system, i.e.

$$U(\mathbf{s}_1, \ldots, \mathbf{s}_{N+1}, L) = U(\mathbf{s}_1, \ldots, \mathbf{s}_N, L) + \Delta U. \tag{6.41}$$

We immediately realize that μ_{ex} then can be rewritten as

$$\mu_{ex} = -k_B T \ln \int_0^1 d\mathbf{s}_{N+1} \langle \exp(-\Delta U/k_B T) \rangle_N, \tag{6.42}$$

where $\langle \ldots \rangle_N$ is a canonical ensemble average over the configuration space of the N-particle system. This average now can be sampled by the conventional Monte Carlo methods. In practice one proceeds as follows: one carries out a standard NVT Monte Carlo simulation of the system of N particles (as described in Section 6.1.1). Often one randomly generates additional coordinates s_{N+1} of the test particle, *uniformly distributed* in the d-dimensional unit cube in order to carry out the remaining integral in Eqn. (6.42). With this value of s_{N+1}, one computes ΔU from Eqn. (6.41) and samples then $\exp(-\Delta U/k_{\mathrm{B}}T)$.

Thus one computes the average of the Boltzmann factor associated with the random insertion of an additional particle in an N-particle system, but actually this insertion is never carried out, because then we would have created an $N+1$-particle system, but we do need an N-particle system for the averaging in Eqn. (6.42).

Care is necessary when applying this method to other ensembles. One can show that (for details see e.g. Frenkel and Smit, 1996) in the NpT ensemble Eqns. (6.40) and (6.42) are replaced by

$$\mu = \mu_{\mathrm{id}}(p) + \mu_{\mathrm{ex}}(p),$$

$$\mu_{\mathrm{id}}(p) = -k_{\mathrm{B}}T\ln(k_{\mathrm{B}}T/p\Lambda^{d}),$$

$$\mu_{\mathrm{ex}}(p) = -k_{\mathrm{B}}T\ln\left\langle\frac{pV}{(N+1)k_{\mathrm{B}}T}\int_{0}^{1}ds_{N+1}\exp(-\Delta U/k_{\mathrm{B}}T)\right\rangle. \tag{6.43}$$

Thus one uses the ideal gas reference state at the same pressure (rather than at the same density as in Eqn. (6.40)) as the investigated system, and the quantity that is sampled is $V\exp(-\Delta U/k_{\mathrm{B}}T)$ rather than $\exp(-\Delta U/k_{\mathrm{B}}T)$.

An obvious extension of the particle insertion method is to binary mixtures (A,B) where one often is interested only in chemical potential differences $\mu_{\mathrm{A}} -\mu_{\mathrm{B}}$ rather than in individual chemical potentials $\mu_{\mathrm{A}}, \mu_{\mathrm{B}}$. Then trial moves can be considered in which one attempts to transform a particle of species A into one of species B (without ever accepting such a transformation, of course).

While the Widom test particle method works well for moderately dense fluids (such as near and below the critical density), it breaks down long before the triple point density of a fluid is reached, simply because the probability $\exp(-\Delta U/k_{\mathrm{B}}T)$ that a random insertion is accepted becomes too small. Even for hard spheres, the insertion probability is down to 4×10^{-5} at a packing fraction of 0.4, long before the freezing transition is reached. Therefore, substantial effort has been devoted to devising schemes for biasing the insertions (rather inserting them 'blindly') as well as implementing 'gradual insertions'. We refer to Allen (1996) for a recent review.

We conclude this section with a caveat: often the chemical potential is computed in a desire to establish phase diagrams (remember that chemical potentials of coexisting phases are equal). Then very good accuracy is needed, and one must carefully pay attention to systematic errors both due to finite size effects and due to the potential cutoff (if the potential is truncated, see Section 6.2.1, one may approximately correct for this truncation by applying so-called 'tail corrections', see Frenkel and Smit, (1996)).

6.2 'SHORT RANGE' INTERACTIONS

6.2.1 Cutoffs

One significant advantage of a potential like Lennard-Jones is that it falls off quite fast, and only those particles within a nearby environment have much effect. As a consequence it is possible to limit, or 'cut off', the maximum range of the interaction at a distance r_c. This effectively introduces a step function into the distance dependence, but the hope is that if the potential has already decayed substantially, this effect will be small. (The situation is perhaps less complex than for molecular dynamics for which this cutoff can introduce a singularity in the force; there the potential is then often 'shifted' so that the force is quite small at r_c.) The choice of cutoff radius is somewhat arbitrary and depends upon the potential used. For Lennard-Jones a convenient choice is often $r_c = 2.5\sigma$. The use of a cutoff dramatically reduces the number of near neighbors which must be included in the calcu-lation of energy of the new trial state, but in order to take advantage of this fact one must use an intelligent data structure. One simple, but very good choice, is discussed in the next sub-section. In general one must balance the desire to speed up the program by using a small cutoff with the concern that the cutoff may change the physics!

6.2.2 Verlet tables and cell structure

A very simple method to reduce the amount of work needed to calculate energy changes is to construct a table of neighbors for each particle which contains only those neighbors which are closer than r_c. This can be further improved by making the following observation: as particles move due to the acceptance of Monte Carlo moves they may leave the 'interaction volume' or new particles may enter this region. The recalculation of the table following each successful move may be avoided by keeping track of all particles within some distance $r_{max} > r_c$ where $(r_{max} - r_c) = n\delta_{max}$ is large enough that no particle can enter the 'interaction volume' in n Monte Carlo steps of max-imum size δ; the table is then only recalculated after every n steps. For very large systems even this occasional recalculation can become very time con-suming, so an additional step can be introduced to further limit the growth in time requirement as the system size increases. The system can be divided into a set of cells of size l which are small compared to the size of the simulation box L but larger than the cutoff radius r_c. The only interacting neighbors must then be found within the same cell or the neighboring cells, so the remainder of the simulation volume need not be searched.

6.2.3 Minimum image convention

Periodic boundary conditions may be easily implemented by simply attaching copies of the system to each 'wall' of the simulation volume. An 'image' of each particle is then replicated in each of the fictitious volumes; only the distance between the nearest neighbor, including one of the 'images' is used in computing the interaction.

6.2.4 Mixed degrees of freedom reconsidered

Often one of the degrees of freedom in the semi-canonical ensemble is continuous. An example that we considered earlier was Si/Ge mixtures for which the choice of atom was determined by a discrete (Ising) variable and a continuous variable was used to determine the movement of the particles. In this case a three-body interaction was included so that the table structure became more complicated. Since the interactions of a 'trimer' needed to be calculated, it was sometimes necessary to calculate the position of the neighbor of a neighbor, i.e. both the nearest neighbor distance as well as the bond angle. This effectively extends the range of the interaction potential substantially.

6.3 TREATMENT OF LONG RANGE FORCES

Long range interactions represent a special challenge for simulation because they cannot be truncated without producing drastic effects. In the following we shall briefly describe several different methods which have been used to study systems with long range interactions (Pollock and Glosli, 1996).

6.3.1 Reaction field method

This approach is taken from the continuum theory of dielectrics and is effective for the study of dipolar systems. We consider a system of N particles each of which has a dipolar moment of magnitude μ. The dipole–dipole interaction between two dipoles i and j is given by

$$v_{dd} = \frac{\mu_i \cdot \mu_j}{r_{ij}^3} - \frac{3(\mu_i \cdot \mathbf{r}_{ij})(\mu_j \cdot \mathbf{r}_{ij})}{r_{ij}^5}, \tag{6.44}$$

and the total energy of a given dipole is determined by summing over all other dipoles. An approximation to the sum may be made by carving out a spherical cavity about the dipole, calculating the sum exactly within that cavity, and treating the remaining volume as a continuum dielectric. In the spirit of dielectric theory, we can describe the volume within the cavity of radius r_c by a homogeneous polarization which in turn induces a 'reaction field' $\mathbf{E}_R$

$$\mathbf{E}_R(i) = \frac{2(\varepsilon - 1)}{r_c^3(2\varepsilon + 1)} \sum_i^N \mu_i, \tag{6.45}$$

which acts on each dipole. The correct choice of the dielectric constant ε is still a matter of some debate. The total dipolar energy of a particle is thus given by the sum of the 'local' part within the cavity and the 'global' part which comes from the reaction field.

6.3.2 Ewald method

The Ewald method is not new; in fact it has long been used to sum the Coulomb energy in ionic crystals in order to calculate the Madelung constant. The implementation to the simulation of a finite system is straightforward with the single modification that one must first periodically replicate the simulation volume to produce an 'infinite' array of image charges. Each cell is identified by the integer n and the vector $\mathbf{r}_n$ is the replication vector. The electrostatic energy is calculated, however, only for those charges in the original cell. The potential at charge q_i is

$$v_i(r) = \sum_j^N \sum_{n=n_i}^{\infty} \frac{q_j}{|\mathbf{r} - \mathbf{r}_j + \mathbf{r}_n|} \qquad \begin{cases} n_i = 0, & j \neq i \\ n_i = 1, & j = i \end{cases} \qquad (6.46)$$

which excludes self-interaction. The trick is to add and subtract a Gaussian charge distribution centered at each site $\mathbf{r}_j$ and separate the potential into two sums, one in real space and one in reciprocal space. The Coulomb potential then becomes

$$v_i(\mathbf{r}) = v_i^r(\mathbf{r}) + v_i^k(\mathbf{r}), \qquad (6.47)$$

with

$$v_i^k(\mathbf{r}) = \sum_{m \neq 0} W(k_m) S(k_m) e^{2\pi k_m \cdot \mathbf{r}} - 2\sqrt{\frac{2}{\pi}} \frac{q_i}{\alpha} \qquad (6.48)$$

where the second term corrects for the self-energy, and

$$W(k_m) = \frac{1}{\pi L^3 k_m^2} e^{-2\pi^2 k_m^2 \alpha^2}, \qquad (6.49a)$$

$$S(k_m) = \sum_j^N q_j e^{-2\pi i k_m \cdot \mathbf{r}_j}. \qquad (6.49b)$$

The width of the Gaussian distribution is α. By proper choice of α the sums in both real space and reciprocal space in Eqn. (6.47) can be truncated.

6.3.3 Fast multipole method

The fast multipole method (Greengard and Rokhlin, 1985) plays a particularly important role in calculating Coulomb interactions in large systems because it exhibits $O(N)$ scaling, where N is the number of particles. The method relies on two expansions which converge for large distances and short distances, respectively. The multipole expansion is

$$V(r) = 4\pi \sum_{l,m}^{l_{max}} \frac{M_{lm}}{(2l+1)} \frac{Y_{lm}(\Omega)}{r^{l+1}} + \cdots \tag{6.50}$$

where $Y_{lm}(r)$ is a spherical harmonic, the multipole moment is

$$M_{lm} = \sum_{i}^{N} q_i r_i^l Y_{lm}^*(\Omega_i) \tag{6.51}$$

and the 'local' expansion is

$$V(r) = 4\pi \sum_{l,m}^{l_{max}} L_{lm} r^l Y_{lm}(\Omega) + \cdots \tag{6.52}$$

where the 'local' moment is

$$L_{lm} = \sum_{l} \frac{q_i}{(2l+1)} \frac{Y_{lm}^*(\Omega_i)}{r_i^{l+1}} + \cdots . \tag{6.53}$$

The algorithm is implemented in the following way:

Fast multipole method

(1) Divide the system into sets of successively smaller sub-cells.

(2) Shift the origin of the multipole expansion and calculate the multipole moments at all sub-cell levels starting from the lowest level.

(3) Shift the origin of the local expansion and calculate the local expansion coefficients starting from the highest level.

(4) Evaluate the potential and fields for each particle using local expansion coefficients for the smallest sub-cell containing the particle.

(5) Add the contributions from other charges in the same cell and near neighbor cells by direct summation.

This procedure becomes increasingly efficient as the number of particles is made larger.

6.4 ADSORBED MONOLAYERS

6.4.1 Smooth substrates

The study of two-dimensional systems of adsorbed atoms has attracted great attention because of the entire question of the nature of two-dimensional melting. In the absence of a periodic substrate potential, the system is free to form an ordered structure determined solely by the interparticle interactions. As the temperature is raised this planar 'solid' is expected to melt, but the nature of the transition is a matter of debate.

6.4.2 Periodic substrate potentials

Extensive experimental data now exist for adsorbed monolayers on various crystalline substrates and there have been a number of different attempts made to carry out simulations which would describe the experimental observations. These fall into two general categories: lattice gas models, and off-lattice models with continuous, position dependent potentials. For certain general features of the phase diagrams lattice gas models offer a simple and exceedingly efficient simulations capability. This approach can describe the general features of order–disorder transitions involving commensurate phases. (For early reviews of such work see Binder and Landau, 1989; Landau, 1991.) An extension of the lattice gas description for the ordering of hydrogen on palladium (100) in the c(2×2) structure has recently been proposed by giving the adatoms translational degrees of freedom within a lattice cell (Presber *et al.*, 1998).

The situation is complicated if one wishes to consider orientational transitions involving adsorbed molecules since continuous degrees of freedom must be used to describe the angular variables. Both quadrupolar and octupolar systems have been simulated. For a more complete description of the properties of adsorbed monolayers it is necessary to allow continuous movement of particles in a periodic potential produced by the underlying substrate. One simplification which is often used is to constrain the system to lie in a two-dimensional plane so that the height of the adatoms above the substrate is fixed. The problem is still difficult computationally since there may be strong competition between ordering due to the adatom–adatom interaction and the substrate potential and incommensurate phases may result. Molecular dynamics has been used extensively for this class of problems but there have been Monte Carlo studies as well. One of the 'classic' adsorbed monolayer systems is Kr on graphite. The substrate has hexagonal symmetry with a lattice constant of 2.46 Å whereas the lattice constant of a compressed two-dimensional krypton solid is 1.9 Å. The 1×1 structure is thus highly unfavorable and instead we find occupation of next-nearest neighbor graphite hexagons leading to a ($\sqrt{3} \times \sqrt{3}$) commensurate structure with lattice constant 4.26 Å. This means, however, that the krypton structure must expand relative to an isolated two-dimensional solid. Thus, there is competition between the length scales set by the Kr–Kr and Kr–graphite interactions. An important question was thus whether or not this competition could lead to an incommensurate phase at low temperatures. This is a situation in which boundary conditions again become an important consideration. If periodic boundary conditions are imposed, they will naturally tend to produce a structure which is periodic with respect to the size of the simulation cell. In this case a more profitable strategy is to use free edges to provide the system with more freedom. The negative aspect of this choice is that finite size effects become even more pronounced. This question has been studied using a Hamiltonian

$$\mathcal{H} = \sum_i V(\mathbf{r}_i) + \frac{1}{2} \sum_{i \neq j} v_{\mathrm{LJ}}(\mathbf{r}_{ij}), \tag{6.54}$$

where the substrate potential is given by

$$V(\mathbf{r}_i) = V_0(z_{\mathrm{eq}}) + 2V_1(z_{\mathrm{eq}})\{\cos(\mathbf{b}_1 \cdot \mathbf{r}_i) + \cos(\mathbf{b}_2 \cdot \mathbf{r}_i) + \cos[(\mathbf{b}_1 + \mathbf{b}_2) \cdot \mathbf{r}_i]\}, \tag{6.55}$$

where $\mathbf{b}_1$ and $\mathbf{b}_2$ are the reciprocal lattice vectors for the graphite basal plane, and v_{LJ} is the Lennard-Jones potential of Eqn. (6.4). The strength of the corrugation potential is given by V_1. The order parameter for the $(\sqrt{3} \times \sqrt{3})$ registered phase is

$$\Phi = \frac{1}{3N} \sum_i \{\cos(\mathbf{b}_1 \cdot \mathbf{r}_i) + \cos(\mathbf{b}_2 \cdot \mathbf{r}_i) + \cos[(\mathbf{b}_1 + \mathbf{b}_2) \cdot \mathbf{r}_i]\}. \tag{6.56}$$

A local order parameter can also be defined using the reciprocal lattice vectors appropriate to each of the three possible sublattices. A canonical Monte Carlo study (Houlrik et al., 1994) showed that there was a first order transition between a low temperature incommensurate phase and a high temperature commensurate $(\sqrt{3} \times \sqrt{3})$ structure. Both the smearing of the transition and the shift in the transition temperature decrease rapidly as the system size increases. At higher temperature still the $(\sqrt{3} \times \sqrt{3})$ ordered structure melts.

Similar potentials as that given in Eqn. (6.55) can also be used when the substrate surface has square or rectangular symmetry as would be appropriate for (100) and (110) faces of cubic crystals (Patrykiejew et al., 1995, 1998). Interesting effects due to competition occur since the adsorbed layer prefers a triangular structure for weak corrugation.

6.5 COMPLEX FLUIDS

By the term 'complex fluids' as opposed to 'simple fluids' one means systems such as colloidal dispersions, surfactant solutions (microemulsions) and their various microphase-separated structures (sponge phases, phases with lamellar superstructure, solutions containing micelles or vesicles, etc.), polymer solutions and melts, liquid crystalline systems with various types of order (nematic, smectic, cholesteric, etc.). Unlike simple atomic fluids (whose basic constituents, the atoms, e.g. fluid Ar, have nothing but their positional degree of freedom) and unlike diatomic molecules (such as N_2, O_2, etc.), whose basic constituents have just positional and orientational degrees of freedom (neglecting the high-frequency small amplitude molecular vibrations, these molecules are just treated as two point particles kept at a rigidly fixed distance), these complex fluids typically have a large number of atomic constituents. Typically they contain several types of atoms and involve different types of interactions, and sometimes they have a large number of internal degrees of freedom. Typical examples are surfactant molecules such as fatty acids that form monomolecular layers (so-called 'Langmuir monolayers', e.g. Gaines (1996)) at the air–water interface (a related system known to the reader from daily life is a thin soap film!) Typically, these surfactant molecules exhibit self-assembly at an interface because of their structure, comprising a hydrophilic head group and a hydrophobic tail (e.g. a short alkane chain, cf. Fig. 6.7). Similar surfactant molecules have important practical applications as detergents, for oil recovery (when small oil droplets are dispersed in water, surfactants are useful that gather at the oil–water interfaces), and also the biological membranes that are the basis of all biological functions in the living cell are formed from similar amphiphilic phospholipid molecules.

Simulation of such systems in full atomistic detail is a very difficult task, since the single molecule is already a rather large object, with complicated interactions which are often only rather incompletely known, and since a common feature of these systems is a tendency to organize themselves in supramolecular structures on mesoscopic length scales, thermal equilibrium is rather hard to obtain. Figure 6.7 indicates one possibility to simplify the model by a kind of coarse-graining procedure: first of all, the water molecules are not considered explicitly (simulation of water and water surfaces is a difficult task itself, see Alejandre *et al.* (1995); note that there is not even a consensus on a good effective potential for water that is 'good' under all physical conditions, because of the tendency of water molecules to form bridging hydrogen bonds). Thus, the air–water interface here is simply idealized as a flat plane at $z = 0$, and it is assumed that the interaction between the hydrophilic head groups and the water substrate is so strong that the head

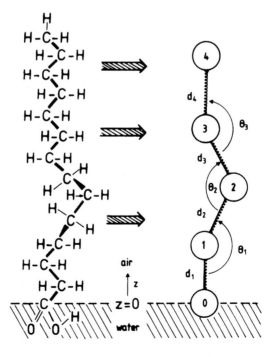

Fig. 6.7 Schematic picture of a fatty acid molecule at the air–water interface and a possible coarse-grained model, where a few successive CH_2 groups are combined into one effective monomer. The effective bonds between these effective monomers are represented by springs, and the stiffness of the chain is controlled by a potential depending on the angle θ_i between the effective bonds. From Haas *et al.* (1996).

groups are also fixed at $z = 0$, they are simply described as point-like particles which interact with a Lennard-Jones-type potential. Similar Lennard-Jones-type potentials are also assumed to act between the effective monomers. In addition, a bond angle potential $V(\theta) \propto (1 - \cos\theta)$ is used. Sometimes one even ignores the internal flexibility of these alkane chains (at low temperatures $V(\theta) \gg T$ apart from the case $\theta = 0$), and treats them as rigid rods with a single orientational degree of freedom (or, more precisely, two polar angles ϑ, φ specifying the orientation of the rod with respect to the z-axis, see Scheringer *et al.* (1992)). While this rigid-rod model clearly is too crude to exhibit much similarity with actual Langmuir monolayers, the model shown in Fig. 6.7 can describe at least qualitatively some of the experimentally observed phases of dense monolayers, such as the untilted structure and phases where the head groups form a regular triangular lattice, while the tails are uniformly tilted towards nearest or next-nearest neighbors, respectively (Schmid *et al.*, 1998). However, at present there exists no model yet that could describe all the experimentally observed phases, that include solid structures with herringbone-type ordering of the CH_2-groups in the xy-plane parallel to the water surface, for instance. However, only for these simplified models has it been possible to study phase changes (applying techniques such as finite size scaling, constant pressure simulations with variable shape of the simulation box, etc.), see Haas *et al.* (1996) and Schmid *et al.* (1998).

While these techniques are straightforwardly generalized from simple to complex fluids, other techniques (such as grand canonical ensemble, Gibbs ensemble, etc.) require special methods, because the particle insertion step for a large surfactant molecule will be rejected in the overwhelming majority of cases. Such special methods (like the 'configurational bias' method) will be discussed later in this chapter.

The situation is similar, as far as the phase behavior of surfactants in bulk solution (rather than at the air–water interface) is concerned. The classic problem is micelle formation in dilute solution (Degiorgio and Corti, 1985). Suppose molecules as shown in Fig. 6.7 are dissolved in a good solvent for alkanes (e.g. benzene or toluol, etc.) while the solvent is a bad solvent for the head group. Then the solution behaves as ideal (i.e. a random, geometrically uncorrelated arrangement of the solute molecules) only at extreme dilution, while for larger concentrations the surfactants cluster together into aggregates, such that the hydrophilic heads form the core of the aggregate, while the tails form the 'corona' of this (star polymer-like) 'micelle'. The transition from the ideal 'gas' of individual surfactant molecules in solution to a 'gas' of micelles occurs relatively sharply at the 'cmc' (critical micelle concentration), although this is not a thermodynamic phase transition. Questions that one likes to answer by simulations concern the precise molecular structure of such micelles, the distribution of their sizes near the cmc, possible transitions between different shapes (spherical vs. cylindrical shape), etc. Again, there is a wide variety of different models that are used in corresponding simulations: fully atomistic models (Karaborni and O'Connell, 1990) are valuable for a description of the detailed structure of a given isolated micelle of *a priori* chosen size, but cannot be used to study the micellar size distribution – there one needs a very large simulation box containing many micelles (to avoid finite size effects) and a very fast simulation algorithm, because in equilibrium many exchanges of surfactant molecules between the different micelles must have occurred. Many different types of coarse-grained models have been used; often it is more realistic to have the hydrophobic and hydrophilic parts comparable in size (unlike the molecule shown in Fig. 6.7), and then one may use symmetric or asymmetric dumbbells (two point-like particles with different Lennard-Jones potential are connected by a spring of finite extensibility (see Rector *et al.*, 1994)) or short flexible chains of type A-A-B-B, where A stands for hydrophilic and B for hydrophobic (von Gottberg *et al.*, 1997; Viduna *et al.*, 1998), etc. In addition, a model where the hydrophilic part is a branched object has also been studied (Smit *et al.*, 1993). Here we cannot review this rapidly developing field, but we only try to convey to the reader the flavor of the questions that one asks and the spirit of the model-building that is both possible and necessary. Due to structure formation on mesoscopic scales, and the large number of mesophases that are possible both at interfaces and in the bulk, this field of 'soft condensed matter' is rapidly growing and still incompletely explored. Since entropy is a dominating factor regarding structure on mesoscopic scales, it is very difficult to develop analytical theories, and hence simulation studies are expected to play a very important rôle. We elaborate on this fact only for one particular example of 'complex fluids', namely polymer solutions and melts, to be described in the next section.

6.6 POLYMERS: AN INTRODUCTION

6.6.1 Length scales and models

Polymers represent an area where computer simulations are providing an ever increasing amount of information about a complex and very important class of physical systems. Before beginning a discussion of the simulation of polymer models, we want to provide a brief background on the special characteristics which are unique to polymers. For systems of small molecules, such as simple fluids containing rare gas atoms, diatomic molecules or water etc., it is possible to treat a small region of matter in full atomistic detail. Since away from the critical point the pair correlation function often exhibits no significant structure on a length scale of 10 Å, such systems may be simulated using a box of linear dimensions 20 Å or thereabouts which contains a few thousand atoms.

For macromolecules the situation is quite different, of course (Binder, 1995). Even a single, flexible, neutral polymer in dilute solution exhibits structure on multiple length scales ranging from that of a single chemical bond (1 Å) to the 'persistence length' ($\approx$10 Å) to the coil radius (100 Å). Note that the persistence length (l_p) describes the length scale over which correlations between the angles formed by subsequent chemical bonds along the 'backbone' of the chain molecule have decayed. Assuming a 'random walk'-like structure is formed by N uncorrelated subunits of length l_p, one concludes that the end-to-end distance R of this 'polymer' should scale like $R \approx l_p \sqrt{N}$ (see Section 3.6). In fact, such a random walk-like structure occurs only in rather special polymer solutions, namely at the so-called 'theta temperature, Θ' where the excluded volume repulsive interaction between the segments of the chains is effectively canceled by an attractive interaction mediated by the solvent (De Gennes, 1979). In 'good solvents', where the excluded volume interactions dominate, the coils are 'swollen' and rather non-trivial correlations in their structure develop. The radius then scales with N according to a non-trivial exponent ν, i.e. $R \propto l_p N^\nu$ with $\nu \approx 0.588$ in $d = 3$ dimensions while $\nu = 3/4$ in $d = 2$ dimensions (De Gennes, 1979). We have already discussed these relations in the context of self-avoiding walks on lattices in Chapter 3.

The above description applies to simple synthetic polymers such as polyethylene $(CH_2)_N$ or polystyrene $(CH_2CH(C_6H_5))_N$. Additional length scales arise for liquid-crystalline polymers, for polymers carrying electrical charges (polyelectrolytes carry charges of one sign only; polyampholytes carry charges of both sign), branched polymers, etc. Such macromolecules are not at all unimportant; a biopolymer such as DNA is an example of a rather stiff polyelectrolyte, and for some biopolymers the understanding of structure formation ('protein folding') is one of the 'grand challenge problems' of modern science!

Nevertheless we shall consider neither polyelectrolytes nor branched polymers further, since they pose special problems for simulations, and thus computer simulation of these systems is much less well developed. For polyelectrolytes the explicit treatment of the long range Coulomb interactions among the charges is a problem for the large length scales that need to be considered (Dünweg *et al.*, 1995). For polymer networks (like rubbery materials) or other branched polymers (randomly branched chains near the gel point, etc.) equilibration is a problem, and one may need special algorithms to move the crosslink points of the network. Since the chemical structure of a network is fixed one also needs to average over many equivalent configurations (Kremer and Grest, 1995). We thus restrict ourselves to flexible neutral polymers. Even then the treatment of full chemical detail is rather difficult, and simplified, coarse-grained models are often the only acceptable choice. We have already encountered the extreme case of coarse-grained models for polymers in Chapter 3 of this book where we dealt with random walks and self-avoiding walks on lattices. Of course, the precise choice of model in a simulation dealing with polymers depends very much on the type of problem that one wishes to clarify. Thus, if one wants to estimate precisely the exponent ν mentioned above or associated 'correction to scaling' exponents, the self-avoiding walk on the lattice is indeed the most appropriate model (Sokal, 1995), since these exponents are 'universal'. On the other hand, if one wants to elucidate where the anomalous anisotropic thermal expansion of crystalline polyethylene comes from, full chemical detail must be kept in the model. In the orthorhombic phase of solid polyethylene there is a contraction of the lattice parameter c in the z-direction (Fig. 6.8c) while the lattice parameters a, b in the x, y directions expand (Fig. 6.8a, b). These experimental trends are qualitatively reproduced by the simulation but there is no quantitative agreement. (i) The simulation is classical Monte Carlo sampling in the NpT ensemble, and hence the temperature derivatives of lattice parameters $da(T)/dT$ etc. remain non-zero as $T \rightarrow 0$, while quantum mechanics requires that $da(T)/dT \rightarrow 0$ as $T \rightarrow 0$, as is also borne out by the data $T < 100$K. (ii) There are uncertainties about the accurate choice of the non-bonded interactions, which typically are chosen of the Lennard-Jones form (suitably truncated and shifted). Even for the chemically simplest polymer, polyethylene, potentials for use in classical Monte Carlo or molecular dynamics work are not perfectly known! As one can see from the simulation data in Fig. 6.8, even in

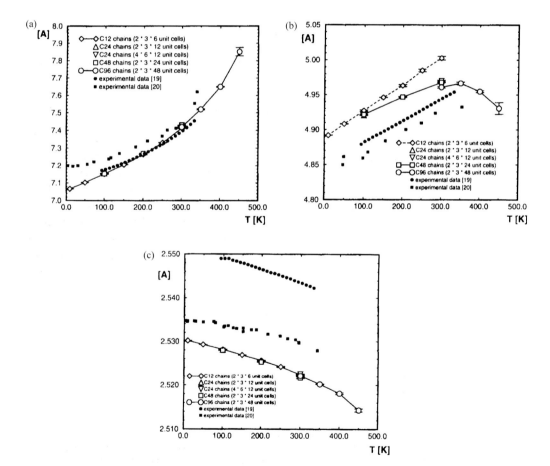

Fig. 6.8 Monte Carlo data for the temperature dependence of the lattice parameters for crystalline polyethylene together with the experimental data of Davis *et al.* (1970) (labeled as [19]) and Dadobaev and Slutsker (1981) (labeled as [20]). Lines are only guides to the eye. From Martonak *et al.* (1997).

the case of polymer crystals there is a need to carefully watch out for finite size effects.

While in a polymer crystal the chain structure is essentially linear, in melts and solutions the chains are coils of random walk or self-avoiding walk type, and their structure needs to be characterized. There are several important quantities which can be used to characterize the behavior of polymer chains. In addition to the mean-square end-to-end distance $\langle R^2 \rangle$, the relative fluctuation of $\langle R^2 \rangle$,

$$\chi(R) = \left(\langle R^4 \rangle - \langle R^2 \rangle^2 \right)/\langle R^2 \rangle^2, \tag{6.57}$$

and the mean-square gyration radius

$$\langle R_g^2 \rangle = \frac{1}{N} \sum \langle (\mathbf{r}_i - \mathbf{r}_j)^2 \rangle, \tag{6.58}$$

where $\mathbf{r}_i$ is the position of the ith monomer, are all important quantities to measure. Similarly the mean-square displacement of the center of mass of the chain,

$$g(t) \equiv \left\langle (\mathbf{r}_{cm}(t) - \mathbf{r}_{cm}(0))^2 \right\rangle \tag{6.59}$$

leads to an estimate of the self-diffusion constant of the chain from the Einstein relation,

$$D_N = \lim_{t \to \infty} \frac{g(t)}{6t}. \tag{6.60}$$

In the simulation of crystalline polyethylene, in principle the problem of large length scales is extremely severe, since the polymer is stretched out in an 'all-trans' zig-zag type linear configuration (Fig. 6.9), i.e. $R \propto N$ rather than $R \propto N^\nu$. This problem is overcome by neglecting the CH_3-groups at the chain ends completely and simply applying a periodic boundary condition in the z-direction. As Fig. 6.8 shows, there are non-trivial finite size effects in one of the other directions if the size of the simulation box is not large enough. In addition, this artificial periodicity prevents a physically reasonable description of the melting transition at high temperature. Significant discre-

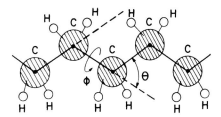

Fig. 6.9 Schematic model for polyethylene: Hydrogen atoms (H) are shown by small white circles, the carbon atoms (C) by larger shaded circles which are connected by harmonic bonds (thick straight lines) of lengths l_i. Segments are labeled consecutively by an index i ($i = 0$ to $N_p - 1$ where N_p is the degree of polymerization). Three successive segments define a bond angle Θ_i, and four successive segments define a torsional angle Φ_i. All the angles $\Phi_i = 0$ in the 'all-trans' configuration.

pancies are seen between the sets of experimental data included in Fig. 6.7; however, since polyethylene single crystals do not occur in nature, and lamellar arrangements separated by amorphous regions may occur in the laboratory, measurements may suffer from unknown systematic errors. The aim of the simulation is to realize an ideal sample of a material that cannot yet be prepared in the laboratory! Technically, a simulation of crystalline polyethylene is rather demanding (Martonak *et al.*, 1996), since the potentials for the lengths l_i of the covalent bonds and the angles θ_i between them are rather stiff, and the scale for the barriers of the torsional potential (Fig. 6.10) is an order of magnitude larger than temperatures of interest ($\sim 10^3$ K). Hence the trial displacements ($\Delta x, \Delta y, \Delta z$) of carbon atoms in the local Monte Carlo moves have to be chosen extremely small, in order to ensure that the acceptance rate of these trial moves is not too low. The relaxation of the (much weaker and slowly varying) non-bonded energy is then very slow. To overcome such problems where the Hamiltonian contains terms with very different energy scales, it is advisable to randomly mix different types of Monte Carlo moves. In the present example, global displacements of chains by amounts $\Delta x_c, \Delta y_c, \Delta z_c$ were chosen, as well as rigid rotations around the *c*-axis, in addition to the standard local moves. In this way a reasonable convergence was achieved. If one is interested in the properties of molten polyethylene at high temperatures (i.e. $T \geq 450$ K), a study of models that include hydrogen atoms explicitly is only possible for rather small N_p (Yoon *et al.*, 1993). An approach which allows the study of larger chains is to model the system using the 'united atom' model where an entire CH_2-monomer is treated as an effective spherical entity. With such models it is still possible to equilibrate polyethylene melts at $T = 500$ K and $N_p = 100$ (Paul *et al.*, 1995). Actually these studies of melts are carried out mostly using molecular dynamics techniques rather than by Monte Carlo, simply because of the lack of efficient Monte Carlo moves for these locally stiff chains. For the study of isolated chains with realistic interactions, however, Monte Carlo techniques are very efficient, and chains as long as $N_p = 4096$ can be simulated (Ryckaert, 1996). However, it is difficult to relate a simulation of such an

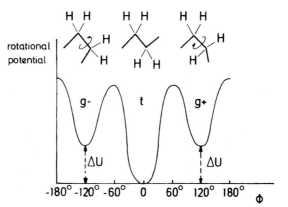

Fig. 6.10 Qualitative sketch of the torsional potential for alkane chains, indicating the three energetically preferred states gauche minus (g−), trans (t), and gauche plus (g+). The minimum of the trans configuration is deeper by an amount ΔU. From Kremer and Binder (1988).

isolated chain in vacuum to a physically meaningful situation (Baschnagel *et al.*, 1992). We shall therefore not discuss such single chain simulations further, although many sophisticated Monte Carlo techniques which have already proven useful for lattice models (Sokal, 1995) are applicable to the off-lattice case as well.

In polymer science, in addition to the explanation of material properties of specific macromolecular substances by simulations, an important goal is the clarification of qualitative questions such as whether polymer chains in a melt 'reptate' (Lodge *et al.*, 1990). By 'reptation' (De Gennes, 1979) one means a snake-like motion of polymer chains along their own contour, since the 'entanglements' with other chains create an effective 'tube' along the contour that constrains the motions. Since this type of motion is a universal phenomenon, it can be studied by coarse-grained models of polymers (Fig. 6.11) where one dispenses with much of the chemical detail such as the torsional potential (Fig. 6.10). Rather one considers models where effective bonds are formed by treating $n \approx 3$–5 successive covalent bonds along the backbone of the chain in one effective subunit. While the chains are generally treated as being completely flexible, i.e. the only potentials considered are bond length potentials and non-bonded forces, a treatment of stiff chains by bond angle potentials is straightforward (Haas *et al.*, 1996). Such models are useful for describing the alkane tails in monolayers of amphiphilic fatty acids at the air–water interface (Haas *et al.*, 1996). In the freely jointed chain (a) rigid links of length l are jointed at beads (shown by dots) and may make arbitrary angles with each other. The stochastic chain conformational changes, that on a microscopic level come about by jumps between the minima of the torsional potential (Fig. 6.10), are modeled by random rotations about the axis connecting the nearest neighbor beads along the chain, as indicated. A new bead position i may be chosen by assigning an angle φ_i, drawn randomly from the interval $[-\Delta\varphi, +\Delta\varphi]$ with $\Delta\varphi \leq \pi$. For the simulation of melts, freely jointed chains are often supplemented by a Lennard-Jones-type potential (Fig. 6.10) between any pairs of beads (Baumgärtner and Binder, 1981). An alternative model is the pearl-necklace model (b), where the beads are at the center of

Fig. 6.11 (a–c) Several off-lattice models for polymer chains; (d) Lennard-Jones potential. For further explanations see text.

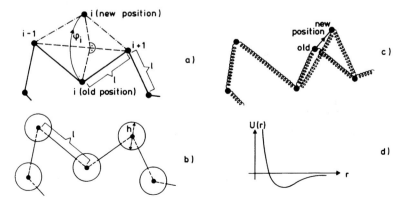

hard spheres of diameter h, which must not intersect each other. By varying the ratio h/l one can to some extent control the persistence length of the polymer chains. With this model studies of rather long chains have been possible (Baumgärtner, 1984). The most popular model, however, is the bead–spring model (c), which is used both for Monte Carlo simulations as indicated (Milchev *et al.*, 1993) and for molecular dynamics simulations (Kremer and Grest, 1995). In both cases the non–bonded interactions are modeled by Lennard–Jones potentials among the beads or by Morse potentials, respectively. These coarse–grained off–lattice models exist in several variants, and defining a model that is optimally suited for the desired application is the first step of a successful Monte Carlo simulation in polymer science.

Problem 6.7 Write a Monte Carlo algorithm that generates recursively freely jointed chains containing N rigid links of length ℓ, i.e. start from the origin and build up a random walk step-by-step. For $N = 10, 20, 30, 40$ and 50 generate a sample of $n = 10\,000$ configurations. Use these configurations to calculate the mean-square end-to-end distance $\langle R^2 \rangle$ and the mean-square gyration radius. Analyze the ratio $\langle R^2 \rangle / \langle R_g^2 \rangle$ as a function of N.

Problem 6.8 Using the algorithm of Problem 6.7 calculate the relative fluctuation of $\langle R^2 \rangle$, i.e. $\chi(R)$, see Eqn. (6.52), as a function of N. How can you interpret the result?

Problem 6.9 Use a configuration generated in Problem 6.7 as the initial state for the algorithm shown in Fig. 6.11a, with $\Delta\phi = \pi/4$. (End bonds may rotate freely by arbitrary angles to a new point on the surface of a sphere of radius ℓ and center at the monomer adjacent to the end.) Calculate the mean-square displacement of the center of mass of the chain. Obtain the self-diffusion constant D_N of the chain from the Einstein relation (Eqn. 6.60). Choose the time unit such that each bead on average is chosen randomly for a move once. Analyze the behavior D_N vs. N on a log–log plot.

Problem 6.10 Use a configuration of Problem 6.7 as a starting configuration for the algorithm in Fig. 6.11a, but with a Lennard-Jones interaction between the beads with $\sigma = 1/2$, $\varepsilon = 3$. Study the relaxation of the end-to-end distance. Analyze $\langle R^2 \rangle$ vs. N on a log–log plot and compare the result to the self-avoiding walk problem.

6.6.2 Asymmetric polymer mixtures: a case study

Many aspects of Monte Carlo simulations of polymeric systems are in fact rather similar to those of simulations of systems composed of atoms or small molecules. This fact will become apparent from the case study treated in this subsection, where we consider a mixture of two polymers (A, B) with different chain lengths, $N_A < N_B$. In other physical properties (shape and size of monomeric units, chain stiffness, etc.) the two types of chains are assumed to be identical, but a choice of pairwise interaction parameters is made which leads to unmixing:

$$\varepsilon_{AB}(\mathbf{r}) = \varepsilon_{AA}(\mathbf{r}) = \varepsilon_{BB}(\mathbf{r}) = \infty, \qquad r < r_{min}, \tag{6.61}$$

$$\varepsilon_{AB}(\mathbf{r}) = -\varepsilon_{AA}(\mathbf{r}) = -\varepsilon_{BB}(\mathbf{r}) = T\varepsilon, \qquad r_{min} \le r \le r_{max}, \tag{6.62}$$

$$\varepsilon_{AB}(\mathbf{r}) = \varepsilon_{AA}(\mathbf{r}) = \varepsilon_{BB}(\mathbf{r}) = 0, \qquad r > r_{max}. \tag{6.63}$$

If, in addition $N_A = N_B$, there would be a symmetry in the problem with respect to the interchange of A and B, and due to that symmetry phase coexistence between unmixed A-rich and B-rich phases could only occur at a chemical potential difference $\Delta\mu = \mu_A - \mu_B = 0$ between the two species. The critical value ρ_c of the concentration ρ of species A, defined in terms of the densities of monomers ρ_A, ρ_B as $\rho = \rho_A/(\rho_A + \rho_B)$, would thus be simply $\rho_c = 1/2$ due to this symmetry A⇔B. In the case of chain length asymmetry, however, this symmetry is destroyed, and then phase coexistence between the A-rich and the B-rich phase occurs along a non-trivial curve $\Delta\mu = \Delta\mu_{coex}(T)$ in the plane of variables (temperature T, chemical potential difference $\Delta\mu$). Also ρ_c now has a non-trivial value. Problems of this sort are of interest in materials science, since polymer blends have many practical applications. As a consequence we would very much like to understand to what extent simple mean-field theories of this problem, such as the Flory–Huggins theory (Binder, 1994), are reliable. These predict the critical point to be at

$$\rho_c = \left(\sqrt{N_A/N_B} + 1\right)^{-1}, \qquad (2z\varepsilon_c)^{-1} = 2\left(1/\sqrt{N_A} + 1/\sqrt{N_B}\right)^{-2}, \tag{6.64}$$

where z is the effective number of monomers within the interaction range specified in Eqn. (6.62) and ε_c is the effective value of ε (see Eqn. (6.62)) at the critical point.

An actual study of this problem has been carried out by Müller and Binder (1995) in the framework of the bond fluctuation lattice model of polymers (see Section 4.7). We nevertheless describe this case study here, because the problem of asymmetric mixtures is rather typical for the off-lattice simulations of binary mixtures in general. For the bond fluctuation model, $r_{min} = 2a$, where a is the lattice spacing, and $r_{max} = \sqrt{6}a$.

We now describe how such a simulation is carried out. The first step consists in choosing an initial, well-equilibrated configuration of an athermal $(T \to \infty)$ polymer melt, consisting purely of B-chains, at the chosen total monomer density $\rho_m = \rho_A + \rho_B$. This part of the simulation is a standard problem for all kinds of polymer simulations of dense polymeric systems, because if we would fill the available volume of the simulation box by putting in simple random-walk type configurations of polymers, the excluded volume interaction, Eqn. (6.61), would not be obeyed. If we put in the chains consecutively, growing them step by step as growing self-avoiding walk type configurations, we would create a bias with subtle correlations in their structure rather than creating the configurations typical for chains in dense melts which do respect excluded volume interactions locally but behave like simple random walks on large length scales, since then the excluded volume interactions are effectively screened out. Thus, whatever procedure one chooses to define the initial configuration, it needs to be carefully relaxed (e.g. by applying the 'slithering snake' algorithm or the 'random hopping' algorithm, cf. Section 4.7 for lattice models of polymers and the previous subsection for off-lattice models). In the case where large boxes containing many short polymers is simulated, one may simply put them into the box until the memory of this ordered initial configuration is completely lost. Of course, this particular choice requires that the linear dimension L of the box exceeds the length of the fully stretched polymer chain.

When dealing with problems of phase coexistence and unmixing criticality of mixtures, it is advisable to work in the semi-grand canonical ensemble, with temperature T and chemical potential difference $\Delta\mu$ being the independent thermodynamic variables. This is exactly analogous to the problem of phase coexistence and criticality in simple fluids, see Section 6.1 of the present chapter, where we have also seen that the grand canonical ensemble is preferable. However, while there it is straightforward to use Monte Carlo moves where particles are inserted or deleted, the analogous move for a mixture (an A-particle transforms into a B-particle, or vice versa) is straightforward to use only for the case of symmetric polymer mixtures (we can take out an A-chain and replace it by a B-chain in the identical conformation: essentially this identity switch is just a relabeling of the chain). Of course, there is no problem in taking out a long B-chain and using part of the emptied volume to insert a shorter A-chain, but the inverse move will hardly ever be successful for a dense polymeric system, because of the excluded volume interaction the acceptance probability for such chain insertions in practice always is zero!

But this problem can be overcome in the special situation $N_B = kN_A$, when k is an integer ($k = 2, 3, 4, \ldots$), by considering the generalized semi-grand canonical moves where a single B-chain is replaced by k A-chains, or vice versa. In the net effect, one has to cut (or insert, respectively) $k - 1$ covalent bonds together with the relabeling step. While the cutting of bonds of a B-chain is unique, the reverse step of bond insertion is non-unique, and hence one must use carefully constructed weighting factors in the acceptance probability of such moves to ensure that the detailed balance principle holds!

We shall not dwell on these weighting factors here further but rather discuss how one can find the chemical potential difference $\Delta\mu_{\text{coex}}(T)$ where phase coexistence occurs, applying such an algorithm. This is done exactly with the same 'equal weight rule' that we have discussed in Section 4.2 in the context of finite size effects at first order transitions: the distribution function $P_L(\rho)$ in the $L \times L \times L$ box (with periodic boundary conditions as usual) will exhibit a double-peak structure for $\Delta\mu$ near $\Delta\mu_{\text{coex}}(T)$, and at $\Delta\mu_{\text{coex}}(T)$ the weights of the two peaks have to be equal. In practice, histogram reweighting techniques are needed (and for T far below T_c, even the application of the 'multicanonical' method is advisable, see Chapter 5) in order to sample $P_L(\rho)$ efficiently. Furthermore, several choices of L need to be studied, in order to check for finite size effects. The analysis of finite size effects is subtle particularly near the critical point, because there the 'field mixing' problem (order parameter density and energy density are coupled for asymmetric systems, see Section 4.2.3) comes into play, too.

Figure 6.12 shows typical results from the finite size scaling analysis applied to this problem. For a given choice of N_A, N_B and the normalized energy $\varepsilon/k_B T$, we have to find $\Delta\mu_{coex}(T)$ such that the second moment $\langle m^2 \rangle$ of the order parameter $m \equiv (\rho_A - \rho_A^{crit})/(\rho_A + \rho_B)$ satisfies the finite size scaling characteristic of first order transitions as long as $T < T_c$, namely $\langle m^2 \rangle$ is a universal function of $L^3(\Delta\mu - \Delta\mu_{coex}(T))$, in $d = 3$ dimensions (Fig. 6.12a). Along the line $\Delta\mu = \Delta\mu_{coex}(T)$ one can then apply the moment analysis as usual, recording ratios such as $\langle m^2 \rangle/\langle |m| \rangle^2$ and $1 - \langle m^4 \rangle/3\langle m^2 \rangle^2$ for different choices of L, in order to locate the critical temperature T_c from the common size-independent intersection point (Fig. 6.12b). The consistency of this Ising-model type finite size scaling description can be checked for $T = T_c$ by analyzing the full order parameter distribution (Fig. 6.12c). We see that the same type of finite size scaling at T_c as discussed in Chapter 4 is again encountered, the order parameter distribution $P(m)$ scales as $P(m) = L^{\beta/\nu}\tilde{P}(mL^{\beta/\nu})$, where $\beta = 0.325$, $\nu = 0.63$ are the Ising model critical exponents of order parameter and correlation length, respectively, and the scaling function $\tilde{P}(\zeta)$ is defined numerically from the 'data collapse' of $P(m)$ as obtained for the different linear dimensions L in the figure. Of course, this data collapse is not perfect – there are various sources of error for a complicated model like the present asymmetric polymer mixture. Neither $\Delta\mu_{coex}(T)$, nor T_c and ρ_A^{crit} ($= 0.57$ here, see Fig. 6.12c) are known without error, there are statistical errors in the simulation data for $P(m)$ and systematic errors due to finite size scaling, etc., but the quality of this data collapse is good enough to make this analysis credible and useful. For the example chosen ($N_A = 40$, $N_B = 80$) one expects from Eqn. (6.64) that $\rho_c \approx 0.586$ and hence the finding $\rho_c = \rho_A^{crit}/(\rho_A + \rho_B) \approx 0.57$ (Fig. 6.12c) deviates from the prediction only slightly.

6.6.3 Applications: dynamics of polymer melts; thin adsorbed polymeric films

The reptation concept alluded to above is only effective if the chain length N far exceeds the chain length N_e between 'entanglements'. For short chains, with $N \approx N_e$ or less, entanglements are believed to be ineffective, and neighboring chains only hinder the motion of a chain by providing 'friction' and random forces acting on the bonds of the chain. In more mathematical terms, this is the content of the 'Rouse model' (Rouse, 1953) of polymer dynamics, where one considers the Langevin equation for a harmonic bead-spring chain exposed to a heat bath. Now it is clear that random motions of beads as

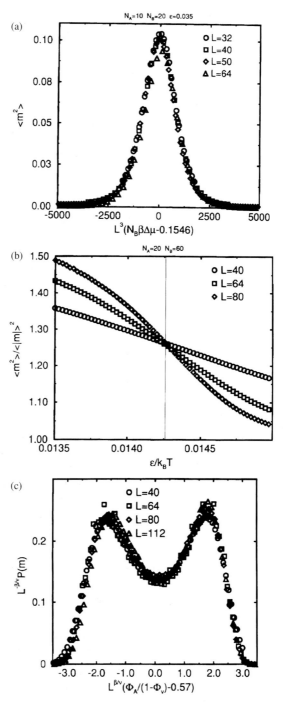

Fig. 6.12 Scaling of the second moment of the order parameter at $\varepsilon/k_B T = 0.035$ (a); locating the critical temperature from the intersection of moment ratios (b); scaled order parameter distribution at criticality (c). In all cases the different symbols indicate different linear dimensions L.

considered in Fig. 6.11c can be considered as discretized realizations of such a stochastic dynamical process described by a Langevin equation. Monte Carlo moves are thus suitable for the modeling of the slow Brownian motion of polymer chains in melts, and since the non-bonded potentials can be chosen such that they have the side effect that no chain intersections can occur in the course of the random motions of the beads, all essential ingredients of the reptation mechanism are included in the Monte Carlo algorithm. As a consequence, various Monte Carlo studies of models shown in Fig. 6.11 have been made to attempt to clarify questions about reptation theory (Baumgärtner, 1984). These simulations supplement molecular dynamics studies (Kremer and Grest, 1990) and Monte Carlo work on lattice models, e.g. Paul et $al.$ (1991). One typical example is the crossover behavior in the self-diffusion of chains. From Fig. 6.11c it is clear that random displacements $\Delta\mathbf{r}$ will lead to a mean-square displacement of the center of mass of a chain of the order $(\Delta\mathbf{r}/N)^2 N \propto l^2/N$ after N moves (the natural unit of time is such that every monomer experiences an attempted displacement on average once, and the mean-square distances between the old and new positions are of the same order as the bond length square, l^2). This shows that the self-diffusion constant of the Rouse model, D_{Rouse}, should scale with chain length like $D_{\text{Rouse}} \propto 1/N$. The characteristic relaxation time, τ_{Rouse}, can be found as the time needed for a chain to diffuse its own sizes $l\sqrt{N}$. Putting D_{Rouse} $\tau_{\text{Rouse}} \propto (l\sqrt{N})^2$ and using $D_{\text{Rouse}} \propto l^2/N$ yields $\tau_{\text{Rouse}} \propto N^2$. This behavior is indeed observed both in single chain simulations at the θ-temperature (Milchev et $al.$, 1993) and for melts of very short chains (Baumgärtner and Binder, 1981)

If we consider instead the motion of very long chains, we can argue that this can be again described by a Rouse-like diffusion but constrained to take place in a tube. During the Rouse time the chain has traveled a distance proportional to $l\sqrt{N}$ along the axis of the tube. However, the axis of the tube follows the random-walk-like contour of the chain, which hence has a length proportional to lN rather than $l\sqrt{N}$. A mean-square distance of order $l^2 N^2$, i.e. the full length of the contour, hence is only traveled at a time of order $\tau_{\text{Rouse}} N \propto N^3$. Hence the characteristic time τ_{Rep} for a chain to 'creep out' of its tube scales like N^3. On the other hand, the distance traveled in the coordinate system of laboratory space (not along the tube contour!) is no more than the chain radius, $R \approx l\sqrt{N}$. Putting again a scaling relation between diffusion constant D_N and relaxation time, $D_N \tau_{\text{Rep}} \propto (l\sqrt{N})^2$ we conclude $D_N \propto N^{-2}$. In general, then, one expects that $D_N \propto N^{-1}$ for $N \ll N_e$ and $D_N \propto N^{-2}$ for $N \gg N_e$, with a smooth crossover for $N \approx N_e$.

Figure 6.13 shows that these expectations indeed are borne out by the simulations. Rescaling D by D_{Rouse} and N by N_e (which can be estimated independently by other means, such as an analysis of the mean-square displacement of inner monomers) one finds that Monte Carlo data for the bond fluctuation model (Paul et $al.$, 1991) and molecular dynamics data for the bead-spring model with purely repulsive Lennard-Jones interaction (Kremer and Grest, 1990) fall on a common curve. The bond

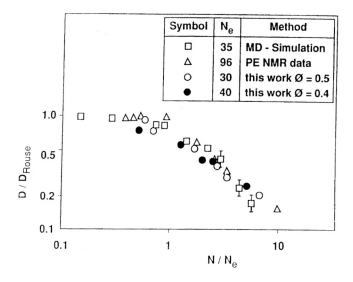

Fig. 6.13 The self-diffusion constant D of polymer melts vs. chain length N, normalized by the diffusion constant in the short chain Rouse limit, D_{Rouse}. The entanglement chain length is N_e. Circles are Monte Carlo results for the bond fluctuation model at two volume fractions Φ of occupied sites, squares are molecular dynamics results of Kremer and Grest (1990), and triangles are experimental data for polyethylene (Pearson et al., 1987). From Paul et al. (1991).

fluctuation model is actually a lattice model, but unlike the self-avoiding walk model of Chapter 3 where a bead is a lattice site and a bond connects two nearest neighbor sites of the lattice, the discretization is rather fine: a bead takes all 8 sites of the elementary cube of the lattice, an effective bond has a length of ~ 3 lattice spacings, and rather than 6 bond vectors connecting nearest neighbor sites on the simple cubic lattice one has 108 bond vectors. The result is a rather close approximation to the properties of continuum models. Although Monte Carlo methods certainly omit many aspects of the dynamics of polymer melts – from bond length vibrations to hydrodynamic flows – they can model the slow Brownian diffusive motion of polymer chains rather well. This is indicated by the agreement with the experimental data on polyethylene (Pearson et al., 1987). Note that there is no inconsistency in the observation that the experimental value of N_e is about three times as large as in the simulation: here the count is simply the degree of polymerization, i.e. number of C–C bonds along the backbone of the chain, while in the simulations each effective bond corresponds to $n \approx$ 3–5 such C–C bonds.

As a final example, we briefly mention thin polymeric films adsorbed on walls. While the adsorption of single chains at walls from dilute solution has been studied for a long time, both in the framework of lattice (Eisenriegler et al., 1982) and continuum models (Milchev and Binder, 1996), the study of many-chain systems at surfaces in equilibrium with surrounding solution has just begun. A particular advantage of the off-lattice models is that from the virial theorem it is straightforward to obtain the components $p_{\alpha\beta}(z)$ of the local pressure tensor as a function of the distance z from the attractive wall (Rao and Berne, 1979)

$$p_{\alpha\beta}(z) = \rho(z)k_{\mathrm{B}}T\delta_{\alpha\beta} - \frac{1}{2A}\sum_{i\neq j}(\mathbf{r}_{ij})_\alpha \frac{\partial U(\mathbf{r}_{ij})}{\partial(\mathbf{r}_{ij})_\beta}\theta[(z-z_i)/z_{ij}]\theta[(z_j-z)/z_{ij}]/|z_{ij}|$$

$$(6.65)$$

where $\rho(z)$ is the local density, A is the surface area of the wall in the simulated system, θ is the step function, and U the total potential. This pressure tensor, which generalizes the expression for the average pressure given in Eqn. (6.9), provides a good criterion for judging whether the simulation box is large enough that bulk behavior in the solution coexisting with the adsorbed layer is actually reached, since in the bulk solution the pressure tensor must be isotropic,

$$p_{xx}(z) = p_{yy}(z) = p_{zz}(z), \qquad (6.66)$$

and independent of z. On the other hand, the anisotropy of the pressure tensor near the wall can be used to obtain interfacial free energies (e.g. Smit, 1988). For a geometry where the wall at $z = 0$ is attractive while the wall at the opposite surface, $z = D$, is purely repulsive, even two different interfacial free energies can be estimated (Pandey *et al.*, 1997; Nijmeijer *et al.*, 1990)

$$\gamma_{\mathrm{I}}^{\mathrm{att}} = \int\limits_0^{D/2} dz\left[p_{zz}(z) - (p_{xx}(z) + p_{yy}(z))/2 - \rho(z)\frac{d}{dz}\phi^{\mathrm{att}}(z)\right], \qquad (6.67)$$

$$\gamma_{\mathrm{I}}^{\mathrm{rep}} = \int\limits_{D/2}^D dz\left[p_{zz}(z) - (p_{xx}(z) + p_{yy}(z))/2 - \rho(z)\frac{d}{dz}\phi^{\mathrm{rep}}(z)\right], \qquad (6.68)$$

if the thickness of the system is large enough such that in the center (near $z = D/2$) the pressure tensor is isotropic. Here $\phi^{\mathrm{att}}(z)$, $\phi^{\mathrm{rep}}(z)$ denote the attractive and repulsive wall potentials.

Of course, understanding the dynamics of chains in these adsorbed layers is a particular challenge (Milchev and Binder, 1996, 1997; Pandey *et al.*, 1997). Also, non-equilibrium phenomena such as 'dewetting' can be observed (Milchev and Binder, 1997): if at time $t = 0$ the strength of the adsorption potential is strongly reduced, a densely adsorbed, very thin polymer film becomes thermodynamically unstable, and it breaks up into small droplets which slowly coarsen as time passes, similar to the coarsening observed in intermediate and late stages of spinodal decomposition of mixtures (Fig. 6.14). While some features of such simulations are qualitatively similar to those found in some experiments, one must consider the possibility that effects due to hydrodynamic flow, which are not included in the Monte Carlo 'dynamics', could be important.

Thus, for simulation of polymers it is particularly important for the reader to consider quite carefully the question of which models and simulation technique are most suitable for the investigation of a particular problem. We have not attempted to give an exhaustive survey but hope that our treatment provides a feeling for the considerations that need to be made.

Fig. 6.14 Snapshots of a system with 64 chains, each containing $N = 32$ beads, in an $L \times L \times D$ box with $L = 32$, $D = 8$. There are periodic boundary conditions in x- and y-directions, while at $z = 0$ and $z = D$ there are impenetrable hard walls; at the bottom wall there is also an attractive square-well potential of strength ε and range $\delta = 1/8$. The chains are described by a bead-spring model with a preferred bond length of 0.7. Note that the springs between the beads are not shown. From Milchev and Binder (1997).

6.7 CONFIGURATIONAL BIAS AND 'SMART MONTE CARLO'

If the trial states generated in attempted Monte Carlo moves are chosen completely 'blindly', without paying particular attention to the state the system is in when the move is attempted, sometimes the acceptance of such a move is very small. An example is the insertion of a rod-like molecule in a nematic liquid crystal, where the molecules have some preferred orientation characterized by the nematic order parameter: if the molecule to be inserted is randomly oriented, it is very likely that the repulsive interaction with the other molecules would be too strong, and hence the trial move would be rejected. Under these circumstances it is an obvious idea to choose an 'orientational bias'. Of course, one has to be very careful that the algorithm that is devised still satisfies detailed balance and provides a distribution with Boltzmann weights in the sampling. In practice, this can be done by a suitable modification of the transition probability $W(\text{o} \to \text{n})$ by which the move from the old (o) to the new (n) configuration is accepted (see Frenkel and Smit (1996) for an extensive discussion). Suppose now the *a priori* transition probability (i.e. without consideration of the Boltzmann factor) depends on the potential energy $U(\text{n})$ of the new configuration through a biasing function $f[U(\mathrm{n})]$, $W_{a\,priori}(\mathrm{n} \to \mathrm{o}) = f[U(\mathrm{n})]$. For the reverse move we would have $W_{a\,priori}(\mathrm{n} \to \mathrm{o}) = f[U(\mathrm{o})]$. Then the proper choice of transition probability is a modified Metropolis criterion

$$W(\mathrm{o} \to \mathrm{n}) = \min\left\{1, \frac{f(U(\mathrm{n}))}{f[U(\mathrm{o})]} \exp\{-[U(\mathrm{n}) - U(\mathrm{o})]/k_B T\}\right\}. \qquad (6.69)$$

This prescription is not only appropriate for the case of rigid molecules where we choose a bias for the trial orientation of a molecule that is inserted, but also holds for other cases too. For example, for flexible chain molecules the insertion of a chain molecule in a multichain system, if it is done blindly, very likely creates a configuration that is 'forbidden' because of the excluded volume interaction. Thus one biases the configuration of the chain that is inserted such that these unfavorable interactions are avoided. We emphasize that the configurational bias Monte Carlo method is not only useful in the off-lattice case, but similarly on lattices as well. In fact, for the lattice case these methods were developed first (Siepmann and Frenkel, 1992). Here the biased configuration of the chain that is inserted is stepwise grown by the Rosenbluth scheme (Rosenbluth and Rosenbluth, 1955). There one 'looks ahead' before a new bond is attached to the existing part of the chain, to see for which directions of the new bond the excluded volume constraint would be satisfied. Only from the subset of these 'allowed' bond directions is the new bond direction then randomly chosen. As has been discussed in the literature elsewhere (Kremer and Binder, 1988), we note that such biased sampling methods have serious problems for very long chains, but for chains of medium length (e.g. less than 100 steps on a lattice) the problem of estimating the statistical errors resulting from such techniques is typically under control. In this stepwise insertion of the polymer chain, one constructs the Rosenbluth weight $W(\mathrm{n})$ of this chain – which is the analog of the biasing function f mentioned above – according to the usual Rosenbluth scheme. In order to be able to introduce the appropriate correction factor $W(\mathrm{n})/W(\mathrm{o})$ in the modified Metropolis criterion, one has to select one of the chains, that are already in the system at random, and retrace it step by step from one end to calculate its Rosenbluth weight. Of course, this type of algorithm can also be extended to the off-lattice case.

Still another type of biased sampling, that sometimes is useful, and can even be applied to simple fluids, is force bias sampling (Ceperley et al., 1977; Pangali et al., 1978): one does not choose the trial move of a chosen particle completely blindly at random, but biases the trial move along the forces and torques acting on the particles. One wishes to choose the transition probability W_{ij} to move from state i to state j proportional to the Boltzmann factor $\exp(U_j/k_B T)$: then detailed balance will be automatically satisfied. Assuming that states i and j are close by in phase space, differing only by center of mass displacements $\mathbf{R}_m(j) - \mathbf{R}_m(i)$ of molecule m and by an angular displacement $\Omega_m(j) - \Omega_m(i)$ (in a formulation suitable for rigid molecules, such as water, for instance). Then one can expand the energy of state j around the energy of state i to first order, which yields

$$W_{ij} = \frac{W_{ij}^M}{Z(i)} \exp\left\{\frac{\lambda}{k_B T}\left(\mathbf{F}_m(i) \cdot [\mathbf{R}_m(j) - \mathbf{R}_m(i)] + \mathbf{N}_m(i) \cdot [\mathbf{\Omega}_m(j) - \mathbf{\Omega}_m(i)]\right)\right\},$$

$$(6.70)$$

where W_{ij}^M is the usual Metropolis acceptance factor $\min\{1, \exp[-(U_j - U_i)/k_B T]\}$, and $\mathbf{F}_m(i)$ is the total force acting on molecule m in state i, and $\mathbf{N}_m(i)$ the corresponding torque. Here $Z(i)$ is a normalization factor and λ is a parameter in the range $0 \le \lambda \le 1$: $\lambda = 0$ would be the unbiased Metropolis algorithm, of course. Note that the displacements have to be limited to fixed (small) domains around the initial values $\mathbf{R}_m(i)$ and $\mathbf{\Omega}_m(i)$.

An alternative force bias scheme proposed by Rossky *et al.* (1978) was inspired by the 'Brownian dynamics' algorithm (Ermak, 1975), where one simulates a Langevin equation. For a point particle of mass m this Langevin equation describes the balance of friction forces, deterministic and random forces:

$$\ddot{\mathbf{r}} = -\dot{\mathbf{r}}\zeta + (\mathbf{F} + \boldsymbol{\eta}(t))/m, \qquad (6.71)$$

where ζ is the friction coefficient, $\mathbf{F} = -\nabla U$ is the force due to the potential, and $\boldsymbol{\eta}(t)$ is a random force, which is linked to ζ in thermal equilibrium by a fluctuation-dissipation relation. A simulation of this Langevin equation could be done by discretizing the time derivatives $\dot{\mathbf{r}} = d\mathbf{r}/dt$ as $\Delta\mathbf{r}/\Delta t$ to find

$$\Delta\mathbf{r} = (D/k_B T)\mathbf{F}\Delta t + \Delta\boldsymbol{\rho}, \qquad (6.72)$$

where $\Delta\mathbf{r}$ is the change of $\mathbf{r}$ in a time step Δt, $\mathbf{F}$ is the force on the particle at the beginning of the step, D is the diffusion constant of the particle in the absence of interparticle interactions, and $\Delta\boldsymbol{\rho}$ is the random displacement corresponding to the random force $\boldsymbol{\eta}(t)$. For a faithful description of the dynamics that would follow from the Langevin equation, Δt and $\Delta\boldsymbol{\rho}$ would have to be very small. However, if we are interested in static equilibrium properties only, we can allow much larger Δt, $\Delta\boldsymbol{\rho}$ and use the corresponding new state obtained from $\mathbf{r}' = \mathbf{r} + \Delta\mathbf{r}$ as a trial move in a Metropolis Monte Carlo sampling. This is the basic idea behind the algorithm proposed by Rossky *et al.* (1978) and called 'smart Monte Carlo'.

A very straightforward type of biased sampling is useful for dilute solutions (Owicki and Scheraga, 1977): one does a preferential sampling of molecules close to a solute molecule. In fact, this idea is similar to the preferential selection of sites near external surfaces or internal interfaces which has already been discussed for lattice models, e.g. in Section 5.7.1.

There are many conditions where such biased Monte Carlo methods produce equilibrium faster than do the standard Monte Carlo methods; but often molecular dynamics (Chapter 12) is then even more efficient! Thus the choice of 'which algorithm and when' remains a subtle problem.

REFERENCES

Alder, B. J. and Wainwright, T. E. (1962), Phys. Rev. **127**, 359.

Alejandre, J., Tildesley, D. J., and Chapela, G. A. (1995), J. Chem. Phys. **102**, 4574.

Allen, M. P. (1996), in *Monte Carlo and Molecular Dynamics of Condensed Matter Systems*, eds. K. Binder and G. Ciccotti (Società Italiana di Fisica, Bologna), p. 255.

Baschnagel, J., Qin, K., Paul, W., and Binder, K. (1992), Macromolecules **25**, 3117.

Baumgärtner, A. (1984), Ann. Rev. Phys. Chem. **35**, 419.

Baumgärtner, A. and Binder, K. (1981), J. Chem. Phys. **75**, 2994.

Binder, K. (1994) Adv. Polymer Sci. **112**, 181.

Binder, K. (1995), (ed.) *Monte Carlo and Molecular Dynamics Simulations in Polymer Science* (Oxford University Press, New York and Oxford).

Binder, K. and Landau, D. P. (1989), in *Advances in Chemical Physics: Molecule-Surface Interaction*, ed. K. P. Lawley (Wiley, NY) p. 91.

Catlow, C. R. A. (1992), (ed.) *Modelling of Structure and Reactivity in Zeolites* (Academic Press, London).

Ceperley, D., Chester, C. V., and Kalos, M. H. (1977), Phys. Rev. B **16**, 3081.

Dadobaev, G. and Slutsker, A. I. (1981), Sov. Phys. Solid State **23**, 1131.

Davies, G. T., Eby, K., and Colson, J. P. (1970), J. Appl. Phys. **41**, 4316.

De Gennes, P. G. (1979), *Scaling Concepts in Polymer Physics* (Cornell University Press, Ithaca).

Degiorgio, V. and Corti, M. (1985) (eds.), *Physics of Amphiphiles: Micelles, Vesicles and Microemulsions* (North-Holland, Amsterdam).

Dünweg, B., Stevens, M., and Kremer, K. (1995), in Binder, K. (1995), p. 125.

Eisenriegler, E., Kremer, K., and Binder, K. (1982), J. Chem. Phys. **77**, 6296.

Ermak, D. L. (1975), J. Chem. Phys. **62**, 4189.

Frenkel, D. and Smit, B. (1996), *Understanding Molecular Simulation: From Algorithms to Applications* (Academic Press, New York).

Gaines, G. L. Jr. (1996), *Insoluble Monolayers at Liquid-Gas Interfaces* (Intersciences, New York).

Greengard, L. and Rokhlin, V. (1987), J. Comp. Phys. **73**, 325.

Haas, F. M., Hilfer, R., and Binder, K. (1996), J. Phys. Chem. **100**, 15290.

Houlrik, J., Landau, D. P., and Knak Jensen, S. (1994), Phys. Rev. E **50**, 2007.

Jaster, A. (1998), Europhys. Lett. **42**, 277.

Karaborni, S. and O'Connell, J. P. (1990), J. Phys. Chem. **94**, 2624.

Kremer, K. and Binder, K. (1988), Computer Phys. Rep. **7**, 259.

Kremer, K. and Grest, G. S. (1990), J. Chem. Phys. **92**, 5057.

Kremer, K. and Grest, G. S. (1995), in Binder, K. (1995), p. 194.

Landau, D. P. (1991), in *Phase Transitions and Surface Films 2*, eds. H. Taub, G. Torzo, H. J. Lauter, and S. C. Fain, Jr., p. 11.

Landau, L. D. and Lifshitz, E. M. (1980), *Statistical Physics* 3rd ed., Part 1 (Pergamon Press, Oxford).

Lodge, T. P., Rotstein, N. A., and Prager, S. (1990), in *Advances in Chemical Physics*, Vol. **79**, eds. I. Prigogine and S. A. Rice, (Wiley, New York) p. 1.

Martonak, R., Paul, W., and Binder, K. (1996), Computer Phys. Commun. **99**, 2.

Martonak, R., Paul, W., and Binder, K. (1997), J. Chem. Phys. **106**, 8918.

McDonald, I. R. (1972), Mol. Phys. **23**, 41.

Metropolis, N. *et al.* (1953), J. Chem. Phys. **21**, 1087.

Milchev, A., and Binder, K. (1996), Macromolecules **29**, 343.

Milchev, A. and Binder, K. (1997), J. Chem. Phys. **106**, 1978.

Milchev, A., Paul, W., and Binder, K. (1993), J. Chem. Phys. **99**, 4786.

Mooij, G. C. A. M., Frenkel, D., and Smit, B. (1992), J. Phys. Condens. Matter **4**, L255.

Müller, M. and Binder, K. (1995), Macromolecules **28**, 1825.

Nelson, D. R. and Halperin, B. I. (1979), Phys. Rev. B **19**, 2457.

Nijmeijer, M. J. P., Bruin, C., Bakker, A. F., and van Leeuwen, M. J. M. (1990), Phys. Rev. A **42**, 6052.

Norman, G. E. and Filinov, V. S. (1969), High Temp. (USSR) **7**, 216.

Owicki, J. C. and Scheraga, H. A. (1977), Chem. Phys. Lett. **47**, 600.

Panagiotopoulos, A. Z. (1987), Molecular Physics **61**, 813.

Panagiotopoulos, A. Z. (1995), in *Observation Prediction and Simulation of Phase Transitions in Complex Fluids*, eds. M. Baus, L. F. Rull and J. P. Ryckaert, (Kluwer Academic Publ., Dordrecht) p. 463.

Pandey, R. B., Milchev, A., and Binder, K. (1997), Macromolecules **30**, 1194.

Pangali, C., Rao, M., and Berne, B. J. (1978), Chem. Phys. Lett. **55**, 413.

Patrykiejew, A., Sokolowski, S., Zientarski, T., and Binder, K. (1995), J. Chem. Phys. **102**, 8221.

Patrykiejew, A., Sokolowski, S., Zientarski, T., and Binder, K. (1998), J. Chem. Phys. **108**, 5068.

Paul, W., Binder, K., Heermann, D. W., and Kremer, K. (1991), J. Phys. II (France) **1**, 37.

Paul, W., Yoon, D. Y., and Smith, G. D. (1995), J. Chem. Phys. **103**, 1702.

Pearson, D. S., Verstrate, G., von Meerwall, E., and Schilling, F. C. (1987), Macromolecules **20**, 1133.

Pollock, E. L. and Glosli, J. (1996), Comput. Phys. Commun. **95**, 93.

Presber, M., Dünweg, B., and Landau, D. P. (1998), Phys. Rev. E **58**, 2616.

Rao, M. and Berne, B. J. (1979), Mol. Phys. **37**, 455.

Rector, D. R., van Swol, F., and Henderson, J. R. (1994), Molecular Physics **82**, 1009.

Rosenbluth, M. N. and Rosenbluth, A. W. (1955), J. Chem. Phys. **23**, 356.

Rossky, P. J., Doll, J. D., and Friedman, H. L. (1978), J. Chem. Phys. **69**, 4628.

Rouse, P. E. (1953), J. Chem. Phys. **21**, 127.

Rovere, M., Heermann, D. W., and Binder, K. (1990), J. Phys. Cond. Matter **2**, 7009.

Ryckaert, J. P. (1996), in *Monte Carlo and Molecular Dynamics of Condensed Matter Systems*, eds. K. Binder and G. Ciccotti (Società Italiana di Fisica, Bologna) p. 725.

Sariban, A. and Binder, K. (1987), J. Chem. Phys. **86**, 5859.

Scheringer, M., Hilfer, R., and Binder, K. (1992), J. Chem. Phys. **96**, 2296.

Schmid, F., Stadler, C., and Lange, H. (1998), in *Computer Simulation Studies in Condensed-Matter Physics X*, eds. D. P. Landau, K. K. Mon, and H.-B. Schüttler (Springer, Berlin) p. 37.

Siepmann, J. I. and Frenkel, D. (1992), Mol. Phys. **75**, 90.

Smit, B. (1988), Phys. Rev. A **37**, 3481.

Smit, B. (1995), J. Phys. Chem. **99**, 5597.

Smit, B., Esselink, K., Hilbers, P. A. J., van Os, N. M., Rupert, L. A. M., and Szleifer, I. (1993), Langmuir **9**, 9.

Sokal, A. D. (1995), in Binder, K. (1995), p. 47.

Viduna, D., Milchev, A., and Binder, K. (1998), Macromol. Theory and Simul. **7**, 649.

Von Gottberg, F. K., Smith, K. A., and Hatton, T. A. (1997), J. Chem. Phys. **106**, 9850.

Weber, H., Marx, D., and Binder, K. (1995), Phys. Rev. B **15**, 14636.

Werner, A., Schmid, F., Müller, M., and Binder, K. (1997), J. Chem. Phys. **107**, 8175.

Widom, B. (1963), J. Chem. Phys. **39**, 2808.

Wilding, N. B. (1997), J. Phys. Condensed Matter **9**, 585.

Wilding, N. B. and Binder, K. (1996), Physica A **231**, 439.

Yoon, D. Y., Smith, G. D., and Matsuda, T. (1993), J. Chem. Phys. **98**, 10037.

Zollweg, J. A. and Chester, G. V. (1992), Phys. Rev. B **46**, 11187.

7   Reweighting methods

7.1 BACKGROUND

7.1.1 Distribution functions

One longstanding limitation on the resolution of Monte Carlo simulations near phase transitions has been the need to perform many runs to precisely characterize peaks in response functions such as the specific heat. Dramatic improvements have become possible with the realization that entire distributions of properties, not just mean values, can be useful; in particular, they can be used to predict the behavior of the system at a temperature other than that at which the simulation was performed. There are several different ways in which this may be done. The reweighting may be done after a simulation is complete or it may become an integral part of the simulation process itself. The fundamental basis for this approach is the realization that the properties of the systems will be determined by a distribution function in an appropriate ensemble. Thus, in the canonical ensemble, for example, the probability of observing a particular state in a simple Ising ferromagnet with interaction constant $\mathcal{J}$ at temperature T is proportional to the Boltzmann weight, $\exp(-KE)$ where we define $K = \mathcal{J}/k_{\mathrm{B}}T$ as the dimensionless coupling. The probability of simultaneously observing the system with total (dimensionless) energy $E = -\sum \sigma_i \sigma_j$ and total magnetization $M = \sum \sigma_i$ is then

$$P_K(E, M) = \frac{W(E, M)}{Z(K)} \exp(-KE), \tag{7.1}$$

where $W(E, M)$ is the number of configurations (density of states) with energy E and magnetization M, and $Z(K)$ is the partition function of the system. Thus, the density of states contains all the relevant information about the systems and the effect of temperature can be straightforwardly included.

7.1.2 Umbrella sampling

In the following discussion we follow Frenkel and Smit (1996) by introducing the 'overlapping distribution method' (Bennett, 1976) for the estimation of the free energy difference ΔF between two systems, labeled 0 and 1, with partition functions Z_0 and Z_1. At this point, we consider off-lattice systems with N particles in a volume V at the same inverse temperature β (but differing in some other property, e.g. systems in different phases, or with some parameter in the Hamiltonian being different). The free energy difference can then be written as ($\beta \equiv 1/k_B T$)

$$\beta \Delta F = -\ln(Z_1/Z_0)$$

$$= -\ln\left(\int d\mathbf{r}^N \exp[-\beta U_1(\mathbf{r}^N)] \bigg/ \int d\mathbf{r}^N \exp[-\beta U_0(\mathbf{r}^N)]\right), \quad (7.2)$$

where $\mathbf{r}^N$ stands symbolically for the set of coordinates $\{\mathbf{r}_1, \mathbf{r}_2, \ldots, \mathbf{r}_N\}$ of the N particles, and U_0, U_1 are the potential energies of the two systems.

Suppose that a Monte Carlo sampling of the configuration space of system 1 is carried out. For every configuration ($\mathbf{r}^N$) of system 1 generated in this process the potential energy $U_0(\mathbf{r}^N)$ of the system 0 can be computed, and hence $\Delta U \equiv U_1(\mathbf{r}^N) - U_0(\mathbf{r}^N)$ can be obtained for every configuration. We use this information to generate a histogram that is proportional to the probability density $p_1(\Delta U)$ that this energy difference ΔU is observed,

$$p_1(\Delta U) = \int d\mathbf{r}^N \exp(-\beta U_1)\delta(U_1 - U_0 - \Delta U)/Z_1. \quad (7.3)$$

Substituting $U_1 = U_0 + \Delta U$ in the argument of the exponential function, we find

$$p_1(\Delta U) = \exp(-\beta \Delta U)\int d\mathbf{r}^N \exp(-\beta U_0)\delta(U_1 - U_0 - \Delta U)/Z_1,$$

$$= \frac{Z_0}{Z_1}\exp(-\beta \Delta U)p_0(\Delta U), \quad (7.4)$$

where

$$p_0(\Delta U) = \int d\mathbf{r}^N \exp(-\beta U_0)\delta(U_1 - U_0 - \Delta U)/Z_0 \quad (7.5)$$

is nothing but the probability density to find the same potential energy difference ΔU between systems 1 and 0 in a Boltzmann sampling of the configurations of system 0. Combining Eqns. (7.2) and (7.4) we readily obtain

$$\ln p_1(\Delta U) = \ln(Z_0/Z_1) - \beta \Delta U + \ln p_0(\Delta U) = \beta(\Delta F - \Delta U) + \ln p_0(\Delta U). \quad (7.6)$$

Thus, if there is a range of values ΔU where both $p_1(\Delta U)$ and $p_0(\Delta U)$ can be estimated from two separate simulations, one for system 0 and one for system 1, one can try to obtain $\beta \Delta F$ from a fit of Eqn. (7.6) to the difference between $\ln p_0(\Delta U)$ and $[\beta \Delta U + \ln p_1(\Delta U)]$.

The sampling of the chemical potential $\mu_{ex} \equiv \mu - \mu_{id}(V)$ ($\mu_{id}(V)$ being the chemical potential of an ideal gas of N particles at temperature T in a volume V) can be understood readily in the following way (see the discussion on particle insertion/removal techniques in Chapter 6). We simply assume that system 1 has N interacting particles while system 0 contains $N - 1$ interacting particles and one non-interacting ideal gas particle. This yields (Shing and Gubbins, 1983)

$$\mu_{ex} = \ln p_1(\Delta U) - \ln p_0(\Delta U) + \beta \Delta U. \tag{7.7}$$

Since Eqn. (7.6) can also be written as

$$p_1(\Delta U) = p_0(\Delta U) \exp[\beta(\Delta F - \Delta U)], \tag{7.8}$$

we conclude that, in principle, knowledge of either $p_1(\Delta U)$ or $p_0(\Delta U)$ suffices to fix ΔF, since these probabilities are normalized, i.e. $\int_{-\infty}^{+\infty} p_1(\Delta U) d(\Delta U) = 1$, $\int_{-\infty}^{+\infty} p_0(\Delta U) \Delta U = 1$. Hence

$$1 = \int_{-\infty}^{+\infty} p_0(\Delta U) \exp[\beta(\Delta F - \Delta U)] d(\Delta U) = \exp(\beta \Delta F) \langle \exp(-\beta \Delta U) \rangle_0.$$

$$\tag{7.9}$$

Thus, in principle 'only' the factor $\exp(-\beta \Delta U)$ in the system 0 needs to be sampled. However, this result already clearly reveals the pitfall of this method: for the 'typical' configurations of system 0 the difference $\Delta U \propto N$ and hence $\exp(-\beta \Delta U)$ is very small, while larger contributions to this average may come from regions of phase space where $p_0(\Delta U)$ is not so small. As a result, the statistical accuracy of any estimate of ΔF based on Eqn. (7.9) can be very poor.

Torrie and Valleau (1977) attempted to cure this problem by a scheme called 'umbrella sampling'. The basic idea is to improve the accuracy of the estimation of the average in Eqn. (7.9) by modifying the Markov chain that is constructed in the sampling in such a way that one samples both the part of configuration space accessible to system 1 and the part accessible to system 0. This is achieved by replacing the Boltzmann factor of the system by a (non-negative) weight function $\pi(\mathbf{r}^N)$. Using such a weight, and remembering that $\Delta U \equiv U_1(\mathbf{r}^N) - U_0(\mathbf{r}^N)$, the desired average can be rewritten as

$$\exp(-\beta \Delta F) = \int_{-\infty}^{+\infty} d\mathbf{r}^N \exp(-\beta U_1) \Big/ \int_{-\infty}^{+\infty} d\mathbf{r}^N \exp(-\beta U_0)$$

$$= \int_{-\infty}^{+\infty} d\mathbf{r}^N \pi(\mathbf{r}^N)[\exp(-\beta U_1)/\pi(\mathbf{r}^N)] \Big/ \int_{-\infty}^{+\infty} d\mathbf{r}^N \pi(\mathbf{r}^N)$$

$$[\exp(-\beta U_0)/\pi(\mathbf{r}^N)]. \tag{7.10}$$

With the notation $\langle \dots \rangle_\pi$ to denote an average over a probability distribution $\pi(\mathbf{r}^N)$ one obtains

$$\exp(-\beta \Delta F) = \langle \exp(-\beta U_1)/\pi \rangle_\pi / \langle \exp(-\beta U_0)/\pi \rangle_\pi. \tag{7.11}$$

The distribution π must have an appreciable overlap with both the regions of configuration space that are sampled by system 0 and by system 1, in order that both the numerator and the denominator in Eqn. (7.11) are meaningful. This 'bridging' property of π is alluded to in the name 'umbrella sampling'.

Of course, a drawback of the method is that π is not known *a priori*; rather one has to construct it using information about the Boltzmann weights of the two systems. It may also be advantageous not to bridge all the way from system 0 to system 1 with a single overlapping distribution, but actually it may be better to perform several 'umbrella sampling' runs in partially over-lapping regions. This formulation of the method actually is closely related in spirit to the 'multicanonical sampling', see Section 7.5.

Umbrella sampling has been used to determine absolute values of the free energy in two-dimensional and three-dimensional Ising models by Mon (1985). In two dimensions the nearest neighbor Ising ferromagnet was con-sidered in two different situations: on a $2N \times 2N$ square lattice with periodic boundary conditions (and Hamiltonian $\mathcal{H}_{2N}$), and with the lattice divided up into four separate $N \times N$ square lattices, each with periodic boundaries (and composite Hamiltonian $\overline{\mathcal{H}}_N$.). The free energy difference is then

$$f_{2N} - f_N = \frac{\ln \langle \exp[-\beta(\overline{\mathcal{H}}_N - \mathcal{H}_{2N})] \rangle_{\mathcal{H}_{2N}}}{4N^2}. \tag{7.12}$$

For three dimensions this difference can then be evaluated by umbrella sampling by simulating a series of systems with Hamiltonian

$$\mathcal{H}' = a\mathcal{H}_{2N} - b\overline{\mathcal{H}}_N, \tag{7.13}$$

where a and b vary from 0 to 1 with $a + b = 1$. The result in two dimensions agrees quite well with the exact value and in three dimensions very precise values were obtained for both simple cubic and body centered cubic models.

7.2 SINGLE HISTOGRAM METHOD: THE ISING MODEL AS A CASE STUDY

The idea of using histograms to extract information from Monte Carlo simu-lations is not new, but has only recently been applied with success to the study of critical phenomena (Ferrenberg and Swendsen, 1988; Ferrenberg, 1991). Here we provide a brief description of the method and show some characteristic analyses.

We first consider a Monte Carlo simulation performed at $T = T_0$ which generates system configurations with a frequency proportional to the Boltzmann weight, $\exp[-K_0 E]$. Because the simulation generates config-urations according to the equilibrium probability distribution, a histogram $H(E, M)$ of energy and magnetization values provides an estimate for the equilibrium probability distribution; this estimate becomes exact in the limit of an infinite-length run. For a real simulation, the histogram will suffer from statistical errors, but $H(E, M)/N$, where N is the number of mea-surements made, still provides an estimate for $P_{K_0}(E, M)$ over the range of E and M values generated during the simulation. Thus

$$H(E, M) = \frac{N}{Z(K_0)} \tilde{W}(E, M) e^{-K_0 E}, \tag{7.14}$$

where $\tilde{W}(E, M)$ is an estimate for the true density of states $W(E, M)$. Knowledge of the exact distribution at one value of K is thus sufficient to determine it for any K. From the histogram $H(E, M)$, we can invert Eqn. (7.13) to determine $\tilde{W}(E, M)$:

$$\tilde{W}(E, M) = \frac{Z(K_o)}{N} H(E, M) e^{K_o E}. \tag{7.15}$$

If we now replace $W(E, M)$ in Eqn. (7.1) with the expression for $\tilde{W}(E, M)$ from Eqn. (7.15), and normalize the distribution, we find that the relationship between the histogram measured at $K = K_o$ and the (estimated) probability distribution for arbitrary K is

$$P_K(E, M) = \frac{H(E, M) e^{\Delta K E}}{\sum H(E, M) e^{\Delta K E}} \tag{7.16}$$

with $\Delta K = (K_0 - K)$. From $P_K(E, M)$, the average value of any function of E and M, denoted $f(E, M)$, can be calculated as a continuous function of K:

$$\langle f(E, M) \rangle_K = \sum f(E, M) P_K(E, M). \tag{7.17}$$

The ability to continuously vary K makes the histogram method ideal for locating peaks, which occur at different locations, in different thermodynamic derivatives, and provides the opportunity to study critical behavior with unprecedented resolution.

As an example of the implementation of this method, we shall now discuss results for the three-dimensional ferromagnetic Ising model. We remind the reader that the Hamiltonian is

$$\mathcal{H} = -\mathcal{J} \sum_{\langle i,j \rangle} \sigma_i \sigma_j, \tag{7.18}$$

where the spins σ_i, σ_j take on the values ± 1 and the sum is over all nearest neighbor pairs. As we saw in Chapter 4, in a finite system the phase transition is rounded and shifted from its infinite lattice location (Eqn. (4.13)). If one looks closely one sees that the difference between the true critical temperature and a 'pseudocritical' temperature of the finite system (estimated e.g. from the specific heat maximum) is not simply given by a power of L but rather includes correction terms as well. Obviously, great resolution is needed if these correction terms are to be included properly. We use this Ising model as an example to demonstrate the manner in which an accurate analysis can be carried out.

A detailed Monte Carlo study was made for $L \times L \times L$ simple cubic lattices with fully periodic boundary conditions (Ferrenberg and Landau, 1991). Most of the simulations were performed at $K_o = 0.221\,654$, an earlier estimate for the critical coupling K_c obtained by a Monte Carlo renormalization group (MCRG) analysis (Pawley *et al.*, 1984) of the kind which will be described in Chapter 9. Data were obtained for lattices with $8 \le L \le 96$, and between 3×10^6 and 1.2×10^7 MCS and measurements were made at intervals of either 5 or 10 MCS after up to 10^5 MCS were discarded for equili-

brium. (For the largest lattice, the total run length was more than 5000 times
the relevant correlation time τ (Wansleben and Landau, 1991), with τ determined as described in Chapter 4.) Error estimates were obtained by dividing
the data from each simulation into a set of between 5 and 11 statistical
samples (bins) and considering the distribution of values obtained from
each bin. Because each histogram is used to determine multiple quantities,
some correlations are expected between the different results; however, these
were found to be smaller than the statistical errors, and the individual errors
could thus be treated as uncorrelated. An analysis was performed for bins of
different sizes choosing the final bin sizes so that systematic errors were
negligible compared to the statistical error.

Sufficiently far from K_o the histogram method yielded values which are
obviously wrong, because in the range of E that is then required the histogram has so few entries that the method has broken down. As K is varied, the
peak in the reweighted distribution moves away from that of the measured
histogram and into the "wings" where the statistical uncertainty is high, thus
leading to unreliable results. This is because of the finite range of E and M
generated in a simulation of finite length as well as the finite precision of the
individual histogram entries. This problem is demonstrated in Fig. 7.1 which
shows the normalized (total) energy histogram for the $L = 16$ lattice measured at $K_o = 0.221\,654$ along with the probability distributions for two
additional couplings ($K = 0.224$ and $K = 0.228$) calculated by reweighting
this histogram. The calculated distribution for $K = 0.224$ is fairly smooth,
although the right side of the distribution, which occurs closer to the peak of
the measured histogram, is clearly smoother than the left side. The 'thicken-
ing' of the distribution on the side in the tail of the measured histogram is an
indication that the statistical errors are becoming amplified and that the
extrapolation is close to its limit of reliability. The distribution calculated
for $K = 0.228$ is clearly unreliable. This limitation in ΔK must always be
kept in mind, particularly for large systems, because the reliable range of K
values decreases as the system size increases!

Fig. 7.1 Probability
distribution of the
dimensionless energy
E for $L = 16$. The
data from the
simulation were
obtained at
$K_o = 0.221\,654$; the
other distributions
come from
reweighting as
described in the text.
From Ferrenberg and
Landau (1991).

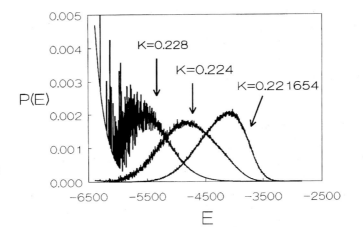

In the critical region a simple histogram covers a finite fraction of the required region in finite size scaling irrespective of size. By performing a small number of additional simulations at different values of K we can guarantee that the results obtained from the single-histogram equation do not suffer from systematic errors. These were done for $L = 32$ at $K_o = 0.2215$, and the location and value of the peaks in the thermodynamic derivatives were determined. Simulations were also performed using two different sets of the random number generator 'magic numbers'. Within the observed statistical errors, no systematic deviations are present. A further test for systematic errors is to use the histogram measured at $K_o = 0.221\,654$ to predict the behavior of the system at $K = 0.2215$ and then compare the results with those obtained directly from the simulation performed at $K_o = 0.2215$. The reweighted results agreed, within the calculated error, with the directly measured results for all quantities except the specific heat (which also agreed to within 2σ).

As described previously (Chapter 4), the critical exponent v can be estimated without any consideration of the critical coupling K_c. For sufficiently large systems it should be possible to ignore the correction term so that linear fits of the logarithm of the derivatives as a function of $\ln L$ provide estimates for $1/v$. In fact, $L_{min} = 24$ was the smallest value that could be used except in the case of the derivative of the magnetization cumulant where linear fits are still satisfactory for $L_{min} = 12$. Combining all three estimates, the analysis yielded $1/v = 1.594(4)$ or $v = 0.627(2)$. By adding a correction term, data from smaller systems can be included. Fits were made of the derivatives to Eqn. (4.12) by fixing the values of v and w, determining the values of a and b which minimize the χ^2 of the fit and then repeating the procedure for different values of v and w. The errors are correlated and the minimum in χ^2 is quite shallow. Scans over a region of (v, w) space for the different quantities revealed the global minimum where $v = 0.6289(8)$.

Once there is an accurate value for v, K_c can be estimated quite accurately. As discussed in Chapter 4, the locations of the maxima of various thermodynamic derivatives provide estimates for effective transition couplings $K_c(L)$ which scale with system size like Eqn. (4.13). These estimates for $K_c(L)$ are plotted as a function of L for $L < 96$ in Fig. 7.2. The solid lines are second order polynomial fits to the data and are drawn to guide the eye. The specific heat peaks (open circles in Fig. 7.2), which occur further from the simulated temperature than any other quantity considered here, fall just *outside the range of validity* of the histogram analysis, especially for $L = 96$. This systematic underestimation of the error, particularly pronounced for $L = 96$, can be compensated for by either increasing the error values, or by removing the $L = 96$ result. In either case, the estimate for K_c is in agreement with that from the other quantities but the error bar is much larger. The result for the derivative of m on the $L = 96$ system is just at the limit of reliability for the histogram analysis. There is noticeable curvature in the lines in Fig. 7.2 indicating that corrections to scaling are important for the smaller systems. If only the results for $L \geq 24$ are analyzed, linear fits to Eqn. (4.12), with no

Fig. 7.2 Size
dependence of the
finite-lattice effective
couplings temperatures
for the three-
dimensional Ising
model. The symbols
represent the data
while the lines (dashed
for the specific heat
and solid for the other
quantities) are fits to
Eqn. (7.3) with $v =$
0.6289 and including
the correction term.
From Ferrenberg and
Landau (1991).

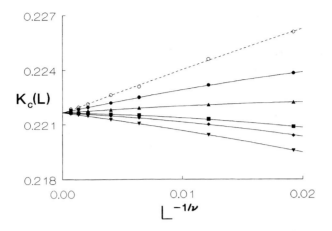

correction terms are obtained; i.e. for sufficiently large L, K_c should extrapolate linearly with $L^{-1/v}$ to K_c. Figure 7.2 shows noticeable curvature for
small system sizes so corrections must be included; these produce estimates
for w and K for each of the quantities which yield a value $K_c = 0.221\,659\,5$
(26). The values of the correction exponent are again consistent with $w = 1$
except for the finite-lattice susceptibility (which has the smallest correction
term). Fits performed by allowing both w and v to vary yield consistent
estimates for v and K_c but with larger errors due to the reduced number
of degrees of freedom of the fit.

The finite size scaling analysis was repeated using corrections to scaling
and the theoretically predicted forms with $w = 1$ and the re-analysis of all
thermodynamic derivatives yielded $v = 0.6294(2)$. While the statistical error
in these values was small, the χ^2 of the fit, as a function of $1/v$, has a broad
shallow minimum so that the actual statistical error, calculated by performing
a true non-linear fit would be larger. Unfortunately, neither the resolution
nor the number of different lattice sizes allows such a fit. With this value of v,
K_c was estimated as $K_c = 0.221\,657\,4(18)$ which is in excellent agreement
with the previous estimate.

Finite size scaling can also be used to estimate other exponents from bulk
properties at K_c. The value of v which was obtained from the derivative of
the magnetization cumulant and the logarithmic derivatives of m and m^2 at K_c
is identical to that obtained by scaling the maximum value of the derivatives.
The scaling behavior of m at K_c yields $\beta/v = 0.518(7)$. (The linear fit for
$L > 24$ yields $\beta/v = 0.505$.) Combining this value for β/v with the estimate
for v, we obtain $\beta = 0.3258(44)$ which agrees with the ε-expansion result
0.3270(15). Estimates for γ/v could be extracted from the scaling behavior
of the finite-lattice susceptibility yielding $\gamma/v = 1.9828(57)$ or $\gamma =1.2470(39)$
or from the true susceptibility at K_c which gave $\gamma/v = 1.970(11)$ or $\gamma =
1.2390(71)$, in excellent agreement with the ε-expansion value of 1.2390(25).

In Fig. 7.3 we show the results of this Monte Carlo study as well as other
high-resolution simultaneous estimates for v and K_c. The boxes present the
quoted error bars in both K_c and v assuming independent errors. To the best

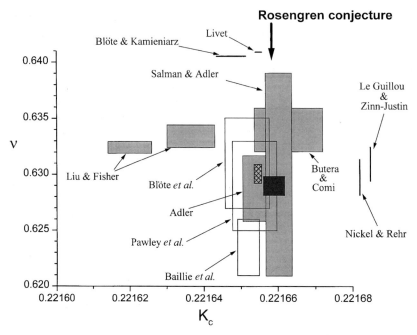

Fig. 7.3 High resolution estimates for K_c and ν for the simple cubic Ising model (boxes show estimates including errors bars; horizontal and vertical lines show the range of independent estimates for only one parameter): series expansions (Adler, 1983; Liu and Fisher, 1989; Nickel and Rehr, 1990; Butera and Comi, 1997; Salman and Adler, 1998), Monte Carlo renormalization group (Pawley *et al.*, 1984; Bloete *et al.*, 1989; Baillie *et al.*, 1992), ε-expansion renormalization group (Le Guillou and Zinn-Justin, 1980), Monte Carlo (Livet, 1991; Bloete and Kamieniarz, 1993). The highest resolution studies, combining Monte Carlo with finite size scaling, are shown by the solid box (Ferrenberg and Landau, 1991) and the cross-hatched box (Bloete *et al.*, 1995). The Rosengren conjecture (Rosengren, 1986) is shown by the vertical arrow.

of our knowledge, all error estimates represent 1 standard deviation. The results from the Monte Carlo study are represented by the filled box and agree well with some MCRG results (Blöte *et al.*, 1989; Pawley *et al.*, 1984), but are outside the error bars of Baillie *et al.* (1992), which in turn have only tenuous overlap with the other MCRG values. The value for ν is also consistent with the ε-expansion result (LeGuillou and Zinn-Justin, 1980) and some of the series expansion results (Nickel and Rehr, 1990; Adler, 1983) but disagrees with others (Liu and Fisher, 1989) which also disagree with the other series values. Transfer matrix Monte Carlo results (Nightingale and Blöte, 1988) yield $\nu = 0.631$ with errors of either 0.006 or 0.002 depending on the range of sizes considered in the analysis and a numerical lower bound (Novotny, 1991), $\nu = 0.6302$ falls within 2σ of the result. Other estimates for K_c (Blöte *et al.*, 1995; Livet, 1991), obtained by assuming fixed values for ν, also lie outside these error bars. The estimate for K_c derived from the maximum slope of m differs substantially from that obtained from the other quantities, although it does agree within 2 standard deviations. If we remove it from the analysis, this estimate for K_c drops to 0.221 657 6(22) which is in even better agreement with the other values presented above. Clearly the question of precise error bar determination remains for all of these numerical methods.

In another high resolution study (Blöte *et al.*, 1995) high statistics runs were made on many, smaller systems and the finite size scaling behavior was carefully examined. Corrections were found beyond those caused by the leading irrelevant scaling field, and with the inclusion of correction exponents from renormalization group theory the critical point was estimated to be at $K_c = 0.221\,654\,6(10)$.

Why do we expend so much effort to locate K_c? In addition to testing the limits of the method, one can also test the validity of a conjectured closed form for K_c (Rosengren, 1986) obtained by attempting to generalize the combinatorial solution of the two-dimensional Ising model to three dimensions:

$$\tanh K_c = (\sqrt{5} - 2)\cos(\pi/8). \tag{7.20}$$

This relation gives $K_c = 0.221\,658\,63$ which agrees rather well with current best estimates. However, Fisher (Fisher, 1995) argued quite convincingly that this conjecture is not unique and most probably not valid.

The combination of high-statistics MC simulations of large systems, careful selection of measured quantities and use of histogram techniques yields results at least as good as those obtained by any other method. All of the analysis techniques used here are applicable if yet higher quality data are obtained and should help define the corrections to scaling. (These same techniques have also been used to provide very high resolution results for a continuous spin model, the three-dimensional classical Heisenberg model (Chen *et al.*, 1993). The size of the error bars on current estimates for K_c indicate that even higher resolution will be required in order to unambiguously test the correctness of the conjectured 'exact' value for K_c. Further improvement will require substantially better data for some of the larger lattice sizes already considered and very high quality data for substantially larger lattices. In addition, since different thermodynamic derivatives have peaks at different temperatures, multiple simulations are indeed needed for each lattice size for the optimal extrapolation of effective critical temperatures to the thermodynamic limit for more than one quantity. Such calculations will be quite demanding of computer memory as well as cpu time and are thus not trivial in scope.

Problem 7.1 Consider an Ising square lattice with nearest neighbor ferromagnetic interactions. Carry out a simple, random sampling Monte Carlo simulation of an 8×8 lattice with p.b.c. at $T = \infty$ and construct a histogram of the resultant energy values. Use this histogram to calculate the specific heat at finite temperature and compare your estimates with data from direct importance sampling Monte Carlo simulation. Estimate the location of the 'effective transition temperature' from the histogram calculation. Then, simulate the system at this temperature, construct a new histogram, and recalculate the specific heat. Compare these new results with those obtained by direct importance sampling Monte Carlo simulation. Estimate the temperature at which you would have to simulate the system to get excellent results near the 'effective phase transition' using the histogram method.

7.3 MULTI-HISTOGRAM METHOD

If data are taken at more than one value of the varying 'field', the resultant histograms may be combined so as to take advantage of the regions where each provides the best estimate for the density of states. The way in which this can be done most efficiently was studied by Ferrenberg and Swendsen (1989). Their approach relies on first determining the characteristic relaxation time τ_j for the jth simulation and using this to produce a weighting factor $g_j = 1 + 2\tau_j$. The overall probability distribution at coupling K obtained from n independent simulations, each with N_j configurations, is then given by

$$P_K(E) = \frac{\left[\sum_{j=1}^{n} g_j^{-1} H_j(E)\right] e^{KE}}{\sum_{j=1}^{n} N_j g_j^{-1} e^{K_j E - f_j}}, \tag{7.21}$$

where $H_j(E)$ is the histogram for the jth simulations and the factors f_j are chosen self-consistently using Eqn. (7.21) and

$$e^{f_j} = \sum_{E} P_{K_j}(E). \tag{7.22}$$

Thermodynamic properties are determined, as before, using this probability distribution, but now the results should be valid over a much wider range of temperature than for any single histogram.

7.4 BROAD HISTOGRAM METHOD

The simulation methods which are generally used to produce the histograms for the methods outlined above tend to yield histograms which become increasingly narrower as the lattice size increases; as we saw in Section 7.2. This can lead to such a narrow range over which the reweighting is valid that the applicability of the method is seriously limited. The broad histogram method (de Oliveira et al., 1996) is an attempt to produce histograms which cover a greater range in energy space and which remain useful for quite large systems. The broad histogram Monte Carlo (BHMC) method produces a histogram which spans a wide energy range and differs from other methods in that the Markov process for the method is based upon a random walk dynamics. Although the original implementation of this method appears to have been flawed, a modified version has proven to be quite effective for the treatment of Potts glasses (Reuhl, 1997). There has been extensive discussion of whether or not the method, in its various forms, completely obeys detailed balance. Thus, until the broad histogram method is examined more intensively it is premature to say if it will be viewed as an interesting case study in statistical sampling methods or a truly useful research tool.

7.5 MULTICANONICAL SAMPLING

7.5.1 The multicanonical approach and its relationship to canonical sampling

In some cases the probability distribution for the states of the system will contain multiple maxima which are widely spaced in configuration space. (Examples include systems near first order phase transitions and spin glasses.) Standard methods may 'flow' towards one of the maxima where they may be easily 'trapped'. Transitions between maxima may occur but, as long as they are infrequent, both the relative weights of the multiple maxima as well as the probability distribution between maxima will be ill determined. One effective approach to such circumstances is to modify the traditional single spin-flip probability to enhance the probability that those 'unlikely' states between the maxima occur. This is not always easy to do and often multiple 'trial runs' must first be made in order to determine what is the best probability to use.

This method reformulates the problem in terms of an effective Hamiltonian:

$$\mathcal{H}_{\mathrm{eff}}(\sigma) = H_{\mathrm{eff}}(\beta \mathcal{H}(\sigma)). \tag{7.23}$$

The probability distribution for the energy can then be written as

$$P(E) = \frac{\exp(S(E) - \mathcal{H}_{\mathrm{eff}})}{\displaystyle\sum_E \exp(S(E) - \mathcal{H}_{\mathrm{eff}})}. \tag{7.24}$$

In the multicanonical algorithm (Berg and Neuhaus, 1991, 1992) the desired form of the probability of states with energy E is determined self-consistently by performing a simulation and using the resultant distribution as a probability estimate for a second simulation, etc. The 'final' probability found is shown in Fig. 7.4, where we show the probability in the canonical ensemble for comparison. Thus, a substantial fraction of the computer resources needed to solve a problem with the multicanonical ensemble may be consumed in the effort to find an optimum probability distribution. The resultant estimate of a thermodynamic average is given by

$$\langle A \rangle_\beta = \frac{\langle A \exp(\mathcal{H}_{\mathrm{eff}} - \mathcal{H}) \rangle}{\langle \exp(\mathcal{H}_{\mathrm{eff}} - \mathcal{H}) \rangle}, \tag{7.25}$$

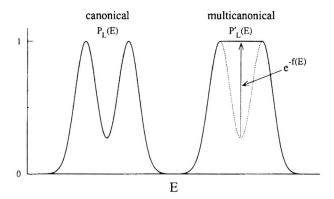

Fig. 7.4 Probability distribution for canonical Monte Carlo sampling for a model with multiple minima compared to that for multicanonical Monte Carlo.

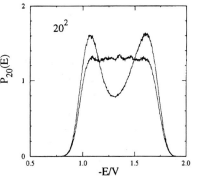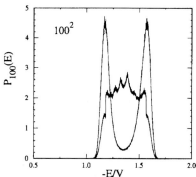

Fig. 7.5 Multicanonical energy distribution $P'_L(E)$ together with the reweighted canonical distribution $P_L(E)$. Both distributions are normalized to unit area. After Janke (1992).

and it is more likely to give correct answers in a situation where the energy landscape is quite complicated than most canonical ensemble methods.

A practical approach to the determination of the effective Hamiltonian is to first determine the probability distribution of states under conditions for which it is easy to measure using a standard Monte Carlo method. Then, use this distribution as an estimate for another run which is made closer to the region of real interest. This process continues all the way to the 'unknown' region where standard sampling methods fail.

As an example of the applicability of the multicanonical algorithm, in Fig. 7.5 we show the results for a $q = 7$ Potts model, a system which has a fairly strong first order transition. The simulations were performed on $L \times L$ square lattices with periodic boundary conditions. For $L = 20$ the multi-canonical distribution is quite flat even though the reweighted, canonical distribution shows two clear peaks. For $L = 100$, it is clearly difficult to find a smooth multicanonical probability, but the resultant canonical distri-bution shows two smooth and very pronounced peaks. Obtaining the relative heights of these two maxima would have been quite difficult using canonical sampling.

7.5.2 Near first order transitions

Having made the above qualitative remarks and shown the example shown in Fig. 7.5, intended to whet the appetite of the reader to learn more about multicanonical sampling, we now proceed to examine the situation near a standard first order transition in greater detail. The systems which we have in mind are the q-state Potts models, which have thermally driven first order transitions in $d = 2$ for $q > 4$, in $d = 3$ for $q \geq 3$, and – even simpler – the transition of the Ising ferromagnet as a function of magnetic field H for $T < T_c$. Remember (see, e.g. Sections 2.1.2.4, 2.3.2, 4.2.3.3, 4.2.5.4) that at $H = 0$ the order parameter (i.e. the magnetization) jumps from a positive value (M_+) to a negative value $(M_- = -M_+,$ cf. Fig. 2.10), and this is accompanied by a dramatic (exponential!) increase of the relaxation time τ_e with lattice size for transitions between states of opposite magnetization in the framework of a simulation with the Metropolis algorithm. Actually this 'ergodic time' τ_e was already roughly estimated in Eqn. (4.61). In the literature (e.g. Berg, 1997) this exponential variation of τ_e with L is sometimes called 'supercritical slowing down'. By the multicanonical method, or its variants, one is able to reduce the correlation time τ to a power law of size dependence, $\tau \propto L^p$. While p is rather large, namely $2d \leq p \leq 5d/2$ where d is the dimension of the lattice (Berg, 1997), the method is clearly useful for large L: while in Fig. 7.5 the minimum and maximum values of $P_L(E)$ differ only by about a factor of 10, there are other examples where maximum and minimum of the distribution differ by astronomically large factors, e.g. in the study of symmetrical polymer mixtures (Müller et $al.$, 1995) the difference was up to a factor of 10^{45} at temperatures far below criticality! Variations of the multicanonical method have also proven to be effective including the 'multimagnetical method' (Berg et $al.$, 1993), where a flat distribution $P_L^1(M)$ of the magnetization M is constructed in between M_- and M_+ in analogy to the flat distribution $P_L^1(E)$ shown in Fig. 7.4, and the 'multibondic algorithm' (Janke and Kappler, 1995), where a combination with cluster algorithms is worked out.

We now consider how to make the step from the canonical distribution $P_L(E)$, in Fig. 7.4, to the multicanonical one, $P_L^1(E)$, which has the property $P_L^1(E) = $ const. for $E_{min} < E < E_{max}$, with $\varepsilon_{min} = E_{min}/L^d < \varepsilon_{max} = E_{max}/L^d$ being constants as $L \to \infty$, by a first-principles approach (following Berg, 1997). This task is achieved by reweighting the canonical distribution $P_L(E)$ with a weight factor $W(E)$ which is related to the spectral density of states $n(E)$ or the (microcanonical) entropy $S(E)$,

$$W(E) = 1/n(E) = \exp[-S(E)] \equiv \exp[-\beta(E)E + \alpha(E)]. \qquad (7.26)$$

In the last step we have introduced the inverse temperature $1/T(E) = \beta(E) = \partial S(E)/\partial E$ and thus the problem is to construct the as yet unknown function $\alpha(E)$ (at least up to an additive constant). This problem in principle can be solved recursively. For a model where the energy spectrum is discrete (such as Ising, Potts models, etc.), there is a minimum spacing between energy levels, which we denote as δE here. Then the discrete analog of the above partial derivative $\beta(E) = \partial S(E)/\partial E$ becomes

$$\beta(E) = [S(E + \delta E) - S(E)]/\delta E, \tag{7.27}$$

and using the identity (from Eqn. (7.26)) $S(E) = \beta(E)E - \alpha(E)$ we can write

$$S(E) - S(E - \delta E) = \beta(E)E - \beta(E - \delta E)(E - \delta E) - [\alpha(E) - \alpha(E - \delta E)]. \tag{7.28}$$

Eliminating now the entropy difference on the left-hand side of Eqn. (7.28) with the help of Eqn. (7.27) we find the recursion

$$\alpha(E - \delta E) = \alpha(E) + [\beta(E - \delta E) - \beta(E)]E, \tag{7.29}$$

where $\alpha(E_{max} = 0)$ is a convenient choice of the additive constant.

In order to use Eqn. (7.29), we would have to do a very accurate set of microcanonical runs in order to sample the relation $\beta = \beta(E)$ from E_{max} to E_{min}, and this requires of the order $L^{d/2}$ different states (which then can be combined into one smooth function by multi-histogram methods, see above). The multicanonical sampling of the flat distribution $P_L^1(E)$ itself (obtained by reweighting with $W(E)$ in Eqn. (7.26), once the weights are estimated) is then a random walk in the energy space, and hence implies a relaxation time $\tau \propto L^{2d}$ since the 'distance' the random walker has to travel scales as $E_{max} - E_{min} \propto L^d$. Actually, in practice the recursion in Eqn. (7.27) may be avoided for a large system, because good enough weights $\alpha(E)$ can often be obtained from a finite size scaling-type extrapolation from results for small systems. Still, the problem remains that τ scales as L^{2d}, a rather large power of L. An alternative to the procedure outlined above involves using the inverse of the histogram obtained between E_{max} and E_{min} at a higher temperature as an estimate for the weighting function. A short multicanonical run is made using this estimate and then the resultant distribution is used to obtain an improved weight factor to be used for longer runs (Janke, 1997).

While the pioneering studies of finite size scaling at first order transitions described in Section 4.2.3.3 used the Metropolis algorithm, and thus clearly suffered from the problem of 'supercritical slowing down', rather accurate studies of Potts models with the multicanonical algorithm are now available (Berg, 1997). Various first order transitions in lattice gauge theory have also been studied successfully with this method (see Berg, 1997 and Chapter 11 of the present book).

7.5.3 Groundstates in complicated energy landscapes

We have encountered complicated energy landscapes in systems with randomly quenched competing interactions, such as spin glasses (Section 5.4.3), and related problems with conflicting constraints (e.g. the 'traveling salesman problem', Section 5.4.3). It is also possible to treat such problems with a variant of multicanonical methods, only the recursion is done slightly differently by starting high up in the disordered phase, where reliable canonical simulations can be performed. In the extreme case E_{max} is chosen such that the corresponding temperature is infinite, $\beta^0(E_{max}) = 0$ and then a recursion is defined as (Berg, 1996, 1997)

$$\beta^{n+1}(E) = (\delta E)^{-1} \ln[H_0^n(E + \delta E)/H_\beta^n(E)], \qquad (7.30)$$

where $H_0^k(E)$ is the (unnormalized) histogram obtained from a simulation at $\beta^k(E)$, while H_β^n contains combined information from all the runs with $\beta^0(E), \ldots, \beta^n(E)$:

$$H_\beta^n(E) = \sum_{k=0}^{n} g_k(E)H_\beta^n(E) \qquad (7.31)$$

and the factors $g_k(E)$ weigh the runs suitably (see Berg, 1996, 1997 for details). With these techniques, it has become possible to estimate rather reliably both groundstate energy and entropy for $\pm J$ nearest neighbor Edwards–Anderson spin glasses in both $d = 2$ and $d = 3$ dimensions. However, the slowing down encountered is very bad ($\tau \propto L^{4d}$ or even worse!) and thus the approach has not been able to finally clarify the controversial aspects about the spin glass transition and the nature of the spin glass order (two-fold degenerate only or a phase space with many 'valleys'?) so far.

At this point we draw attention to a related method, namely the method of expanded ensembles (Lyubartsev et al., 1992), where one enlarges the configuration space by introducing new dynamical variables such as the inverse temperature (this method then is also called 'simulated tempering', see Marinari and Parisi (1992)). A discrete set of weight factors is introduced

$$w_k = \exp(-\beta_k E + \alpha_k), \qquad k = 1, \ldots, n, \qquad \beta_1 < \beta_2 < \cdots < \beta_{n-1} < \beta_n. \qquad (7.32)$$

The transitions $(\beta_k, \alpha_k) \to (\beta_{k-1}, \alpha_{k-1})$ or $(\beta_{k+1}, \alpha_{k+1})$ are now added to the usual $E \to E'$ transitions. Particularly attractive is the feature that this method can be efficiently parallelized on n processors ('parallel tempering', Hukusima and Nemoto, 1996).

Just as the multicanonical averaging can estimate the groundstate energy of spin glass models, it also can find the minimum of cost functions in optimization problems. Lee and Choi (1994) have studied the traveling salesman problem with up to $N = 10\,000$ cities with this method.

7.5.4 Interface free energy estimation

Returning to the magnetization distribution $P_L(M)$ of the Ising model for $T < T_c$, we remember (as already discussed in Section 4.2.5.4) that the minimum of $P_L(M)$ which occurs for $M \approx 0$ is realized for a domain configuration, where two domain walls (of area L^{d-1} each) run parallel to each other through the (hyper-cubic) simulation box, such that one half of the volume L^d is in a domain with magnetization M_+, and the other half of the volume forms the domain with magnetization $M_- = -M_+$. Thus, the free energy cost of this configuration relative to a state with uniform magnetization M_+ or M_-, respectively) is estimated as $2\sigma L^{d-1}$, σ being the interfacial tension. Hence one predicts (Binder, 1982) that $P_L(M = 0)/P_L(M_+) = \exp(-2\beta\sigma L^{d-1})$. Since this ratio, however, is nothing but the weight $W(M)$ needed to convert $P_L(M)$ to the flat distribution $P_L^1(M)$, it follows that we can estimate σ if we know this weight:

$$\sigma = -\frac{1}{2\beta L^{d-1}} \lim[P_L(M = 0)/P_L(M_+)]. \tag{7.33}$$

While the first application of this idea for the Ising model (Binder, 1982) using the Metropolis algorithm failed to obtain accurate results, combination with multicanonical methods did produce very good accuracy (Berg et al., 1993). Meanwhile these techniques have been extended to estimate interfacial tensions between the ordered and disordered phases of Potts models (Berg, 1997), coexisting phases in polymer mixtures (Müller et al., 1995), and various models of lattice gauge theory (see Chapter 11).

At this point, we emphasize that the reweighting techniques described in this section are still a rather recent development and form an active area of research; thus we have not attempted to describe the algorithms in full detail but rather give the flavor of the various approaches.

Problem 7.2 Use the multicanonical sampling method to determine the energy histogram for a 16×16 Ising square lattice at $k_B T/J = 2.0$. From these data determine the canonical ensemble distribution and compare with the distribution obtained from Metropolis Monte Carlo simulation.

7.6 A CASE STUDY: THE CASIMIR EFFECT IN CRITICAL SYSTEMS

Before ending this chapter, we wish to briefly review a Monte Carlo study which could not have been successful without use of the combination of advanced sampling techniques discussed in Chapter 5 together with the reweighting methods presented in this chapter. If a critical system is confined between two walls, critical fluctuations of the order parameter generate effective long range interactions which are reminiscent of those due to zero point fluctuations of the electromagnetic spectrum for a system of two closely spaced magnetic plates. This phenomenon, known as the Casimir effect, can be described in terms of universal amplitudes which determine the strength of the contribution to the effective interface potential due to a term proportional to $\Delta l^{-(d-1)}$ where l is the thickness of the film and Δ is known as the Casimir amplitude. The direct determination of the Casimir amplitudes is quite difficult since it demands the very careful measurement of the small free energy difference between two systems with different boundary conditions. A careful study of the Casimir amplitudes of two-dimensional and

Fig. 7.6 (top)
Histograms for the
$q = 4$, 320×40 Potts
model with periodic
boundary conditions;
(bottom) Casimir
amplitude Δ_{per} for q-
state Potts models
with fixed aspect ratio
of 1/8. After Krech
and Landau (1996).

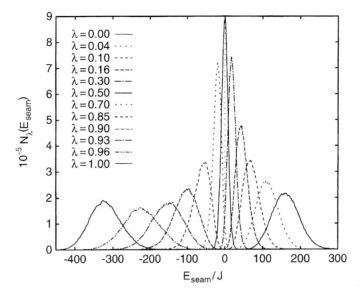

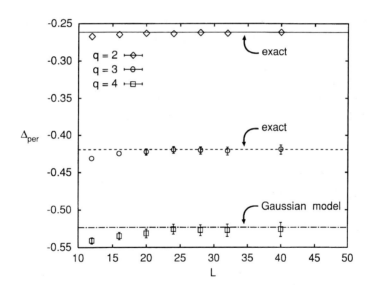

three-dimensional Potts models with different boundary conditions was performed by Krech and Landau (1996). The system was divided into two
pieces, e.g. in two dimensions an $L \times M$ system was divided into two strips
of width $L/2$ coupled through a seam Hamiltonian so that

$$\mathcal{H}_\lambda = \mathcal{H} + \lambda \mathcal{H}_{\text{seam}}. \tag{7.34}$$

They used a hybrid Monte Carlo sampling algorithm which combined
Metropolis and Wolff steps and umbrella sampling to simulate $L \times M$ square
lattices. The difference in free energy with and without the seam gave the
combination of different Casimir amplitudes as $L, M \to \infty$ but with fixed

aspect ratio s. In Fig. 7.6 the histograms produced by simulations for different values of λ show just how little overlap there is between curves unless their λ values are quite close together. Even with the improved sampling algorithm, extensive sampling was needed and 7.2×10^5 hybrid steps were used to produce each of the histograms shown in Fig. 7.6. Note that the spacing of the histograms changes with λ and it is important to choose the values of λ which produce adequate overlap of the histograms! On the right in this figure the results for three different Potts models are compared with the exact answers. Other Casimir amplitudes were measured including some for which the answer is not known.

REFERENCES

Adler, J. (1983), J. Phys. A **16**, 3585.

Baillie, C. F., Gupta, R., Hawick, K. A., and Pawley, G. S. (1992), Phys. Rev. B **45**, 10438.

Bennett, C. H. (1976), J. Comput. Phys. **22**, 245.

Berg, B. and Neuhaus, T. (1991), Phys. Lett. B **267**, 249.

Berg, B. and Neuhaus, T. (1992), Phys. Rev. Lett. **69**, 9.

Berg, B. A. (1996), J. Stat. Phys. **82**, 343.

Berg, B. A. (1997), in *Proceedings of the International Conference on Multiscale Phenomena and Their Simulations* (Bielefeld, Oct. 1996), eds. F. Karsch, B. Monien and H. Satz (World Scientific, Singapore).

Berg, B. A., Hansmann, U., and Neuhaus, T. (1993), Phys. Rev. B **47**, 497.

Binder, K. (1982), Phys. Rev. A **25**, 1699.

Blöte, H. W. J. and Kamieniarz, G. (1993), Physica A **196**, 455.

Blöte, H. W. J., de Bruin, J., Compagner, A., Croockewit, J. H., Fonk, Y. T. J. C., Heringa, J. R., Hoogland, A., and van Willigen, A. L. (1989), Europhys. Lett. **10**, 105.

Blöte, H. W. J., Compagner, A., Croockewit, J. H., Fonk, Y. T. J. C., Heringa, J. R., Hoogland, A., Smit, T. S., and van Willigen, A. L. (1989), Physica A **161**, 1.

Blöte, H. W. J., Luijten, E., and Heringa, J. R. (1995), J. Phys. A **28**, 6289.

Butera, P. and Comi, M. (1997), Phys. Rev. B **56**, 8212.

Chen, K., Ferrenberg, A. M., and Landau, D. P. (1993), Phys. Rev. B **48**, 239.

de Oliveira, P. M. C., Penna, T. J. P., and Herrmann, H. J. (1996), Braz. J. Phys. **26**, 677.

Ferrenberg, A. M. (1991), in *Computer Simulation Studies in Condensed Matter Physics III*, eds. D. P. Landau, K. K. Mon and H.-B. Schüttler (Springer-Verlag, Heidelberg).

Ferrenberg, A. M. and Landau, D. P. (1991), Phys. Rev. B **44**, 5081.

Ferrenberg, A. M. and Swendsen, R. H. (1988), Phys. Rev. Lett. **61**, 2635.

Ferrenberg, A. M. and Swendsen, R. H. (1989), Phys. Rev. Lett. **63**, 1195.

Fisher, M. E. (1995), J. Phys. A **28**, 6323.

Frenkel, D. and Smit, B. (1996), *Understanding Molecular Simulation: From Algorithms to Applications* (Academic Press, New York).

Hukusima, K. and Nemoto, K. (1996), J. Phys. Soc. Japan **65**, 1604.

Janke, W. (1992), in *Dynamics of First Order Transitions*, eds. H. J.

Herrmann, W. Janke, and F. Karsch (World Scientific, Singapore).

Janke, W. (1997), in *New Directions in Statistical Physics*, eds. C.-K. Hu and K.-T. Leung (Elsevier, Amsterdam) p. 164.

Janke, W. and Kappler, S. (1995), Phys. Rev. Lett. **74**, 212.

Krech, M. and Landau, D. P. (1996), Phys. Rev. E **53**, 4414.

Lee, Y. and Choi, M. Y. (1994), Phys. Rev. E **50**, 4420.

LeGuillou, J.-C. and Zinn-Justin, J. (1980), Phys. Rev. B **21**, 3976.

Liu A. J. and Fisher, M. E. (1989), Physica A **156**, 35.

Liu, A. and Fisher, M. E. (1990), J. Stat. Phys. **58**, 431.

Livet, F. (1991), Europhys. Lett. **16**, 139.

Lyubartsev, A. P., Martsinovski, A. A., Shevkunov, S. V., and Vorentsov-Velyaminov, P. N. (1992), J. Chem. Phys. **96**, 1776.

Marinari, E. and Parisi, G. (1992), Europhys. Lett. **19**, 451.

Mon, K. K. (1985), Phys. Rev. Lett. **54**, 2671.

Müller, M., Binder, K., and Oed, W. (1995), J. Chem. Soc. Faraday Trans. **28**, 8639.

Nickel B. G. (1982), in *Phase Transitions: Cargese 1980*, eds. M. Levy, J.-C. Le Guillou, and J. Zinn-Justin (Plenum, New York) p. 291.

Nickel, B. G. and Rehr, J. J. (1990), J. Stat. Phys. **61**, 11.

Nightingale, M. P. and Blöte, H. W. J. (1988), Phys. Rev. Lett. **60**, 1662.

Novotny, M. A. (1991), Nucl. Phys. B Proc. Suppl. **20**, 122.

Pawley, G. S., Swendsen, R. H., Wallace, D. J., and Wilson, K. G. (1984), Phys. Rev. B **29**, 4030.

Reuhl, M. (1997), Diplomarbeit (University of Mainz, unpublished).

Rosengren, A. (1986), J. Phys. A **19**, 1709.

Salman, Z. and Adler, J. (1998), Int. J. Mod. Phys. C **9**, 195.

Shing, K. S. and Gubbins, K. E. (1983), Mol. Phys. **49**, 1121.

Torrie, G. M. and Valleau, J. P. (1977), J. Comput. Phys. **23**, 187.

Wansleben, S. and Landau, D. P. (1991), Phys. Rev. B **43**, 6006.

8 Quantum Monte Carlo methods

8.1 INTRODUCTION

In most of the discussion presented so far in this book, the quantum character of atoms and electrons has been ignored. The Ising spin models have been an exception, but since the Ising Hamiltonian is diagonal (in the absence of a transverse magnetic field!), all energy eigenvalues are known and the Monte Carlo sampling can be carried out just as in the case of classical statistical mechanics. Furthermore, the physical properties are in accord with the third law of thermodynamics for Ising-type Hamiltonians (e.g. entropy S and specific heat vanish for temperature $T \to 0$, etc.) in contrast to the other truly classical models dealt with in previous chapters (e.g. classical Heisenberg spin models, classical fluids and solids, etc.) which have many unphysical low temperature properties. A case in point is a classical solid for which the specific heat follows the Dulong–Petit law, $C = 3Nk_B$, as $T \to 0$, and the entropy has unphysical behavior since $S \to -\infty$. Also, thermal expansion coefficients tend to non-vanishing constants for $T \to 0$ while the third law implies that they must be zero. While the position and momentum of a particle can be specified precisely in classical mechanics, and hence the groundstate of a solid is a perfectly rigid crystal lattice (motionless particles localized at the lattice points), in reality the Heisenberg uncertainty principle forbids such a perfect rigid crystal, even at $T \to 0$, due to zero point motions which 'smear out' the particles over some region around these lattice points. This delocalization of quantum-mechanical particles increases as the atomic mass is reduced; therefore, these quantum effects are most pronounced for light atoms like hydrogen in metals, or liquid helium. Spectacular phenomena like superfluidity are a consequence of the quantum nature of the particles and have no classical counterpart at all. Even for heavier atoms, which do not show superfluidity because the fluid–solid transition intervenes before a transition from normal fluid to superfluid could occur, there are genuine effects of quantum nature. Examples include the isotope effects (remember that in classical statistical mechanics the kinetic energy part of the Boltzmann factor cancels out from all averages, and thus in thermal equilibrium no property depends explicitly on the mass of the particles).

The quantum character of electrons is particularly important, of course, since the mass of the electron is only about $1/2000$ of the mass of a proton, and phenomena like itinerant magnetism, metallic conductivity and superconductivity completely escape treatment within the framework of classical statistical mechanics. Of course, electrons also play a role for many problems of 'chemical physics' such as formation of hydrogen bonds in liquid water, formation of solvation shells around ions, charge transfer in molten oxides, etc. While some degrees of freedom in such problems can already be treated classically, others would still need a quantum treatment. Similarly, for many magnetic crystals it may be permissible to treat the positions of these ions classically, but the quantum character of the spins is essential. Note, for example, in low-dimensional quantum antiferromagnets the Néel state is not the groundstate, and even understanding the groundstate of such quantum spin systems may be a challenging problem.

There is no unique extension of the Monte Carlo method as applied in classical statistical mechanics to quantum statistical mechanics that could deal well with all these problems. Instead, different schemes have been developed for different purposes: e.g. the path integral Monte Carlo (PIMC) technique works well for atoms with masses which are not too small at temperatures which are not too low, but it is not the method of choice if groundstate properties are the target of the investigation. Variational Monte Carlo (VMC), projector Monte Carlo (PMC) and Green's function Monte Carlo (GFMC) are all schemes for the study of properties of many-body systems at zero temperature. Many of these schemes exist in versions appropriate to both off-lattice problems and for lattice Hamiltonians. We emphasize at the outset, however, that important aspects are still not yet satisfactorily solved, most notably the famous 'minus sign problem' which appears for many quantum problems such as fermions on a lattice. Thus many problems involving the quantum statistical mechanics of condensed matter exist, that cannot yet be studied by simulational methods, and the further development of more powerful variants of quantum Monte Carlo methods is still an active area of research. (Indeed we are rather lucky that we can carry out specific quantum Monte Carlo studies, such as path integral simulations described in the next section, at all!) The recent literature is voluminous and has filled several books (e.g. Kalos, 1984; Suzuki, 1986; Doll and Gubernatis, 1990; Suzuki, 1992), and review articles (Ceperley and Kalos, 1979; Schmidt and Kalos, 1984; De Raedt and Lagendijk, 1985; Berne and Thirumalai, 1986; Schmidt and Ceperley, 1992; Gillan and Christodoulos, 1993; Ceperley, 1995, 1996; Nielaba, 1997). Thus in this chapter we can by no means attempt an exhaustive coverage of this rapidly developing field. Instead we present a tutorial introduction to some basic aspects and then describe some simple applications.

8.2 FEYNMAN PATH INTEGRAL FORMULATION

8.2.1 Off-lattice problems: low-temperature properties of crystals

We begin with the problem of evaluating thermal averages in the framework of quantum statistical mechanics. The expectation value for some quantum mechanical operator $\hat{A}$ corresponding to the physical observable A, for a system of N quantum particles in a volume V is given by

$$\langle \hat{A} \rangle = Z^{-1} \mathrm{Tr}\, \exp(-\mathcal{H}/k_B T)\hat{A} = Z^{-1} \sum_n \langle n| \exp(-\mathcal{H}/k_B T)\hat{A}|n \rangle, \quad (8.1)$$

with

$$Z = \mathrm{Tr}\, \exp(-\mathcal{H}/k_B T) = \sum_n \langle n| \exp(-\mathcal{H}/k_B T)|n \rangle, \quad (8.2)$$

where $\mathcal{H}$ is the Hamiltonian, and the states $|n\rangle$ form a complete, orthonormal basis set. In general, the eigenvalues E_α of the Hamiltonian ($\mathcal{H}|\alpha\rangle = E_\alpha|\alpha\rangle$ with eigenstate $|\alpha\rangle$) are not known, and we wish to evaluate the traces in Eqns. (8.1) and (8.2) without attempting to diagonalize the Hamiltonian. This task is possible with the Feynman path integral approach (Feynman and Hibbs, 1965). The basic idea of this method can be explained for a single particle of mass m in a potential $V(x)$, for which the Hamiltonian (in position representation) reads

$$\mathcal{H} = \hat{E}_{\mathrm{kin}} + \hat{V} = -\frac{\hbar^2}{2m}\frac{d^2}{dx^2} + V(x), \quad (8.3)$$

and using the states $|x\rangle$ as a basis set the trace Z becomes

$$Z = \int dx \langle x| \exp(-\mathcal{H}/k_B T)|x\rangle = \int dx \langle x| \exp[-(\hat{E}_{\mathrm{kin}} + \hat{V})/k_B T]|x\rangle. \quad (8.4)$$

If $\hat{E}_{\mathrm{kin}}$ and $\hat{V}$ commuted, we could replace $\exp[-(\hat{E}_{\mathrm{kin}} + \hat{V})/k_B T]$ by $\exp(-\hat{E}_{\mathrm{kin}}/k_B T)\exp(-\hat{V}/k_B T)$ and, by inserting the identity $\hat{1} = \int dx'|x'\rangle\langle x'|$, we would have solved the problem, since $\langle x'| \exp[-\hat{V}(x)/k_B T]x\rangle = \exp$ $[-V(x)/k_B T]\delta(x - x')$ and $\langle x| \exp(-\hat{E}_{\mathrm{kin}}/k_B T)|x'\rangle$ amounts to dealing with the quantum mechanical propagator of a free particle. However, by neglecting the non-commutativity of $\hat{E}_{\mathrm{kin}}$ and $\hat{V}$, we reduce the problem back to the realm of classical statistical mechanics, all quantum effects would be lost.

A related recipe is provided by the exact Trotter product formula (Trotter, 1959; Suzuki, 1971) for two non-commuting operators $\hat{A}$ and $\hat{B}$:

$$\exp(\hat{A} + \hat{B}) \underset{P \to \infty}{\to} [\exp(\hat{A}/P)\exp(\hat{B}/P)]^P, \quad (8.5)$$

where P is an integer. In the specific case of a single particle moving in a potential, the Trotter formula becomes

$$\exp[-(\hat{E}_{\mathrm{kin}} + \hat{V})/k_B T] = \lim_{P \to \infty} \{\exp(-\hat{E}_{\mathrm{kin}}/k_B TP)\exp(-\hat{V}/k_B TP)\}^P. \quad (8.6)$$

As a result, we can rewrite the partition function Z as follows

$$Z = \lim_{P \to \infty} \int dx_1 \int dx_2 \ldots \int dx_P \langle x_1 | \exp(-\hat{E}_{\text{kin}}/k_B TP) \exp(-\hat{V}/k_B TP)|x_2\rangle$$
$$\langle x_2 | \exp(-\hat{E}_{\text{kin}}/k_B TP) \exp(-\hat{V}/k_B TP)|x_3\rangle \langle x_3| \ldots |x_P\rangle$$
$$\langle x_P | \exp(-\hat{E}_{\text{kin}}/k_B TP) \exp(-\hat{V}/k_B TP)|x_1\rangle. \tag{8.7}$$

In practice, it will suffice to work with a large but finite P, and since the matrix elements can be worked out as follows

$$\langle x| \exp(-\hat{E}_{\text{kin}}/k_B TP) \exp(-\hat{V}/k_B TP)|x'\rangle$$
$$= \left(\frac{mk_B TP}{2\pi\hbar^2}\right)^{1/2} \exp\left[-\frac{mk_B TP}{2\hbar^2}(x - x')^2\right] \exp\left[-\frac{V(x) + V(x')}{2k_B TP}\right], \tag{8.8}$$

we obtain the following approximate result for the partition function:

$$Z \approx \left(\frac{mk_B TP}{2\pi\hbar^2}\right)^{P/2} \int dx_1 \ldots dx_P \exp\left\{-\frac{1}{k_B T}\left[\frac{\kappa}{2}\sum_{s=1}^{P}(x_s - x_{s+1})^2\right.\right.$$
$$\left.\left. + \frac{1}{P}\sum_{s=1}^{P} V(x_s)\right]\right\}, \tag{8.9}$$

where the boundary condition $x_{P+1} = x_1$ holds and the effective spring constant is

$$\kappa = mP(k_B T)^2/\hbar^2. \tag{8.10}$$

Equation (8.9) is equivalent to the classical configurational partition function of P classical particles coupled with a harmonic potential $V(x)$, in a kind of 'ring polymer'. When one generalizes this to N particles interacting with a pair potential in d dimensions,

$$\mathcal{H} = \sum_{i=1}^{N}\left(-\frac{\hbar^2}{2m}\nabla_i^2\right) + \sum_{i<j} V(|\mathbf{r}_i - \mathbf{r}_j|), \tag{8.11}$$

one finds that the resulting 'melt' of cyclic polymers has somewhat unusual properties, since monomer–monomer interactions occur only if the 'Trotter index' is the same. Thus the partition function becomes ($\mathbf{r}_i^{(s)}$ is the coordinate of the ith particle in the sth slice of the imaginary time variable)

$$Z = \left(\frac{mk_B TP}{2\pi\hbar^2}\right)^{dNP/2} \int d\mathbf{r}_1^{(1)} \ldots \int d\mathbf{r}_N^{(P)} \exp\left\{-\frac{1}{k_B T}\left[\frac{\kappa}{2}\sum_{i=1}^{N}\sum_{s=1}^{P}(\mathbf{r}_i^{(s)} - \mathbf{r}_i^{(s+1)})^2\right.\right.$$
$$\left.\left. + \frac{1}{P}\sum_{i<j}\sum_{s=1}^{P} V\left(|\mathbf{r}_i^{(s)} - \mathbf{r}_j^{(s)}|\right)\right]\right\}$$
$$= \left(\frac{mk_B T}{2\pi\hbar^2}\right)^{dNP/2} \int d\mathbf{r}_1^{(1)} \ldots \int d\mathbf{r}_N^{(P)} \exp\left\{-\mathcal{H}_{\text{eff}}^{(P)}/k_B T\right\}. \tag{8.12}$$

This 'ring polymer' is shown schematically in Fig. 8.1. If the effect of the potential V could be neglected, we could simply conclude from the equipartition theorem (since Eqns. (8.9 and 8.12) can be viewed as a problem in classical statistical mechanics, this theorem applies), that the potential energy carried by each spring is $(d/2)k_B T = (\kappa/2)\langle(\mathbf{r}_i^{(s)} - \mathbf{r}_i^{(s+1)})^2\rangle$, i.e. the typical interparticle mean-square displacement of two neighboring particles along the chain is $\ell^2 = \langle(\mathbf{r}_i^{(s)} - \mathbf{r}_i^{(s+1)})^2\rangle = dk_B T/\kappa = \hbar^2 d/(mk_B TP)$. Now the gyration radius of a ring polymer containing P monomers is $\langle R_g^2\rangle = \ell^2 P/12 = (d/12)(\hbar^2/mk_B T)$. Thus we see that the diameter $2\sqrt{\langle R_g^2\rangle} = \hbar\sqrt{d/(3mk_B T)}$ is of the same order as the thermal de Broglie wavelength $\lambda_T = h/\sqrt{2\pi m T}$ of a particle. This formalism brings out in a very direct fashion the fact that in quantum mechanics the uncertainty principle forbids the simultaneous precise specification of both momenta and positions of the particles; and for free particles, integrating out the momenta then leaves the particles delocalized in space in 'cells' of linear dimension λ_T. The advantage of the formalism written in Eqns. (8.9–12) is, of course, that it remains fully valid in the presence of the potential $V(|\mathbf{r}_i - \mathbf{r}_j|)$ – then the linear dimension of the delocalization no longer is simply given by λ_T, but depends on the potential V as well. This fact is well known for harmonic crystals, of course: the delocalization of an atom in a harmonic crystal can be expressed in terms of the harmonic oscillator groundstate wave functions, summed over all eigenfrequencies ω_q of the crystal. In other words, the mean-square displacement of an atom around the position in the ideal rigid lattice for $T = 0$ is $\langle \mathbf{r}_i^2\rangle = (1/2Nm)\sum_q(\hbar\omega_q)^{-1}$. On the other hand, one knows that the harmonic approximation for crystals has many deficiencies, e.g. it does not describe thermal expansion. As an example, Fig. 8.2 compares the lattice constant $a(T)$ of orthorhombic solid polyethylene, as deduced from a PIMC calculation (Martonak *et al.*, 1998), with the corresponding classical results and with

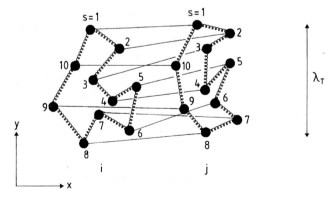

Fig. 8.1 Schematic representation of two interacting quantum particles i, j in two dimensions: each particle (i) represented by a 'ring polymer' composed of $P = 10$ effective monomers $\mathbf{r}_i^{(s)}$, with $s = 1, \ldots, P$. Harmonic springs (of strength κ) only connect 'monomers' in the same 'polymer', while interatomic forces join different monomers with the same Trotter index s, indicated by the thin straight lines. In the absence of such interactions, the size of such a ring polymer coil would be given by the thermal de Broglie wavelength, $\lambda_i = h/\sqrt{2\pi m k_B T}$, where h is Planck's constant.

Fig. 8.2 Temperature dependence of the lattice constant for orthorhombic polyethylene. Results of a PIMC calculation are compared with the value for a classical system and with experiment. After Martonak *et al.* (1998).

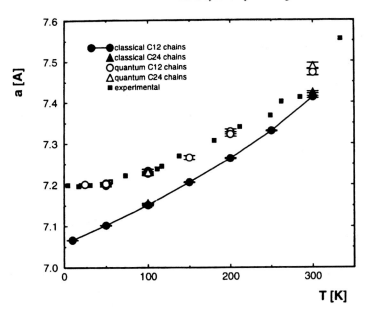

experiment (Dadobaev and Slutsker, 1981). Clearly the classical Monte Carlo result underestimates $a(T)$ systematically at all temperatures from $T = 0\,K$ to room temperature, and yields a constant thermal expansion coefficient $\alpha = a^{-1}da/dT$ as $T \to 0$, in contrast to the result $\alpha(T \to 0) \to 0$ required by the third law of thermodynamics. The PIMC results are clearly in accord with this law, as they should be, and even reproduce the experimental data perfectly, although such good agreement is to some extent fortuitous in view of the uncertainties about the potentials to be used for this polymer.

Now it is well known that one can go somewhat beyond the harmonic approximation in the theory of the dynamics of crystal lattices, e.g. by taking entropy into account via the quasi-harmonic approximation that uses a quadratic expansion around the minimum of the free energy rather than the potential energy, as is done in the standard harmonic approximation. In fact, such a quasi-harmonic lattice dynamics study of orthorhombic polyethylene has also been carried out (Rutledge *et al.*, 1998), and the comparison with the PIMC results shows that the two approaches do agree very nicely at temperatures below room temperature. However, only the PIMC approach in this example is reliable at room temperature and above, up to the melting temperature, where quantum effects gradually die out and the system starts to behave classically. Also, the PIMC method yields information on local properties involving more than two atoms in a very convenient way, e.g. the mean-square fluctuation of the bond angle θ_{CCC} between two successive carbon-carbon bonds along the backbone of the C_nH_{2n+2} chain (Fig. 8.3), which would be rather cumbersome to obtain by lattice dynamics methods. While according to classical statistical mechanics such a bond angle fluctuation vanishes as $T \to 0$, i.e. $\sqrt{\langle(\delta\theta_{CCC})^2\rangle} \propto \sqrt{T}$, so that in the groundstate $(T = 0)$ a perfectly rigid zig-zag structure (Fig. 6.8) remains, this is not

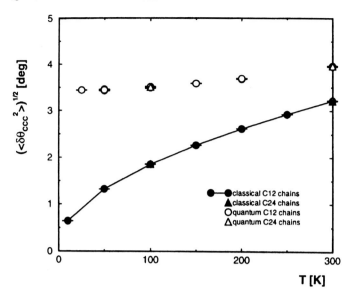

Fig. 8.3 Temperature dependence of the average fluctuation $\sqrt{\langle(\delta\theta_{CCC})^2\rangle}$ of the $C-C-C$ bond angle, according to the classical Monte Carlo calculation (full dots) and according to PIMC simulations (open symbols), for two choices of chain length n ($n = 12$ and $n = 24$, respectively). From Martoňák *et al.* (1998).

true when one considers quantum mechanics and bond angles then fluctuate by around 3 degrees! Even at room temperature the classical calculation underestimates this fluctuation still by about 20%.

Now one point which deserves comment is the proper choice of the Trotter dimension P. According to Eqn. (8.6), the method is only exact in the limit $P \to \infty$. This presents a serious problem as does the extrapolation to the thermodynamic limit, $N \to \infty$. Just as one often wishes to work with as small N as possible, for the sake of an economical use of computer resources, one also does not wish to choose P unnecessarily large. However, since the distance between points along the ring polymer in Fig. 8.1 scales as $\ell^2 \propto (TP)^{-1}$, as argued above, and we have to keep this distance small in comparison to the length scales characterizing the potential, it is obvious that the product TP must be kept fixed so that ℓ is fixed. As the temperature T is lowered, P must be chosen to be larger. Noting that for operators $\hat{A}$, $\hat{B}$ whose commutator is a complex number c, i.e. $[\hat{A}, \hat{B}] = c$, we have the formula

$$\exp[\hat{A} + \hat{B}] = \exp(\hat{A})\exp(\hat{B})\exp\left(-\tfrac{1}{2}[\hat{A}, \hat{B}]\right), \qquad (8.13)$$

we conclude that for large P the error in replacing $\exp[-(\hat{E}_{kin} + \hat{V})/Pk_BT]$ by $\exp(\hat{E}_{kin}/Pk_BT)\exp(\hat{V}/Pk_BT)$ is of order $1/P^2$. This observation suggests that simulations should be tried for several values of P and the data extrapolated versus $1/P^2$. In favorable cases the asymptotic region of this 'Trotter scaling' is indeed reached, as Fig. 8.4 demonstrates. This figure also shows that PIMC is able to identify typical quantum mechanical effects such as 'isotope effects': the two isotopes ^{20}Ne and ^{22}Ne of the Lennard-Jones system neon differ only by their mass, and in classical statistical mechanics there would be no difference in static properties whatsoever. However, as Fig. 8.4 shows, there is a clear distinction between the lattice constants of the

Fig. 8.4 Trotter
scaling plot for the
lattice parameter a of
solid neon. The upper
curve corresponds to
^{20}Ne at $T = 16$ K.
From Müser *et al.*
(1995).

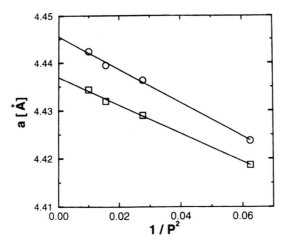

two isotopes, and the difference observed in the simulation in fact is rather close to the value found in the experiment (Batchelder *et al.*, 1968). The examples shown should not leave the reader in a too optimistic mood, however, since there are also examples in the literature where even Trotter numbers as large as $P = 100$ are insufficient to reach this Trotter scaling limit. Indeed, not all quantities are equally well suited for such an extrapolation. Particularly cumbersome, for instance, is the specific heat for an insulating crystal which is expected to vary like $C \propto T^d$ at low temperatures in d dimensions (Debye law). However, the theory of lattice dynamics shows that this behavior results from long wavelength acoustic phonons, with frequency $\omega_q = c_s|\mathbf{q}|$ where c_s is the speed of sound and $\mathbf{q}$ their wavevector. In a finite cubic crystal of size $L \times L \times L$ with periodic boundary conditions the smallest $|\mathbf{q}|$ that fits is of order $2\pi/L$, and hence the phonon spectrum is cut off at a minimum frequency $\omega_{\min} \propto c_s/L$. Due to this gap in the phonon spectrum at low enough temperatures ($k_B T < \hbar\omega_{\min}$) the specific heat does not comply with the Debye law, but rather behaves as $C \propto \exp(-\hbar\omega_{\min}/k_B T)$. In order to deal with such problems, Müser *et al.* (1995) proposed a combined Trotter and finite size scaling. In this context, we also emphasize that the specific heat *cannot* be found from computing fluctuations of the effective Hamiltonian $\mathcal{H}_{\text{eff}}^{(P)}$, Eqn. (8.10), $\langle\mathcal{H}_{\text{eff}}^{(P)2}\rangle - \langle\mathcal{H}_{\text{eff}}^{(P)}\rangle^2$. The reason is that the spring constant κ, Eqn. (8.8), is temperature-dependent, and this fact invalidates the standard derivation of the fluctuation formula. For suitable estimators of the specific heat and other response functions in Monte Carlo calculations we refer to the more specialized literature quoted in Chapter 1, Introduction.

Problem 8.1 Consider a single particle in a harmonic potential well with characteristic frequency of $\omega = (k/m)^{1/2}$. Perform a path integral Monte Carlo simulation for $P = 1$, $P = 2$, and $P = 8$ at an inverse temperature of $\beta = 2.5$. Carry out multiple runs for 10 000 MC steps and determine statistical error bars. Repeat the calculation for runs of 10^6 MC steps. Compare the results and comment.

8.2.2 Bose statistics and superfluidity

We now mention another important problem: in making the jump from the one-particle problem, Eqn. (8.6), to the N-particle problem, Eqn. (8.12), we have disregarded the statistics of the particles (Bose–Einstein vs. Fermi–Dirac statistics) and have treated them as distinguishable. For crystals of not too light atoms, this approximation is acceptable, but it fails for quantum crystals such as solid ^{3}He and ^{4}He, as well as for quantum fluids (Ceperley, 1995). For Bose systems, only totally symmetric eigenfunctions contribute to the density matrix, and hence if we write symbolically $\mathbf{R} = (\mathbf{r}_1, \mathbf{r}_2, \ldots, \mathbf{r}_N)$ and we define a permutation of particle labels by $\hat{P}\mathbf{R}$ where $\hat{P}$ is the permutation operator, we have for any eigenfunction $\phi_\alpha(\mathbf{R})$

$$\hat{P}\phi_\alpha(\mathbf{R}) = \frac{1}{N!} \sum_P \phi(\hat{P}\mathbf{R}), \qquad (8.14)$$

where the sum is over all permutations of particle labels. The partition function for a Bose system therefore takes the form (Ceperley, 1995)

$$Z_B = \left(\frac{mk_B TP}{2\pi\hbar^2}\right)^{dNP/2} \frac{1}{N!} \int d\mathbf{r}_1^{(1)} \ldots \int d\mathbf{r}_N^{(P)} \exp\left\{-\mathcal{H}_{\text{eff}}^{(P)}/k_B T\right\}, \qquad (8.15)$$

where now the boundary condition is not $\mathbf{r}_i^{(P+1)} = \mathbf{r}_i^{(1)}$ as in Eqn. (8.10), but rather $\hat{P}\mathbf{R}^{(P+1)} = \mathbf{R}^{(1)}$. This means that paths are allowed to close on any permutation of their starting positions, and contributions from all $N!$ closures are contained in the partition function. At high temperatures the identity permutation yields the dominating contribution, while at zero temperature all permutations have equal weight. In the classical isomorphic system, this means that 'crosslinks' form and open up again in the system of ring polymers. (Of course, such behavior should not be confused with the actual chemical kinetics of polymerization and crosslinking processes of real polymers!) A two-atom system with P effective monomers can be in two possible permutation states: either two separate ring polymers, each with P springs (as shown in Fig. 8.1), or one larger ring polymer with $2P$ springs.

At this point, it is illuminating to ask what superfluidity (such as actually occurs in ^{4}He) implies in this formalism (Feynman, 1953): a macroscopic polymer is formed which involves on the order of N atoms and stretches over the entire system. From Fig. 8.1, it is clear that this 'crosslinking' among ring polymers can set it only when the linear dimension of a ring polymer coil becomes of the same order as the 'interpolymer spacing': in this way one can get an order of magnitude estimate of the superfluid transition temperature T_λ, by putting the thermal de Broglie wavelength $\lambda_T = h/\sqrt{2\pi m k_B T}$ equal to the 'interpolymer spacing', $\rho^{-1/d}$, where ρ is the density of the d-dimensional system. The 'degeneracy temperature' T_D found from $\lambda_T = \rho^{-1/d}$, i.e. $T_D = \rho^{2/d} h^2/(2\pi k_B m)$, sets the temperature scale on which important quantum effects occur.

In practice, use of Eqns. (8.12) and (8.15) would not work for the study of superfluidity in ^{4}He – although the formalism is exact in principle, values of

P which are unreasonably large would be required for satisfactory results. An alternative approach is to use what is called an 'improved action' rather than the 'primitive action' $\mathcal{H}_{eff}/k_B T$ given in Eqn. (8.12). However, we shall not go into any detail here but rather refer the reader to the original literature (e.g. Ceperley, 1995).

The treatment of fermions is even more cumbersome. The straightforward application of PIMC to fermions means that odd permutations subtract from the sum: this is an expression of the 'minus sign problem' that hampers all Monte Carlo work on fermions. In fact, PIMC for fermions in practice requires additional approximations and is less useful than for bosons or for 'Boltzmannons' (i.e. cases where the statistics of the particles can be neglected altogether, as for the behavior of slightly anharmonic crystals formed from rather heavy particles, as discussed in the beginning of this section). We refer the reader to Ceperley (1996) for a recent review of this problem.

8.2.3 Path integral formulation for rotational degrees of freedom

So far the discussion has tacitly assumed point-like particles and the kinetic energy operator $\hat{E}_{kin}$ (Eqns. (8.3,4)) was meant to describe their translational motion; however, rather than dealing with the effects due to non-commutativity of position operator (**x**) and momentum operator (**p**), $[\hat{x}_\alpha, \hat{p}_\beta] = i\hbar\delta_{\alpha\beta}$, we may also consider effects due to the non-commutativity of the components of the angular momentum operator, $\hat{L}_\alpha$. Such effects are encountered for example in the description of molecular crystals, where the essential degrees of freedom that one wishes to consider are the polar angles (θ_i, φ_i) describing the orientation of a molecule (Müser, 1996). Here we discuss only the simple special case where the rotation of the molecules is confined to a particular plane. For example, in monolayers of N_2 adsorbed on graphite in the commensurate $\sqrt{3} \times \sqrt{3}$ structure (Marx and Wiechert, 1996), one can ignore both the translational degree of freedom of the N_2 molecules and the out-of-plane rotation, i.e. the angle $\theta_i = \pi/2$ is not fluctuating, the only degree of freedom that one wishes to consider is the angle φ_i describing the orientation in the xy-plane, parallel to the graphite substrate. The Hamiltonian hence is (I is the moment of inertia of the molecules, and $\hat{V}$ the intermolecular potential)

$$\mathcal{H} = \sum_{j=1}^{N} \frac{\hat{L}_{jZ}^2}{2I} + \sum_{i\neq j} \hat{V}(\varphi_i, \varphi_j), \tag{8.16}$$

since the commutation relation $[\hat{L}_{jZ}, \hat{\varphi}_i] = -i\hbar\delta_{j,i}$ is analogous to that of momentum and position operator, one might think that the generalization of the PIMC formalism (Eqns. (8.9–12)) to the present case is trivial, but this is not true due to the rotation symmetry $\varphi_j = \varphi_j + n_j 2\pi$; with n_j integer: if we write the partition function as path integral we obtain (Marx and Nielaba, 1992)

$$Z = \left(\frac{Ik_B TP}{2\pi\hbar^2}\right)^{NP/2} \prod_{j=1}^{N}\left\{\sum_{n_j=-\infty}^{+\infty}\int_{0}^{2\pi} d\varphi_i^{(1)} \prod_{s=2}^{P}\int_{-\infty}^{+\infty} d\varphi_j^{(s)}\right\} \exp[-\mathcal{H}_{\text{eff}}^{(P)}/k_B T],$$

(8.17)

with

$$\mathcal{H}_{\text{eff}}^{(P)} = \sum_{s=1}^{P}\left\{\sum_{j=1}^{N}\frac{IPk_B^2 T^2}{2\hbar^2}\left[\varphi_j^{(s)} - \varphi_j^{(s+1)} + 2\pi n_j \delta_{S,P}\right]^2 + \sum_{\langle i,j\rangle}^{N}\frac{1}{P}V(\varphi_i^{(s)}, \varphi_j^{(s)})\right\}.$$

(8.18)

Thus each quantum mechanical rotational degree of freedom is represented in this path integral representation by P classical rotators, which form closed loops and interact via harmonic type interactions. In addition there is the potential $V(\varphi_i^{(s)}, \varphi_j^{(s)})$ denoting the pair potential evaluated separately for the configuration at each imaginary-time slice $s = 1, \ldots, P$. However, in contrast to path integrals for translational degrees of freedom, the loops need not be closed using periodic boundary conditions, but only modulo 2π: the classical angles are not confined to $[0, 2\pi]$ but are allowed on the whole interval $[-\infty, +\infty]$. The resulting mismatch n_j is called the 'winding number' of the jth path and Eqns. (8.17) and (8.18) yield the 'winding number representation' of the partition function. Only the Boltzmann-weighted summation over all possible winding numbers in addition to the integration over all paths having a certain winding number yields the correct quantum partition function in the Trotter limit $P \to \infty$. Thus the Monte Carlo algorithm has to include both moves that update the angular degrees of freedom $\{\varphi_j^{(s)} \to \varphi_j^{(s)'}\}$ and moves that attempt to change the winding number, $n_j \to n_j'$.

As an example of problems that can be tackled with such techniques, Fig. 8.5 shows the order parameter ϕ of a model for N_2 on graphite. This order parameter describes the ordering of the so-called herringbone structure, and

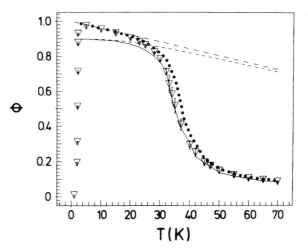

Fig. 8.5 Herringbone structure order parameter for a model of N_2 plotted vs. temperature. Quantum simulation, full line; classical simulation, dotted line; quasi-harmonic theory, dashed line; Feynman–Hibbs quasi-classical approximation, triangles. From Marx et al. (1993).

is calculated from the three order parameter components ϕ_α as $\phi = \left\langle \left[\sum_{\alpha=1}^{3} \phi_\alpha^2 \right]^{1/2} \right\rangle$, with

$$\phi_\alpha = \frac{1}{N} \frac{1}{P} \sum_{j=1}^{N} \sum_{s=1}^{P} \sin(2\varphi_j^{(s)} - 2\eta_\alpha) \exp[\mathbf{Q}_\alpha \cdot \mathbf{R}_j], \qquad (8.19)$$

where $\mathbf{R}_j$ is the center of mass position of the jth molecule, the $\mathbf{Q}_\alpha$ are wavevectors characteristic for the ordering $\{\mathbf{Q}_1 = \pi(0, 2/\sqrt{3});\ \mathbf{Q}_2 = \pi(-1, -1/\sqrt{3});\ \mathbf{Q}_3 = \pi(1, -1/\sqrt{3})\}$ and the phases η_i are $\eta_1 = 0$, $\eta_2 = 2\pi/3$ and $\eta_3 = 4\pi/3$. Using $N = 900$ rotators, even for $T > T_c$ we have the characteristic 'finite size tail' in both the classical and in the quantum calculations. The critical temperature T_c of the classical model has been estimated as 38 K. While at high temperatures classical and quantum calculations merge, near T_c the quantum mechanical result deviates from the classical one, since in this model the quantum fluctuations reduce T_c by about 10%. Furthermore, one can infer that the quantum system does not reach the maximum herringbone ordering ($\phi = 1$) even at $T \to 0$: the quantum librations depress the saturation value by 10%. In Fig. 8.5 the order parameter, as obtained from the full quantum simulation, is compared with two approximate treatments valid at low and high temperatures: quasi-harmonic theory can account for the data for $T > 10\,\mathrm{K}$ but fails completely near the phase transition; the Feynman–Hibbs quasi-classical approximation (based on a quadratic expansion of the effective Hamiltonian around the classical path) works very well at high temperatures, but it starts to deviate from the correct curve just below T_c and completely breaks down as $T \to 0$. We see that all these approximate treatments are uncontrolled, their accuracy can only be judged a posteriori; only the PIMC simulation yields correct results over the whole temperature range from the classical to the quantum regime.

8.3 LATTICE PROBLEMS

8.3.1 The Ising model in a transverse field

The general idea that one follows to develop a useful path integral formulation of quantum models on lattices is again the strategy to decompose the Hamiltonian $\mathcal{H}$ of the interacting many-body system into sums of operators that can be diagonalized separately. The Trotter formula can be then used in analogy with Eqn. (8.6), for $\mathcal{H} = \mathcal{H}_1 + \mathcal{H}_2$ (Trotter, 1959; Suzuki, 1971)

$$\exp[-(\mathcal{H}_1 + \mathcal{H}_2)/k_B T] = \lim_{P \to \infty} \{\exp(-\mathcal{H}_1/k_B TP) \exp(-\mathcal{H}_2/k_B TP)\}^P.$$
$$(8.20)$$

Note that there is no general recipe for how this division of $\mathcal{H}$ into parts should be done - what is appropriate depends on the nature of the model. Therefore, there are many different variants of calculations possible for certain models, and generalizations of Eqn. (8.20), where the error is not of order

$1/P^2$ but of even higher inverse order in P, have also been considered (Suzuki, 1976, 1992).

To illustrate the general principles of the approach we consider a model for which all calculations can be carried out exactly, namely the one-dimensional Ising model in a transverse field. We take (de Raedt and Lagendijk, 1985)

$$\mathcal{H}_1 = -\mathcal{J} \sum_{i=1}^{N} \hat{\sigma}_i^z \hat{\sigma}_{i+1}^z, \qquad \mathcal{H}_2 = -h \sum_{i=1}^{N} \hat{\sigma}_i^x, \qquad (8.21)$$

where $\hat{\sigma}_i^\alpha (\alpha = x, y, z)$ denote the Pauli spin matrices at site i. We assume periodic boundary conditions, $\hat{\sigma}_{N+1}^\alpha = \hat{\sigma}_1^\alpha$. For the representation we choose the eigenstates of $\hat{\sigma}^z$ and label them by Ising spin variables, $S = \pm 1$, i.e. $\hat{\sigma}^z|S\rangle = S|S\rangle$. Of course, $\mathcal{H}_1$ is diagonal in this representation. We then find for the Pth approximant to the partition function

$$Z_p = \text{Tr}[\exp(-\mathcal{H}_1/k_B TP)\exp(-\mathcal{H}_2/k_B TP)]^P$$

$$= \sum_{\{S_i^{(k)}\}} \prod_{k=1}^{P} \prod_{i=1}^{N} \exp\left[\frac{\mathcal{J}}{k_B TP} S_i^{(k)} S_{i+1}^{(k)}\right] \left\langle S_i^{(k)} \left| \exp\left(\frac{h\hat{\sigma}_i^x}{k_B TP}\right) \right| S_i^{(k+1)} \right\rangle. \qquad (8.22)$$

In this trace we have to take periodic boundary conditions in the imaginary time direction as well, $S_i^{(k)} = S_i^{(k+P)}$. Now the matrix element in Eqn. (8.22) is evaluated as follows,

$$\langle S| \exp(a\hat{\sigma}^x)|S'\rangle = \left(\tfrac{1}{2}\sinh 2a\right)^{1/2} \exp\left(\tfrac{1}{2}\ln \coth a\right) SS'. \qquad (8.23)$$

Substituting Eqn. (8.23) in Eqn. (8.22), we see that Z_p looks like the partition function of an anisotropic two-dimensional Ising model,

$$Z_p = C_p \sum_{\{S_i^{(k)}\}} \exp\left[\sum_{k=1}^{P} \sum_{i=1}^{N} \left(K_p S_i^{(k)} S_i^{(k+1)} + \frac{\mathcal{J}}{k_B TP} S_i^{(k)} S_{i+1}^{(k)}\right)\right], \qquad (8.24)$$

with

$$C_p = \left[\tfrac{1}{2}\sinh(2h/k_B TP)\right]^{PN/2}, \qquad K_p = \tfrac{1}{2}\ln \coth(h/k_B TP). \qquad (8.25)$$

At this point we can use the rigorous solution of the finite two-dimensional Ising model (Onsager, 1944). Thus the one-dimensional quantum problem could be mapped onto an (anisotropic) two-dimensional classical problem, and this mapping extends to higher dimensions, as well. However, it is important to note that the couplings depend on the linear dimension P in the 'Trotter direction' and in this direction they also are temperature dependent (analogous to the spring constant κ in the polymer formalism derived above).

8.3.2 Anisotropic Heisenberg chain

A more complex and more illuminating application of the Trotter formula to a simple lattice model is to the spin-$\frac{1}{2}$ anisotropic Heisenberg chain,

$$\mathcal{H} = \sum_i (\mathcal{J}_x \hat{S}_i^x \hat{S}_{i+1}^x + \mathcal{J}_y \hat{S}_i^y \hat{S}_{i+1}^y + \mathcal{J}_z \hat{S}_i^z \hat{S}_{i+1}^z). \tag{8.26}$$

For $\mathcal{J}_x = \mathcal{J}_y = \mathcal{J}_z$ this model is merely a simple quantum Heisenberg chain, and for $\mathcal{J}_x = \mathcal{J}_y$ and $\mathcal{J}_z = 0$ it becomes the quantum XY-chain. There are now several different ways in which the quantum Hamiltonian may be split up. The procedure first suggested by Suzuki (1976b) and Barma and Shastry (1978) was to divide the Hamiltonian by spin component, i.e.

$$\mathcal{H} = \mathcal{H}_0 + V_A + V_B \tag{8.27a}$$

where

$$\mathcal{H}_0 = -\sum_{i=1}^{N} \mathcal{J}_z \hat{S}_i^z \hat{S}_{i+1}^z, \tag{8.27b}$$

$$V_A = \sum_{i\,\text{odd}} V_i, \tag{8.27c}$$

$$V_B = \sum_{i\,\text{even}} V_i, \tag{8.27d}$$

$$V_i = -(\mathcal{J}_x \hat{S}_i^x \hat{S}_{i+1}^x + \mathcal{J}_y \hat{S}_i^y \hat{S}_{i+1}^y). \tag{8.27e}$$

Applying Trotter's formula to the partition function we obtain

$$Z = \lim_{P \to \infty} Z^{(P)} \tag{8.28}$$

with

$$Z^{(P)} = \text{Tr}\left(e^{-\beta\mathcal{H}_0/2P} e^{-\beta V_A/P} e^{-\beta\mathcal{H}_0/2P} e^{-\beta V_B/P}\right)^P, \tag{8.29}$$

where the limit $P \to \infty$ and the trace have been interchanged. Introducing $2P$ complete sets of eigenstates of $\mathcal{H}_0$ (the Ising part) so that there is one complete set between each exponential we obtain

$$Z^{(P)} = \sum_{\alpha_1 \alpha_2 \ldots \alpha_{2P}} \exp\left(\frac{-\beta}{2P}\sum_{r=1}^{2P}\mathcal{H}_{0r} - \beta\sum_{i \ni A}\sum_{r=1}^{2P} h(i, r) - \beta\sum_{i \ni B}\sum_{r=1}^{2P} h(i, r)\right) \tag{8.30}$$

where

$$e^{-\beta h(i,r)} = \langle S_{ir} S_{(i+1)r} | e^{-\beta V_i/P} | S_{i(r+1)} S_{(i+1)(r+1)}\rangle \tag{8.31}$$

and $S_{ir} = \pm 1/2$. Equation (8.30) can be interpreted as describing an $N \times 2P$ lattice with periodic boundary conditions and with two-spin interactions in the real space direction and temperature-dependent four-spin coupling on alternating elementary plaquettes, which couple neighboring sites in both real space and the Trotter direction, as shown in Fig. 8.6. Evaluation of the matrix elements in Eqn. (8.31) shows that only those plaquettes which have an even

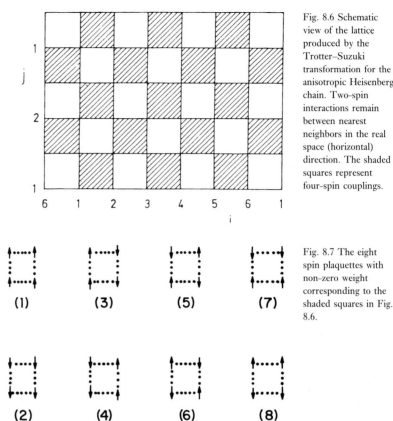

Fig. 8.6 Schematic view of the lattice produced by the Trotter–Suzuki transformation for the anisotropic Heisenberg chain. Two-spin interactions remain between nearest neighbors in the real space (horizontal) direction. The shaded squares represent four-spin couplings.

Fig. 8.7 The eight spin plaquettes with non-zero weight corresponding to the shaded squares in Fig. 8.6.

number of spins in each direction have non-zero weight, and these are enumerated in Fig. 8.7. (This result means that the classical model which results from the general, anisotropic Heisenberg chain is equivalent to an 8-vertex model; moreover, if $J_x = J_y$ it reduces further to a 6-vertex model.) Only those spin-flips which overturn an even number of spins are allowed, to insure that the trial state has non-zero weight, and the simplest possible such moves are either overturning all spins along a vertical line in the Trotter direction or those spins around a 'local' loop as shown in Fig. 8.8. We note further that if $J_x = J_y$ all allowed extensive flips change the magnetization of the system whereas the local flips do not. There is one additional complication that needs to be mentioned: because of the temperature dependent interactions, the usual measures of the thermal properties are no longer corrrect. Thus, for example, the Pth approximant to the thermal average of the internal energy $E^{(P)}$ is

$$E^{(P)} = -\frac{\partial}{\partial \beta} \ln Z^{(P)}$$
$$= \frac{1}{Z^{(P)}} \sum_j F_j^{(P)} \exp(-\beta E_j^{(P)}), \qquad (8.32)$$

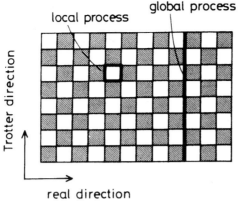

Fig. 8.8 Allowed spin-flip patterns (bold lines) for the lattice shown in Fig. 8.6.

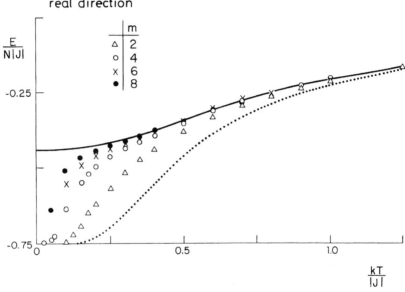

Fig. 8.9 Internal energy for the $S = \frac{1}{2}$ antiferromagnetic Heisenberg model. The solid line is the calculation of Bonner and Fisher (1964) and the dotted line is the exact $P = 1$ result from Suzuki (1966). From Cullen and Landau (1983).

where the sum is over all states and the 'energy function' F_j is now non-trivial. Similarly the calculation of the specific heat has an explicit contribution from the temperature dependence of the energy levels. Results for the antiferromagnetic Heisenberg chain, shown in Fig. 8.9, clearly indicate how the result for a fixed value of P approximates the quantum result only down to some temperature below which the data quickly descend to the classical value. This procedure has been vectorized by Okabe and Kikuchi (1986) who assigned a plaquette number to each four-spin plaquette and noted that a simple XOR operation could be used to effect the spin plaquette flips. When this process was vectorized in an optimal fashion a speed of 34 million spin-flip trials per second could be achieved, a very impressive performance for the computers of that time.

Before we leave this section we wish to return to the question of how the Hamiltonian should be divided up before applying the Trotter transformation. An alternative to the decomposition used in the above discussion would have been to divide the system into two sets of non-interacting dimers, i.e.

$$\mathcal{H} = \mathcal{H}_1 + \mathcal{H}_2, \tag{8.33a}$$

where

$$\mathcal{H}_1 = -\sum_{i \ni \text{odd}} (\mathcal{J}_x \hat{S}_i^x \hat{S}_{i+1}^x + \mathcal{J}_y \hat{S}_i^y \hat{S}_{i+1}^y + \mathcal{J}_z \hat{S}_i^z \hat{S}_{i+1}^z), \tag{8.33b}$$

$$\mathcal{H}_2 = -\sum_{i \ni \text{even}} (\mathcal{J}_x \hat{S}_i^x \hat{S}_{i+1}^x + \mathcal{J}_y \hat{S}_i^y \hat{S}_{i+1}^y + \mathcal{J}_z \hat{S}_i^z \hat{S}_{i+1}^z). \tag{8.33c}$$

When the same process is repeated for this decomposition an $N \times 2P$ lattice is generated but the four-spin interactions have a different geometrical connectivity, as is shown in Fig. 8.10. In general then some thought needs to be given as to the best possible decomposition since there may be a number of different possibilities which present themselves. This approach can be readily extended to higher dimensions and, in general, a d-dimensional quantum spin lattice will be transformed into a $(d + 1)$-dimensional lattice with both two-spin couplings in the real space directions and four-spin interactions which connect different 'rows' in the Trotter direction.

8.3.3 Fermions on a lattice

The one-dimensional spin models considered in the previous section provide the opportunity to use the Trotter–Suzuki decomposition to help us understand concepts, to check the convergence as $P \to \infty$ and to test various refinements. New, non-trivial problems quickly arise when considering other relatively simple models such as spinless fermions in one dimension, where the Hamiltonian $\mathcal{H} = \mathcal{H}_1 + \mathcal{H}_2$ is written as

$$\mathcal{H} = -t \sum_{i=1}^{N} (\hat{c}_i^+ \hat{c}_{i+1} + \hat{c}_{i+1}^+ \hat{c}_i) + v_1 \sum_{i=1}^{N} \hat{n}_i \hat{n}_{i+1}. \tag{8.34}$$

The fermion operator $\hat{c}_i^+ (\hat{c}_i)$ creates (annihilates) a particle at site i, and $\hat{n}_i \equiv \hat{c}_i^+ \hat{c}_i$ is the particle number operator, $\mathcal{N} = \sum_{i=1}^{N} n_i$ being the total number of

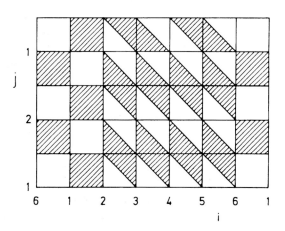

Fig. 8.10 Lattice produced by the alternate decomposition, given in Eqns. (8.33) for the $S = 1/2$ antiferromagnetic Heisenberg model.

particles ($\rho \equiv \mathcal{N}/N$ then is the particle density). The hopping energy t is chosen to be unity, having the strength v_1 of the nearest neighbor interaction as a non-trivial energy scale in the model.

One of the standard tricks for dealing with quantum problems is to make use of clever transformations that make the problem more tractable. In the present situation, we first use Pauli matrices $\hat{\sigma}_i^\alpha (\alpha = x, y, z)$ to define spin-raising and spin-lowering operators by $\hat{\sigma}_\ell^+ = (\hat{\sigma}_\ell^x + i\hat{\sigma}_\ell^y)/2$ and $\hat{\sigma}_\ell^- = (\hat{\sigma}_\ell^x - i\hat{\sigma}_\ell^y)/2$, respectively, and express the $\hat{c}_\ell^+, \hat{c}_\ell$ in terms of the $\hat{\sigma}_\ell$ operators by a Jordan–Wigner transformation, which has a non-local character

$$\hat{c}_\ell^+ = \hat{\sigma}_\ell^+ \exp\left[\frac{i\pi}{2}\sum_{p=1}^{\ell-1}(1+\hat{\sigma}_p^z)\right], \qquad \hat{c}_\ell = \hat{\sigma}_\ell^- \exp\left[-\frac{i\pi}{2}\sum_{p=1}^{\ell-1}(1+\hat{\sigma}_p^z)\right].$$

$$(8.35)$$

With this transformation the spinless fermion model, Eqn. (8.34), can be mapped exactly onto a spin-$\frac{1}{2}$ model, and neglecting boundary terms which are unimportant for $N \to \infty$,

$$\mathcal{H} = -\frac{t}{2}\sum_{i=1}^N\left(\hat{\sigma}_i^x\hat{\sigma}_{i+1}^x + \hat{\sigma}_i^y\hat{\sigma}_{i+1}^y - \frac{v_1}{2t}\hat{\sigma}_i^z\hat{\sigma}_{i+1}^z - \frac{v_1}{t}\hat{\sigma}_i^z - \frac{v_1}{2t}\right). \qquad (8.36)$$

Since the invention of the Bethe ansatz (Bethe, 1931), a huge number of analytical treatments of the model Eqns. (8.34) and (8.36) and its generalization have appeared so that the groundstate properties are rather well known. Here we discuss only the structure factor (a is the lattice spacing)

$$S_T(q) = \sum_{j=1}^N\left(\langle\hat{n}_i\hat{n}_{i+j}\rangle_T - \langle\hat{n}_i\rangle_T\langle\hat{n}_j\rangle_T\right)\cos(jqa) \qquad (8.37)$$

for $T = 0$ and half filling ($\rho = 1/2$). At $v_1 = 2t$ a metal–insulator transition occurs (Ovchinnikov, 1973): for $v_1 < 2t$ there is no energy gap between the groundstate energy and the first excited states, and the system is a metal; $S(q)$ then has a peak at $q = \pi/a$ with finite width. If $v_1 > 2t$ there is a gap and the groundstate has long range order, which implies that $S(q)$ has a delta function (for $N \to \infty$) at $q = \pi/a$. For $v_1 \to \infty$ the groundstate approaches simply that of the classical model where every second lattice site is occupied and every other lattice site is empty. A related quantity of interest is the static wavevector-dependent 'susceptibility' ($\hat{\varphi}_q$ is the Fourier component of the density operator $\hat{n}_i$)

$$\chi(q) = \frac{1}{\hbar}\int_0^{\hbar/k_BT} dx\left[\langle e^{x\mathcal{H}}\hat{\varphi}_q e^{-x\mathcal{H}}\hat{\varphi}_{-q}\rangle_T - \langle\hat{\varphi}_{q=0}\rangle_T^2\right]. \qquad (8.38)$$

If $[\mathcal{H}, \hat{\rho}_q] = 0$, we would simply recover the classical fluctuation relation $\chi(q) = S_T(q)/k_BT$ since $S_T(q) = \langle\hat{\rho}_q\hat{\rho}_{-q}\rangle_T - \langle\hat{\rho}_{q=0}\rangle_T^2$. Thus, in calculating response functions ($\chi(q)$ describes the response of the density to a wave-vector-dependent 'field' coupling linearly to the density) one must carefully consider the appropriate quantum mechanical generalizations of fluctuation formulae, such as Eqn. (8.38).

In order to bring the problem, Eqn. (8.34) or Eqn. (8.36), into a form where the application of the Trotter formula, Eqn. (8.20), is useful, we have to find a suitable decomposition of $\mathcal{H}$ into $\mathcal{H}_1$ and $\mathcal{H}_2$. When we wish to describe the states in the occupation number representation (or the corresponding spin representation: $|S_1 \dots S_i \dots S_N\rangle$ means that $S_i = 1(-1)$ if the site i is occupied (empty)), we have the problem that the non-diagonal first term in Eqn. (8.38) couples different sites. Thus, one uses a decomposition where one introduces two sublattices, $\mathcal{H}_{i,j} = -t(\hat{c}_i^+ \hat{c}_j + \hat{c}_j^+ \hat{c}_i) + v_1 \hat{n}_i n_j$, following Barma and Shastry (1977)

$$\mathcal{H}_1 = \sum_{i=1}^{N/2} \mathcal{H}_{2i-1,2i}, \qquad \mathcal{H}_2 = \sum_{i=1}^{N/2} \mathcal{H}_{2i,2i+1}. \tag{8.39}$$

Of course, here we require N to be even (only then does the system admit an antiferromagnetic ground state with no domain wall in the limit $v_1 \to \infty$). The idea of this partitioning of the Hamiltonian is that now the terms in $\mathcal{H}_1$ all commute with each other as do the terms in $\mathcal{H}_2$, due to the local character of the Hamiltonian,

$$[\mathcal{H}_{2i-1,2i}, \mathcal{H}_{2j-1,2j}] = [\mathcal{H}_{2i,2i+1}, \mathcal{H}_{2j,2j+1}] = 0, \qquad \text{all } i,j. \tag{8.40}$$

Therefore the corresponding Trotter approximation reads

$$Z_P = \mathrm{Tr}\big[\exp(-\mathcal{H}_{1,2}/k_B TP)\exp(-\mathcal{H}_{3,4}/k_B TP)\dots \exp(-\mathcal{H}_{N-1,N}/k_B TP)$$
$$\exp(-\mathcal{H}_{2,3}/k_B TP)\dots \exp(-\mathcal{H}_{N-2,N-1}/k_B TP)\exp(-\mathcal{H}_{N,1}/k_B TP)\big]^P \tag{8.41}$$

since Eqn. (8.40) implies that $\exp(-x\hat{H}_1) = \prod_{i=1}^{N/2} \exp(-x\mathcal{H}_{i,i+1})$ and similarly for $\mathcal{H}_2$, for arbitrary x. Introducing the representation mentioned above, we need to evaluate the matrix elements

$$T(S_i, S_j; \tilde{S}_i, \tilde{S}_j) \equiv \langle S_i, S_j | \exp(-\mathcal{H}_{i,j}/k_B TP)|\tilde{S}_i, \tilde{S}_j\rangle, \tag{8.42}$$

which yields

$$T(S_i, S_j; \tilde{S}_i, \tilde{S}_j)$$
$$= \begin{pmatrix} 1 & 0 & 0 & 0 \\ 0 & \cosh(t/Pk_B T) & \sinh(t/Pk_B T) & 0 \\ 0 & \sinh(t/Pk_B T) & \cosh(t/Pk_B T) & 0 \\ 0 & 0 & 0 & \exp(-v_1/Pk_B T) \end{pmatrix} \tag{8.43}$$

where the lines of the matrix are ordered according to the states $|-1, -1\rangle$, $|1, -1\rangle$, $|-1, 1\rangle$ and $|1, 1\rangle$, from above to below, respectively. Then the Trotter approximation for the partition function becomes (de Raedt and Lagendijk, 1985)

$$Z_P = \sideset{}{'}\sum_{\{S_i^{(S)}\}} \sideset{}{'}\sum_{\{\tilde{S}_i^{(S)}\}} \prod_{s=1}^{P} T\left(S_1^{(s)}, S_2^{(s)}; \tilde{S}_1^{(s)}, \tilde{S}_2^{(s)}\right) \cdots$$

$$T\left(S_{N-1}^{(s)}, S_N^{(s)}; \tilde{S}_{N-1}^{(s)}, \tilde{S}_N^{(s)}\right) \times T\left(\tilde{S}_2^{(s)}, \tilde{S}_3^{(s)}; \tilde{S}_2^{(s+1)}, \tilde{S}_3^{(s+1)}\right) \cdots$$

$$T\left(\tilde{S}_{N-2}^{(s)}, \tilde{S}_{N-1}^{(s)}; \tilde{S}_{N-2}^{(s+1)}, \tilde{S}_{N-1}^{(s+1)}\right)$$

$$\times\, T\left(\tilde{S}_N^{(s)} \tilde{S}_1^{(s)}; \tilde{S}_N^{(s+1)} \tilde{S}_1^{(s+1)}\right)\left(1 - |S_1^{(s)} - S_N^{(s)}|\right)^{N+1}. \tag{8.44}$$

The primes on the summation signs in Eqn. (8.44) mean that the sums over the variables S and $\tilde{S}$ are restricted, because the total number $\mathcal{N}$ of fermions is fixed, i.e. $\sum_{i=1}^{N} S_i^{(s)} = \sum_{i=1}^{N} \tilde{S}_i^{(s)} = 2\mathcal{N} - N$ for all s. The last line in Eqn. (8.44) represents the physical situation in which a particle moves from site 1 to site N and vice versa. Such moves destroy the ordering in which the fermions have been created from the vacuum state. Therefore the last factor is a correction term which results from reordering the fermion operators, taking into account the anticommutation rules. Obviously, there are only negative contributions to Z_P if $\mathcal{N}$ is even, and no minus signs would be present if there were free boundary conditions, because then the entire last line of Eqn. (8.34) would be missing.

8.3.4 An intermezzo: the minus sign problem

For an interpretation of Z_P as the trace of an equivalent classical Hamiltonian, $Z_P = \text{Tr} \exp(-\mathcal{H}_{\text{eff}}^{(P)}/k_B T)$, it is clearly necessary that all terms that contribute to this partition sum are non-negative, because for a real $\mathcal{H}_{\text{eff}}^{(P)}$ the term $\exp(-\mathcal{H}_{\text{eff}}^{(P)}/k_B T)$ is never negative. The anticommutation rule of fermion operators leads to negative terms, as they occur in Eqn. (8.44) for even $\mathcal{N}$, and this problem hampers quantum Monte Carlo calculations, in a very severe way. Of course, the same problem would occur if we simply tried to work with the Fermi equivalent of Z_B in Eqn. (8.15), since then the eigenfunctions $\phi_\alpha(\mathbf{R})$ are antisymmetric under the permutation of particles,

$$\hat{A}\phi_\alpha(\mathbf{R}) = \frac{1}{N!}\sum_{P}(-1)^P \phi(\hat{P}\mathbf{R}), \tag{8.45}$$

where $(-1)^P$ is negative if the permutation is odd, while Eqn. (8.14) did not lead to any such sign problems.

Now it is possible to generalize the Metropolis importance sampling method to cases where a quantity $\rho(x)$ in an average (x stands here symbolically for a high-dimensional phase space)

$$\langle \hat{A} \rangle = \int A(x)\rho(x)dx \bigg/ \int \rho(x)dx \tag{8.46}$$

is not positive semi-definite, and hence does not qualify for an interpretation as a probability density. The standard trick (de Raedt and Lagendijk, 1981) amounts to working with $\tilde{\rho}(x) = |\rho(x)|/\int|\rho(x)|dx$ as probability density for which one can do importance sampling, and to absorb the sign of $\rho(x)$ in the quantity that is sampled. Thus

$$\langle \hat{A} \rangle = \frac{\int A(x)\mathrm{sign}(\rho(x))\tilde{\rho}(x)dx}{\int \mathrm{sign}(\rho(x))\tilde{\rho}(x)dx} = \frac{\langle \hat{A}\hat{s} \rangle}{\langle \hat{s} \rangle}, \qquad (8.47)$$

where $\hat{s}$ is the sign operator that corresponds to the function sign $(\rho(x))$. While Eqn. (8.47) seems like a general solution to this so-called 'minus sign problem', in practice it is useful only for very small particle number N. The problem is that *all* regions of phase space are important but have contributions which tend to cancel each other. In practice this leads to the problem that $\langle \hat{s} \rangle$ is extremely small, huge cancellations occur in both $\langle \hat{A}\hat{s} \rangle$ and $\langle \hat{s} \rangle$, the statistical fluctuations then will render an accurate estimation of $\langle \hat{A} \rangle$ almost impossible. The reader may obtain some insight into this situation by examining a much simpler problem which presents the same difficulty, namely the evaluation of the integral

$$F(\alpha, x) = \int_{-\infty}^{\infty} e^{-x^2} \cos(\alpha x)dx \qquad (8.48)$$

in the limit that $\alpha \to \infty$. The argument of this integral oscillates rapidly for large α and the determination of the value by Monte Carlo methods, see Chapter 3, becomes problematical. In the example given below we show how the determination of the value of the integral becomes increasingly imprecise as α increases. For $\alpha = 0$ the estimate after 10^7 samples is good to better than 0.03% whereas for $\alpha = 4$ the fluctuations with increasing sampling are of the order of 1%. For larger values of α the quality of the result deteriorates still more.

Example

Use simple sampling Monte Carlo to estimate $F(\alpha, x)$ for $x = 0, 1.0, 2.0, 4.0$:

Number of points	α			
	0	1.0	2.0	4.0
10 000 000	0.885 717 0	0.688 967 0	0.325 109 0	0.016 047 0
20 000 000	0.886 325 5	0.689 699 0	0.325 827 5	0.016 436 0
30 000 000	0.886 106 0	0.689 621 0	0.325 403 7	0.015 922 3
40 000 000	0.885 929 5	0.689 765 3	0.325 628 3	0.016 033 8
50 000 000	0.886 271 8	0.690 020 8	0.325 923 2	0.016 206 6
60 000 000	0.886 562 6	0.690 204 5	0.325 984 2	0.016 246 7
70 000 000	0.886 407 4	0.690 090 4	0.325 772 1	0.016 037 0
80 000 000	0.886 347 0	0.690 090 5	0.325 786 0	0.015 939 5
90 000 000	0.886 206 9	0.689 879 8	0.325 688 7	0.015 898 8
100 000 000	0.886 201 2	0.689 889 0	0.325 741 1	0.016 048 8
Exact	0.886 226 6	0.690 194 0	0.326 024 5	0.016 231 8

Another very important quantum problem in which progress has been limited because of the minus sign problem is the Hubbard Hamiltonian (Hubbard, 1963),

$$\mathcal{H}_{\text{Hubbard}} = t \sum_{\langle i,j \rangle} \left(\hat{c}_{i,\sigma}^+ \hat{c}_{j,\sigma} + \hat{c}_{j,\sigma}^+ \hat{c}_{i,\sigma} \right) + U \sum_i \hat{n}_{i\downarrow} \hat{n}_{i\uparrow} \tag{8.49}$$

where $\hat{c}_{i,\sigma}^+ (\hat{c}_{i,\sigma})$ creates (annihilates) a fermion of spin $\sigma = \uparrow, \downarrow$ at site i, t is the hopping matrix element analogously to Eqn. (8.34), while U represents the on-site Coulomb interaction strength. The minus sign problem has been studied in detail, and it was found that (Loh *et al.*, 1990)

$$\langle \hat{s} \rangle \propto \exp(-\gamma NU/k_B T), \tag{8.50}$$

where γ is a constant that depends strongly on the filling of the band. It is obvious that the minus sign problem gets worse as N increases and as the temperature is lowered. Finding methods to avoid this problem (or at least to make γ very small) is still an active area of research.

8.3.5 Spinless fermions revisited

While the minus sign problem is also a severe problem for the Hamiltonian Eqn. (8.34) in $d = 2$ and 3 dimensions, for $d = 1$ the only remnant of this problem is the last factor on the right-hand side of Eqn. (8.34), and this is clearly not a big problem (note that this term would be completely absent for the choice of free boundary conditions).

The first step in dealing with Eqn. (8.44) is the elimination of the $\tilde{S}_i^{(s)}$ variables, which can be done analytically. Note that $T(S_i, S_j; \tilde{S}_i, \tilde{S}_j)$ from Eqn. (8.43) can be rewritten as

$$T(S_i, S_j; \tilde{S}_i, \tilde{S}_j) = \delta_{S_i S_j \tilde{S}_i \tilde{S}_j} T_{S_i, S_j}(S_i, S_i \tilde{S}_i S_j) \tag{8.51}$$

where the remaining (2×2) matrices $T_1(S, \overline{S})$ and $T_{-1}(S, \overline{S})$ are

$$T_1(S, \overline{S}) \equiv \begin{pmatrix} 1 & \cosh(t/k_B TP) \\ \cosh(t/k_B TP) & \exp(-v_1/k_B TP) \end{pmatrix} \tag{8.52}$$

where the upper line refers to state $|-1\rangle$ and the lower line to state $|1\rangle$, and

$$T_{-1}(S, \overline{S}) \equiv \delta_{S,\overline{S}} \sinh(t/k_B TP). \tag{8.53}$$

Summing over the $\tilde{S}_i^{(s)}$ in Eqn. (8.44) then yields

$$Z_P = \sum_{\{S_i^{(s)}\}} \sum_{\{\sigma_j\}} \prod_{j=1}^{P} T_{\sigma_j \phi_1^{(j)}} \left(S_1^{(j)}, \sigma_j \phi_1^{(j)} S_2^{(j)} \right) T_{\sigma_j \phi_2^{(j)}} \left(\sigma_j \phi_2^{(j)} S_2^{(j)}, S_3^{(j+1)} \right) \cdots$$

$$T_{\sigma_j \phi_{N-1}^{(j)}} \left(S_{N-1}^{(j)}, \sigma_j \phi_{N-1}^{(j)} S_N^{(j)} \right) T_{\sigma_j \phi_N^{(j)}} \left(\sigma_j \phi_{N,j} S_N^{(j)}, S_1^{(j+1)} \right) \sigma_j^{N+1} \delta_{\phi_N^{(j)},1}, \tag{8.54}$$

where the $\{\phi_\ell^{(j)}\}$ are string-like variables formed from the $\{S_i^{(s)}\}$,

$$\phi_\ell^{(j)} = \prod_{i=1}^{\ell} S_i^{(j)} S_i^{(j+1)}. \tag{8.55}$$

Therefore, the effective lattice model, $\mathcal{H}_{\text{eff}}^{(P)}$ that results from Eqn. (8.54), $Z_P \equiv \text{Tr} \exp(-\mathcal{H}_{\text{eff}}^{(P)}/k_B T)$, contains non-local interactions both along the chain and in the Trotter imaginary time direction, unlike the Ising model in a transverse field, that had non-local interactions in the Trotter direction only. The total number of variables in Eqn. (8.54) is $P(N+1)$, namely the PN spins $\{S_i^{(s)}\}$ and P variables $\sigma_j = \pm 1$. The extra sum over the latter is a consequence of the use of periodic boundary conditions. If we work with free boundary conditions, this sum can be omitted in Eqn. (8.54) and we can put $\sigma_j \equiv 1$ there and no negative terms occur. Even then a Monte Carlo process that produces states proportional to the Boltzmann weight $\exp(-\mathcal{H}_{\text{eff}}^{(P)}/k_B T)$ is difficult to construct. To avoid the non-local interaction in the spatial direction generated in Eqn. (8.54), one can rather attempt to construct a Monte Carlo scheme that realizes the Boltzmann weight for Eqn. (8.44), at the expense that one has twice as many variables ($S_i^{(s)}$ and $\tilde{S}_i^{(s)}$, respectively). However, the zero matrix elements in Eqn. (8.43) imply that many states generated would have exactly zero weight if one chose trial configurations of the $\{S_i^{(s)}, \tilde{S}_i^{(s)}\}$ at random: rather the Monte Carlo moves have to be constructed such that the Kronecker delta in Eqn. (8.51) is never zero. This constraint can be realized by two-particle moves in the checkerboard representation, Fig. 8.11, as proposed by Hirsch et al. (1982). Figure 8.12 shows the type of results that can be obtained from this method. One can see from

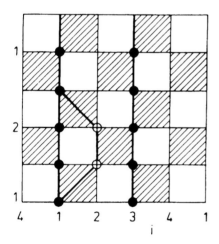

Fig. 8.11 Example of the elementary two-particle jump procedure for the checkerboard lattice, for a chain of four sites. Each shaded square represents a T-matrix and determines which particles can interact with each other (only particles that sit on the corners of the same shaded square). The variables $S_i^{(j)}$ are defined on the rows $j = 1, 2$ whereas the variables $\tilde{S}_i^{(j)}$ are defined on the rows between the $j = 1$ and $j = 2$ rows (note we have chosen $P = 2$ here, and we *must* impose periodic boundary conditions in the Trotter direction because of the trace operation; the figure implies also the choice of periodic boundary conditions in the spatial direction as well). The black dots indicate a state of the lattice with non-zero weight, representing particles present in the occupation number representation (the thick lines connecting them are the so-called 'world lines'). A trial state is generated by moving two particles from one vertical edge of an *unshaded* square to the other. From de Raedt and Lagendijk (1985).

Fig. 8.12 (a) Points
showing Monte Carlo
data for the structure
factor for a 40-site
lattice containing 20
non–interacting
electrons ($t = 1$,
$V = 0$) at low
temperature,
$1/k_{\rm B}T = 4$. Solid line
is the analytical
solution for this
system. (b) Monte
Carlo results for the
structure factor for
$t = 1$, and $V = 2$ at
$1/k_{\rm B}T = 4$. Note the
difference in scale
between parts (a) and
(b). (c) Structure
factor $S(q = \pi)$ for the
half-filled case with
$v/2t = 1$ vs. the lattice
size. From Hirsch *et
al.* (1982).

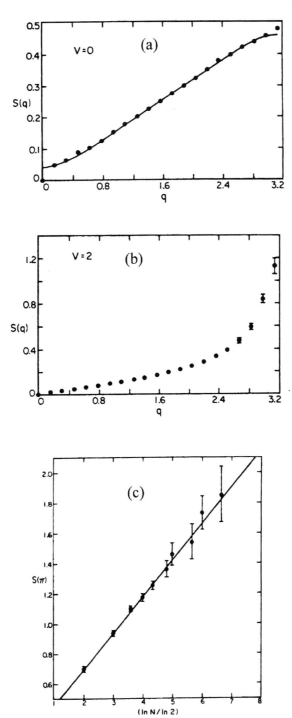

Fig. 8.12 that non-trivial results for this fermion model in $d = 1$ dimensions have been obtained, but even in this case it is difficult to go to large N (the largest size included in Fig. 8.12 is $N = 100$), and statistical errors are considerable at low temperatures. Nevertheless Fig. 8.12 gives reasonable evidence for the quite non-trivial scaling dependence $S(\pi) \propto \ln N$.

This case of fermions in $d = 1$ has again shown that the PIMC methods always need some thought about how best to split the Hamiltonian into parts so that, with the help of the Trotter formalism, one can derive a tractable $\mathcal{H}_{\text{eff}}$. Finding efficient Monte Carlo moves also is a non-trivial problem. Of course, since the steps described in the present section the subject has been pushed much further. We direct the interested reader to the reviews quoted in the introduction for more recent work and details about specialized directions.

8.3.6 Cluster methods for quantum lattice models

In Chapter 5 we saw that for many kinds of classical models there were some specialized techniques that could be used to effectively reduce the correlation times between configurations which have been generated. The constraints on direct application of these methods to the classical models which result from the Trotter–Suzuki transformation arise due to the special constraints on which spins may be overturned. Evertz and coworkers (1993) have introduced a form of the cluster algorithm, known as the 'loop algorithm' which addresses these difficulties. It is basically a worldline formulation that employs non-local changes. We have already mentioned that the transformed spin models are equivalent to vertex models in which every bond contains an arrow which points parallel or anti-parallel to a direction along the bond. Thus, in two dimensions each vertex is the intersection of four arrows which obey the constraints that there must be an even number of arrows flowing into or out of a vertex and that they cannot all point either towards or away from the vertex. A 'loop' is then an oriented, closed, non-branching path of bonds all of which contain arrows which point in the same direction. This path may be self-intersecting. A 'flip' then reverses all arrows along the loop. How are the loops chosen? One begins with a randomly chosen bond and looks to the vertex to which it points. There will be two outgoing arrows and one then needs to decide which arrow the loop will follow; this depends upon the model in question. An example of a possible loop configuration is shown in Fig. 8.13. For some models it is also possible to define improved estimators in terms of the 'cluster' properties just as was done for simple classical spin models.

The method was further generalized to arbitrary spin value by Kawashima and Gubernatis (1995) and we refer the reader there (or to Gubernatis and Kawashima (1996), Kawashima (1997)) for more details.

Fig. 8.13 Possible loop structure for the lattice produced using the Trotter formula for a one-dimensional $S = \frac{1}{2}$ Heisenberg model. Note that there are periodic boundary conditions applied in both the real space and Trotter directions. From Gubernatis and Kawashima (1996).

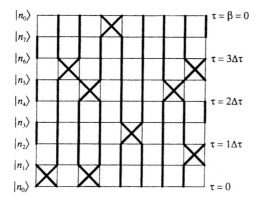

8.3.7 Decoupled cell method

A different approach was proposed by Homma *et al.* (1984, 1986). The system is divided into a set of 'cells' consisting of a center spin i and a symmetric set of surrounding neighbors. The energies of the different states of the cell are solved for as an eigenvalue problem of the cell portion of the Hamiltonian and then Monte Carlo sampling is carried out, i.e. spin-flipping, using relative probabilities of these cell states. The size of the cell is then systematically increased to allow extrapolation to the full lattice.

To examine this method more formally we begin by expressing s_i as the state of the central spin i in a cell, S_i as the state of all other spins in the cell, and $\overline{S}_i$ as the state of all spins outside of the cell. The transition probability between state $S = (s_i, S_i, \overline{S}_i)$ and $S' = (-s_i, S_i, \overline{S}_i)$ is

$$q(S) = \frac{P(S)}{P(S')} = \frac{\langle S| \exp(-\beta \mathcal{H})|S\rangle}{\langle S'| \exp(-\beta \mathcal{H})|S'\rangle} \tag{8.56}$$

and this is then approximated by

$$q^{(v)}(S) = \frac{\langle s_i S_i| \exp(-\beta \mathcal{H}(v, i)|s_i S_i\rangle}{\langle -s_i S_i| \exp(-\beta \mathcal{H}(v, i)| - s_i S_i\rangle} \tag{8.57}$$

where $\mathcal{H}(v, i)$ is the cell Hamiltonian for a cell of size v. The transition probability is then simply

$$W_{DC}(-s_i \rightarrow s_i) = \max[1, q^{(v)}(S_i)]. \tag{8.58}$$

This procedure has been used successfully for a number of different quantum spin systems, but at very low temperatures detailed balance begins to break down and the specific heat becomes negative. A modified version of the decoupled cell method was introduced by Miyazawa *et al.* (1993) to remedy this problem. The improvement consists of dividing the system into overlapping cells such that every spin is at the center of some cell and then using all cells which contain spin i to calculate the flipping probability instead of just one cell in which the ith spin was the center. Miyazawa and Homma (1995) provide a nice overview of the enhanced method and describe a study of the J_1–J_2 model using this approach The decoupled cell method was also used to study the quantum XY-model on a triangular lattice using systems as large as 45×45 and seven-spin cells. Typically 10^4 Monte Carlo steps were used for equilibration and between 10^4 and 8×10^4 were used for averaging. Both groundstate properties and temperature-dependent thermal properties were studied.

8.3.8 Handscomb's method

An alternative method with a completely different philosophy was suggested by Handscomb (1962, 1964). Although it has been used for a rather limited range of problems, we mention it here for completeness. For simplicity, we describe this approach in terms of a simple linear $S = \frac{1}{2}$ Heisenberg chain of N spins whose Hamiltonian we re-express in terms of permutation operators $E(i,j) = (1 + \hat{S}_i \cdot \hat{S}_j)/2$

$$\mathcal{H} = -\mathcal{J} \sum_{i=1}^{N} E(i, i+1) + \tfrac{1}{2}\mathcal{J}N. \tag{8.59}$$

The exponential in the partition function is then expanded in a power series in $\beta\mathcal{H}$ to yield

$$Z = \sum_{n=0}^{\infty} \mathrm{Tr}\{(\beta\mathcal{H})^n\}$$

$$= \sum_{n=0}^{\infty} \sum_{C_n} \frac{1}{n!} \mathrm{Tr}\{\mathcal{H}_{i_1} \ldots \mathcal{H}_{i_n}\} \tag{8.60}$$

$$= \sum_{n=0}^{\infty} \sum_{C_n} \frac{K^n}{n!} \mathrm{Tr}\{P(C_n)\}$$

where $K = \beta\mathcal{J}$ and the second sum is over all possible products $P(C_n)$ with n operators $E(i, i+1)$. The distribution function can then be expressed as

$$\pi(C_n) = \frac{K^n}{n!} \mathrm{Tr}\{P(C_n)\}. \tag{8.61}$$

The Monte Carlo process then begins with an arbitrary sequence of permutation operators. A trial step then consists of either adding an operator to a randomly chosen place in the sequence or deleting a randomly chosen operator from the sequence subject to the condition of detailed balance,

$$P(C_{n+1} \to C_n)\pi(C_{n+1}) = P(C_n \to C_{n+1})\pi(C_n), \tag{8.62}$$

where P_i is the probability of choosing an operator. This approach has been successfully applied to several quantum Heisenberg models by Lyklema (1982), Lee *et al.* (1984), Gomez–Santos *et al.* (1989) and Manousakis and Salvador (1989). Studies of Heisenberg chains used 2×10^5 Monte Carlo steps for equilibration and as many as 5×10^6 Monte Carlo steps for statistical averaging. Lee *et al.* (1984) have modified the approach by shifting the zero of the energy with the result that only terms with an even number of operators give non-zero trace, a modification which helps to largely overcome the minus sign problem in antiferromagnetic quantum Heisenberg models studied by this method. Note that this approach does not make the problem trivial; the study of 32×32 square lattice systems still required 6×10^6 Monte Carlo steps! Sandvik and Kurkijärvi (1991) later introduced a further generalization which is applicable to any spin length.

8.3.9 Fermion determinants

Since it is so hard to deal with fermionic degrees of freedom in quantum Monte Carlo calculations directly, it is tempting to seek methods where one integrates over fermionic degrees of freedom analytically, at the expense of having to simulate a problem with a much more complicated Hamiltonian (Blanckenbecler *et al.*, 1981). This route is, for instance, followed in simulations dealing with lattice gauge theory, see Chapter 11, where one has to deal with a partition function

$$Z = \int \mathcal{D}A_\mu \mathcal{D}\bar{\Psi}\mathcal{D}\Psi \exp[S(A_\mu, \bar{\Psi}, \Psi)] \tag{8.63}$$

where A_μ (μ denotes Cartesian coordinates in the four-dimensional Minkowski space) denote the gauge fields, and $\bar{\Psi}$, Ψ stand for the particle fields (indices $f = 1, \ldots, n_f$ for the 'flavors' and $c = 1, \ldots, n_c$ for the 'colors' of these quarks are suppressed). Now quantum chromodynamics (QCD) implies that the action S is bilinear in $\bar{\Psi}$, Ψ and hence can be written as ($\hat{M}$ is an operator that need not be specified here)

$$S(A_\mu, \bar{\Psi}, \Psi) = \frac{1}{k_B T} \mathcal{H}_0(A_\mu) - \sum_{i=1}^{n_f} \bar{\Psi}\hat{M}\Psi. \tag{8.64}$$

Here we have written the part of the action that depends on gauge fields only as $(1/k_B T)\mathcal{H}_0$, to make the analogy of QCD with statistical mechanics explicit. Note that this formulation is already approximate, since one uses one-component fields (so-called 'staggered fermions' rather than four-component Dirac spinors) here. Now it is well known that the path integration over the fermionic fields (remember these are anticommuting variables) can be integrated out to yield

$$Z = \int \mathcal{D}A_\mu (\det \hat{M})^{n_F} \exp[-\mathcal{H}_0(A_\mu)/k_B T]$$

$$= \int \mathcal{D}A_\mu \exp[-\mathcal{H}_{\text{eff}}(A_\mu)/k_B T]$$

$$\mathcal{H}_{\text{eff}}(A_\mu) = \mathcal{H}_0(A_\mu) - \frac{n_f}{2} \ln[\det(\hat{M}^+\hat{M})]. \tag{8.65}$$

In the last step $(\det \hat{M})$ was replaced by $[\det(\hat{M}^+\hat{M})]^{1/2}$, provided the determinant is positive-definite. Unfortunately, this condition is satisfied only in special cases with 'particle-hole' symmetry, e.g. QCD in a vacuum or the simplest Hubbard model at half-filling! While $\mathcal{H}_0(A_\mu)$ in the lattice formulation of QCD is local, see Chapter 11, the above determinant introduces a non-local interaction among the A_μs.

In condensed matter problems such as the Hubbard Hamiltonian this method does not work directly, since in addition to the bilinear term in the fermion operators $t\hat{c}^+_{i\sigma}\hat{c}_{j\sigma}$ (describing hopping of an electron with spin $\sigma = \uparrow, \downarrow$ from site i to site j) one also has the on-site interaction $U\hat{n}_{i\uparrow}\hat{n}_{i\downarrow} = U\hat{c}^+_{i\uparrow}\hat{c}_{i\uparrow}\hat{c}^+_{i\downarrow}\hat{c}_{i\downarrow}$. However, it is still possible to eliminate the fermionic degrees of freedom from the partition function by introducing auxiliary (bosonic) fields. The key element of this step is the relation

$$\int\limits_{-\infty}^{+\infty} e^{-a\phi^2 - b\phi}\,d\phi = \sqrt{\frac{\pi}{a}}e^{-b^2/4a}, \qquad a > 0. \tag{8.66}$$

Thus a variable b appearing quadratic in the argument of an exponential can be reduced to a linear term (the term $b\phi$ on the left-hand side of the above equation) but on the expense of an integration over the auxiliary variable ϕ. This trick then yields for the on-site interaction of the Hubbard model for $U > 0$

$$\exp\left(-\frac{U}{k_\mathrm{B}TP}\sum_{\ell=1}^{N}\hat{n}_{\ell\uparrow}\hat{n}_{\ell\downarrow}\right)$$

$$\propto \prod_{\ell=1}^{N}\int\limits_{-\infty}^{+\infty} d\phi_\ell \, \exp\left[-\frac{Pk_\mathrm{B}T\phi_\ell^2}{2U} - \phi_\ell(\hat{n}_{\ell\uparrow} - \hat{n}_{\ell\downarrow}) - \frac{U(\hat{n}_{\ell\uparrow} + \hat{n}_{\ell\downarrow})}{2k_\mathrm{B}TP}\right]. \tag{8.67}$$

Using this expression in the framework of the Trotter decomposition, one then can carry out the trace over the fermionic degrees of freedom and again obtain a determinant contribution to the effective Hamiltonian that is formulated in terms of the $\{\phi_\ell\}$, the auxiliary boson fields.

Of course, these remarks are only intended to give readers the flavor of the approach, and direct them to the original literature or more thorough reviews (e.g. de Raedt and von der Linden, 1992) for details.

8.4 MONTE CARLO METHODS FOR THE STUDY OF GROUNDSTATE PROPERTIES

For some quantum mechanical many-body problems even understanding the groundstate is a challenge. A famous example (which is of interest for the understanding of high-T_c superconductivity in the CuO_2 planes of these perovskitic materials) is the groundstate of the spin-$\frac{1}{2}$ Heisenberg antiferromagnet on the square lattice. While for the Ising model the problem is trivial – with a nearest neighbor interaction the lattice simply is split in two ferromagnetic sublattices in a checkerboard fashion, on one sublattice spins are up, on the other they are down. This so-called Néel state is not a groundstate of the Heisenberg antiferromagnet.

Various methods have been devised to deal with problems of this kind, e.g. variational Monte Carlo (VMC) methods, Green's function Monte Carlo (GFMC), and projector quantum Monte Carlo (PQMC). In the following, we only sketch some of the basic ideas, following de Raedt and von der Linden (1992).

8.4.1 Variational Monte Carlo (VMC)

The starting point of any VMC calculation is a suitable trial wave function, $|\Psi^T\{m\}\rangle$, which depends on a set of variational parameters $\{m\}$. Using the fact that the problem of Heisenberg antiferromagnets can be related to the hard-core Boson problem, we describe the approach for the latter case. We write (de Raedt and von der Linden, 1992)

$$|\Psi\rangle_{\text{trial}} = \sum_{\Gamma} \exp\left\{-\sum_{ij} \eta_{ij}\Gamma_i\Gamma_j\right\}|\Gamma\rangle, \qquad (8.68)$$

where the summation extends over all real space configurations Γ, with $\Gamma_i = 1$ if site i is occupied and $\Gamma_i = 0$ otherwise. The expectation value for an arbitrary operator $\hat{O}$ is then

$$\langle\hat{O}\rangle = \frac{_{\text{trial}}\langle\Psi|\hat{O}|\Psi\rangle_{\text{trial}}}{_{\text{trial}}\langle\Psi|\Psi\rangle_{\text{trial}}} = \sum_{\Gamma} P(\Gamma)O(\Gamma) = \frac{1}{M}\sum_{\ell=1}^{M} O(\Gamma^{(\ell)}), \qquad (8.69)$$

with

$$O(\Gamma) = \sum_{\Gamma}\langle\Gamma|\hat{O}|\Gamma'\rangle \exp\left\{-\sum_{ij}\eta_{ij}(\Gamma_i'\Gamma_j' - \Gamma_i\Gamma_j)\right\}. \qquad (8.70)$$

The Markov chain of real space configurations is denoted in Eqn. (8.69) as $\Gamma^{(1)}, \Gamma^{(2)}, \ldots, \Gamma^{(M)}$, M being the total number of configurations over which is sampled. Thus one can use an importance sampling method here, not with a thermal probability density $Z^{-1}\exp(-\mathcal{H}^{\text{eff}}/k_B T)$ but with a probability density $P(\Gamma)$ given as

$$P(\Gamma) = (Z')^{-1}\exp\left(-2\sum_{ij}\eta_{ij}\Gamma_i\Gamma_j\right), \qquad Z' = \sum_{\Gamma}\exp\left(-2\sum_{ij}\eta_{ij}\Gamma_i'\Gamma_j'\right). \qquad (8.71)$$

The energy is calculated using Eqn. (8.69) for the Hamiltonian $\mathcal{H}$ and it is minimized upon the variational parameters $\{\eta_{ij}\}$. Of course, in order that this scheme is tractable, one needs a clever ansatz with as few such parameters as possible. A short range interaction (SR) corresponds to a wave function proposed a long time ago by Hulthén (1938), Kasteleijn (1952) and Marschall (1955):

$$\eta_{ij}^{SR} = \begin{cases} \infty, & \text{if } i = j \text{ (hard-core on-site interaction)} \\ \eta, & \text{if } i, j \text{ are nearest neighbors} \\ 0, & \text{otherwise.} \end{cases} \tag{8.72}$$

The variational principle of quantum mechanics implies $\langle \mathcal{H} \rangle \geq E_0$, the groundstate energy, as is well known. Therefore, the lower energy a trial wave function $|\Psi\rangle_{\text{trial}}$ yields the closer one can presumably approximate the true groundstate. It turns out that lower energies are found when one replaces the 'zero' in the last line by a long range part (Horsch and von der Linden, 1988; Huse and Elser, 1988),

$$\eta_{ij}^{LR} = \alpha |\mathbf{r}_i - \mathbf{r}_j|^{-\beta}, \tag{8.73}$$

if i, j are more distant than nearest neighbors, and α, β then are additional varational parameters. All these trial wave functions lead to long range order for the two-dimensional Heisenberg antiferromagnet which is more complicated than the simple Néel state, namely the so-called 'off-diagonal long range order' (ODLRO). Another famous trial function, the 'resonant valence bond' state (RVB) originally proposed by Liang et al. (1988), corresponds to the choice (Doniach et al. (1988); p is another variational parameter)

$$\eta_{ij} = p \ln(|\mathbf{r}_i - \mathbf{r}_j|) \tag{8.74}$$

in the case where (incomplete) long range order of Néel type is admitted. Also other types of RVB trial functions exist (Liang et al. (1988)) which lead only to a groundstate of 'quantum liquid' type with truly short range antiferromagnetic order.

Problem 8.2 Show that the order parameter $\hat{\mathbf{M}} = \sum_i \hat{\mathbf{S}}_i$ of a quantum Heisenberg ferromagnet ($\mathcal{H} = -J \sum_{\langle i,j \rangle} \hat{\mathbf{S}}_i \cdot \hat{\mathbf{S}}_j$, for spin quantum number $S = \frac{1}{2}$, $J > 0$) commutes with the Hamiltonian. Show that the staggered magnetization (order parameter of the Néel state) does not commute with the Hamiltonian of the corresponding antiferromagnet ($J < 0$). Interpret the physical consequences of these results.

Problem 8.3 Transform the Heisenberg antiferromagnet on the square lattice, $\mathcal{H} = J \sum_{\langle i,j \rangle} \mathbf{S}_i \cdot \mathbf{S}_j$, into the hard-core boson Hamiltonian, $\mathcal{H} = -J \sum_{\langle i,j \rangle} \hat{b}_i^+ \hat{b}_j + J \sum_{\langle i,j \rangle} \hat{n}_i \hat{n}_j + E_0$, by using the transformations $\hat{S}_i^+ = \hat{S}_i^x + i\hat{S}_i^y = \hat{b}_i^+$, $\hat{S}_i^- = \hat{S}_i^x - i\hat{S}_i^y = \hat{b}_i$, and $\hat{S}_i^z = \frac{1}{2} - \hat{b}_i^+ \hat{b}_i$, with the hard-core constraint $\hat{b}_i^{+2} = 0$ and $\hat{b}_i = e_i \hat{b}_i$, with $e_i = 1$ on sublattice 1, $e_i = -1$ on sublattice 2, $\hat{n}_i = \hat{b}_i^+ \hat{b}_i$. Show that $E_0 = -J(N - N_b)$, where N is the number of spins and N_b is related to the z-component of the total magnetization, $N_b = N/2 - S_0^z$ ($N_b = \sum_i \langle \hat{n}_i \rangle$ is the total number of bosons).

8.4.2 Green's function Monte Carlo methods (GFMC)

The basic idea of GFMC (originally used to study the groundstate of the interacting electron gas by Ceperley and Alder (1980); it has also been extended to study the two-dimensional Heisenberg antiferromagnet, Trivedi and Ceperley (1989)) is the repeated application of the Hamiltonian $\mathcal{H}$ to an almost arbitrary state of the system, in order to 'filter out' the groundstate component. To do this, one carries out an iterative procedure

$$|\Psi^{(n+1)}\rangle = [1 - \tau(\mathcal{H} - \hbar\omega)]|\Psi^{(n)}\rangle = \hat{G}|\Psi^{(n)}\rangle, \qquad (8.75)$$

where we have written down the nth step of the iteration, and $\hbar\omega$ is a guess for the groundstate energy. Since $\hat{G}$ can be viewed as the series expansion of the imaginary time evolution operator $\exp[-\tau(\mathcal{H} - \hbar\omega)]$ or of the propagator $[1 + \tau(\mathcal{H} - \hbar\omega)]^{-1}$ for small steps of imaginary time τ, the notion of a Green's function for $\hat{G}$ becomes plausible.

Now the iteration converges to the groundstate only if $\tau < 2/(E_{max} - \hbar\omega)$, E_{max} being the highest energy eigenvalue of $\mathcal{H}$, which shows that GFMC is applicable only if the spectrum of energy eigenvalues is bounded. In addition, this condition implies that τ has to decrease as $1/N$ because $E_{max} - E_o \propto N$. Therefore, one needs a large number of iterations with increasing system size.

In order to realize Eqn. (8.75), one expands the many-body wave function $|\Psi\rangle$ in a suitable set of many-body basis states $|R\rangle$,

$$|\Psi\rangle = \sum_R \Psi(R)|R\rangle \qquad (8.76)$$

which must be chosen such that the coefficients $\Psi(R)$ are real and non-negative, so that they can be regarded as probability densities. In the hard-core boson problem described above (Problem 8.3), one can write explicitly

$$|R\rangle = \prod_{\ell=1}^{N_b} \hat{b}_{\mathbf{r}_\ell}^+|0\rangle \qquad (8.77)$$

where $|0\rangle$ is a state with no bosons, while $\hat{b}_{\mathbf{r}_\ell}^+$ creates a boson at site $\mathbf{r}_\ell$. Thus R stands symbolically for the set $\{\mathbf{r}_\ell\}$ of lattice sites occupied by bosons. In this representation, the iteration Eqn. (8.75) reads

$$\Psi^{(n+1)}(R) = \sum_{R'} G(R, R')\phi^{(n)}(R'), \qquad (8.78)$$

where $G(R, R')$ are the matrix elements of $\hat{G}$ propagating configuration R' to R,

$$G(R, R') = \langle R|[1 - \tau(\mathcal{H} - \hbar\omega)]|R'\rangle = \begin{cases} 1 - \tau[U(R) - \hbar\omega] & \text{if } R = R' \\ \tau\mathcal{J}/2 & \text{if } R \in N(R') \\ 0 & \text{otherwise.} \end{cases}$$

$$(8.79)$$

Here $U(R) = \langle R|\mathcal{H}_{pot}|R\rangle$ is the expectation value of the potential energy of this hard-core boson Hamiltonian, and the set $N(R')$ contains all those configurations that can be obtained from R' by moving one of the bosons to any of the available nearest neighbor positions.

In order to introduce Monte Carlo sampling techniques into this iteration scheme, one decomposes $G(R, R')$ into a matrix $P(R, R')$ and a residual weight $W(R')$, $G(R, R') = P(R, R')W(R')$ such that

$$\sum_R P(R, R') = 1 \text{ and } P(R, R') \geq 0. \tag{8.80}$$

Starting with an initial state $|\phi^{(0)}\rangle$, the probability density after n iterations becomes

$$\begin{aligned}
\phi^{(n)}(R) &= \langle R|\hat{G}^n|\phi^{(0)}\rangle \\
&= \sum_{R_0, R_1 \ldots R_n} \delta_{R,R_n} W(R_{n-1})W(R_{n-2})\ldots W(R_0) \\
&\quad \times P(R_n, R_{n-1})P(R_{n-1}, R_{n-2})\ldots P(R_1, R_0)\phi^{(0)}(R_0). \tag{8.81}
\end{aligned}$$

One defines an n-step random walk on the possible configurations R. With probability $\phi^{(0)}(R_0)$ the Markov chain begins with configuration R_0 and the random walk proceeds as $R_0 \to R_1 \to R_2 \to \ldots R_n$. The transition probability for the move $R_\ell \to R_{\ell+1}$ is given by $P(R_{\ell+1}, R_\ell)$. For each walk the cumulated weight is

$$W^{(n)} = \prod_{\ell=0}^{n-1} W(R_\ell). \tag{8.82}$$

Since the probability for one specific walk is $\prod_{\ell=1}^{n} P(R_\ell, R_{\ell-1})\phi^{(0)}(R_0)$, one finds that the desired wave function can be constructed as the mean value of the weights $W_k^{(n)}$ averaged over M independent walks labeled by index k,

$$\phi^{(n)}(R) = \lim_{M\to\infty} \frac{1}{M} \sum_{k=1}^{M} W_k^{(n)} \delta_{R,R_{n,k}}. \tag{8.83}$$

As it stands, the algorithm is not very practical since the variance of the estimates increases exponentially with the number of iterations n. However, one can reduce the variance by modifying the scheme through the introduction of a 'guiding wave function' $|\Psi_G\rangle$ (Schmidt and Kalos, 1984) which leads to a sort of importance sampling in the iteration process. However, this technique as well as other techniques to reduce the variance (Trivedi and Ceperley, 1989), are too specialized to be treated here.

We conclude this section by comparing the results for the order parameter m of the nearest neighbor Heisenberg antiferromagnet on the square lattice (in a normalization where $m = \frac{1}{2}$ for the Néel state): while Eqn. (8.60) yields $m = 0.42$ (Huse and Elser, 1988), Eqn. (8.50) yields $0.32 \leq m \leq 0.36$ (Horsch and von der Linden, 1988; Huse and Elser, 1988; Trivedi and Ceperley, 1989), GFMC yields $0.31 \leq m \leq 0.37$ (Trivedi and Ceperley, 1989), while grand canonical 'worldline' quantum Monte Carlo (which is based on the Trotter formulation, similar as described in the previous section, and in the end uses an extrapolation to $T \to 0$) yields $m = 0.31$ (Reger and Young, 1988).

8.5 CONCLUDING REMARKS

In this chapter, we could not even attempt to cover the field exhaustively but rather tried to convey to the reader the flavor of what can be accomplished and how it is done. Of course, many recent variations of the technique have not been described at all, though they are quite important to deal with more and more problems of solid state physics (such as lattice dynamics beyond the harmonic approximation, electron–phonon coupling, spin–phonon coupling, magnetism, superconductivity, magnetic impurities in metals, hydrogen and other light interstitials in metals, tunneling phenomena in solids, hydrogen-bonded crystals like ice, HF, HCl etc.). One particularly interesting recent development has not been dealt with at all, namely the study of quantum dynamical information. As is well known, Monte Carlo sampling readily yields correlations in the 'Trotter direction', i.e. in imaginary time, $\langle \hat{A}(0)\hat{A}(\tau)\rangle$. If we could undo the Wick rotation in the complex plane ($it/\hbar \rightarrow \tau$) the propagator $\exp(-\tau\mathcal{H})$ would again become the quantum mechanical time evolution operator $\exp(-it\mathcal{H}/\hbar)$. If exact information on $\langle \hat{A}(0)\hat{A}(\tau)\rangle$ were available, one could find $\langle \hat{A}(0)\hat{A}(t)\rangle$ by analytic continuation; however, in practice this is extremely difficult to do directly because of statistical errors. Gubernatis $et\ al.$ (1991) have shown that using quantum Monte Carlo in conjunction with the maximum entropy method (Skilling, 1989) one can find $\langle \hat{A}(0)\hat{A}(t)\rangle$ from $\langle \hat{A}(0)\hat{A}(\tau)\rangle$ in favorable cases.

REFERENCES

Barma, M. and Shastry, B. S. (1977), Phys. Lett. **61A**, 15.

Barma, M. and Shastry, B. S. (1978), Phys. Rev. B **18**, 3351.

Batchelder, D. N., Losee, D. L. and Simmons, R. O. (1968), Phys. Rev. **73**, 873.

Berne, B. J. and Thirumalai, D. (1986), Ann. Rev. Phys. Chem. **37**, 401.

Bethe, H. A. (1931), Z. Phys. **71**, 205.

Blanckenbecler, R., Scalapino, D. J. and Sugar, R. L. (1981), Phys. Rev. D **24**, 2278.

Bonner, J. C. and Fisher, M. E. (1964), Phys. Rev. **135**, A640.

Ceperley, D. M. (1995), Rev. Mod. Phys. **67**, 279.

Ceperley, D. M. (1996), in *Monte Carlo and Molecular Dynamics of Condensed Matter Systems*, eds. K. Binder and

G. Ciccotti (Societa Italiana di Fisica, Bologna) p. 445.

Ceperley, D. M. and Alder, B. J. (1980), Phys. Rev. Lett. **45**, 566.

Ceperley, D. M. and Kalos, M. H. (1979), *Monte Carlo Methods in Statistical Physics*, ed. K. Binder (Springer, Berlin) p. 145.

Cullen, J. J. and Landau, D. P. (1983), Phys. Rev. B. **27**, 297.

Dadobaev, G. and Slutsker, A. I. (1981), Soviet Phys. Solid State **23**, 1131.

de Raedt, H. and Lagendijk, A. (1981), Phys. Rev. Lett. **46**, 77.

de Raedt, H. and Lagendijk, A. (1985), Phys. Rep. **127**, 233.

de Raedt, H. and von der Linden, W. (1992), in *The Monte Carlo Method in Condensed Matter Physics*, ed. K. Binder (Springer, Berlin) p. 249.

Doll, J. D. and Gubernatis, J. E. (1990), (eds.) *Quantum Simulations* (World Scientific, Singapore).

Doniach, S., Iuni, M., Kalmeyer, V., and Gabay, M. (1988), Europhys. Lett. **6**, 663.

Evertz, H. G., Lana, G. and Marcu, M. (1993), Phys. Rev. Lett. **70**, 875.

Evertz, H. G. and Marcu, M. (1993), in *Computer Simulations Studies in Condensed Matter Physics VI*, eds. D. P. Landau, K. K. Mon and H.-B. Schüttler (Springer, Heidelberg).

Feynman, R. P. (1953), Phys. Rev. **90**, 1116; **91**, 1291; **91**, 1301.

Feynman, R. P. and Hibbs, A. R. (1965), *Quantum Mechanics and Path Integrals* (McGraw Hill, New York).

Gillan, M. J. and Christodoulos, F. (1993), Int. J. Mod. Phys. **C4**, 287.

Gomez-Santos, G. Joannopoulos, J. D., and Negele, J. W. (1989), Phys. Rev. B **39**, 4435.

Gubernatis, J. E. and Kawashima, N. (1996), in *Monte Carlo and Molecular Dynamics of Condensed Matter Systems*, eds. K. Binder and G. Ciccotti (Societa Italiana di Fisica, Bologna) p. 519.

Gubernatis, J. E., Jarrell, M., Silver, R. N. and Sivia, D. S. (1991), Phys. Rev. B **44**, 6011.

Handscomb, D. C. (1962), Proc. Cambridge Philos. Soc. **58**, 594.

Handscomb, D. C. (1964), Proc. Cambridge Philos. Soc. **60**, 115.

Hirsch, J. E., Sugar, R. L., Scalapino, D. J. and Blancenbecler, R. (1982), Phys. Rev. B **26**, 5033.

Homma, S. Matsuda, H. and Ogita, N. (1984), Prog. Theor. Phys. **72**, 1245.

Homma, S. Matsuda, H. and Ogita, N. (1986), Prog. Theor. Phys. **75**, 1058.

Horsch, R. and von der Linden, W. (1988), Z. Phys. B **72**, 181.

Hubbard, J. (1963), Proc. Roy. Soc. London A **276**, 238.

Hulthén, J. L. (1938), Ark. Mat. Astron. Fjs. A **26**, 1.

Huse, D. A. and Elser, V. (1988), Phys. Rev. Lett. **60**, 2531.

Kalos, M. H. (1984), (ed.) *Monte Carlo Methods in Quantum Problems* (D. Reidel, Dordrecht).

Kasteleijn, P. W. (1952), Physica **18**, 104.

Kawashima, N. (1997), in *Computer Simulations Studies in Condensed Matter Physics IX*, eds. D. P. Landau, K. K. Mon and H.-B. Schüttler (Springer, Heidelberg).

Kawashima, N. and Gubernatis, J. E. (1995), Phys. Rev. E **51**, 1547.

Lee, D. H., Joannopoulos, J. D., and Negele, J. W. (1984), Phys. Rev. B **30**, 1599.

Liang, S., Doucot, B., and Anderson, P. W. (1988), Phys. Rev. Lett. **61**, 365.

Loh, E. Y., Gubernatis, J. E., Scalettar, R. T., White, S. R., Scalapino, D. J., and Sugar, R. L. (1990), Phys. Rev. B **41**, 9301.

Lyklema, J. W. (1982), Phys. Rev. Lett. **49**, 88.

Manousakis, E. and Salvador, R. (1989), Phys. Rev. B **39**, 575.

Marshall, W. (1955), Proc. Roy. Soc. London A **232**, 64.

Martonak, P., Paul, W. and Binder, K. (1998), Phys. Rev. E **57**, 2425.

Marx, D. and Nielaba, P. (1992), Phys. Rev. A **45**, 8968.

Marx, D. and Wiechert, H. (1996), Adv. Chem. Phys. **95**, 213.

Marx, D., Opitz, O., Nielaba, P., and Binder, K. (1993), Phys. Rev. Lett. **70**, 2908.

Miyazawa, S. and Homma, S. (1995), in *Computer Simulations Studies in Condensed Matter Physics VIII*, eds. D. P. Landau, K. K. Mon, and H.-B. Schüttler (Springer, Heidelberg).

Miyazawa, S., Miyashita, S., Makivic, M. S., and Homma, S. (1993), Prog. Theor. Phys. **89**, 1167.

Müser, M. H. (1996), Mol. Simulation **17**, 131.

Müser, M. H. , Nielaba, P., and Binder, K. (1995), Phys. Rev. B **51**, 2723.

Nielaba, P. (1997), in *Annul Reviews of Computational Physics V*, ed. D. Stauffer (World Scientific, Singapore) p. 137.

Okabe, Y. and Kikuchi, M. (1986), Phys. Rev. B **34**, 7896.

Onsager, L. (1944), Phys. Rev. **65**, 117.

Ovchinnikov, A. A. (1973), Sov. Phys. JETP **37**, 176.

Reger, J. D. and Young, A. P. (1988), Phys. Rev. B **37**, 5978.

Rutledge, G. C., Lacks, D. J., Martonak, R. and Binder, K. (1998), J. Chem. Phys. **108**, 10274.

Sandvik, A. W. and Kurkijärvi, J. (1991), Phys. Rev. B **43**, 5950.

Schmidt, K. E. and Ceperley, D. M. (1992) in *The Monte Carlo Method in Condensed Matter Physics*, ed. K. Binder (Springer, Berlin) p. 205.

Schmidt, K. E. and Kalos, M. H. (1984) in *Applications of the Monte Carlo Method in Statistical Physics*, ed. K. Binder (Springer, Berlin) p. 125.

Skilling, J. (1989), (ed.) *Maximum Entropy and Bayesian Methods* (Kluwer, Dordrecht).

Suzuki, M. (1966), J. Phys. Soc. Jpn. **21**, 2274.

Suzuki, M. (1971), Prog. Theor. Phys. **46**, 1337.

Suzuki, M. (1976a), Commun. Math. Phys. **51**, 183.

Suzuki, M. (1976b), Prog. Theor. Phys. **56**, 1454.

Suzuki, M. (1986), (ed.) *Quantum Monte Carlo Methods* (Springer, Berlin).

Suzuki, M. (1992), (ed.) *Quantum Monte Carlo Methods in Condensed Matter Physics* (World Scientific, Singapore).

Trivedi, N. and Ceperley, D. M. (1989), Phys. Rev. B **40**, 2737.

Trotter, H. F. (1959), Proc. Am. Math. Soc. **10**, 545.

9   Monte Carlo renormalization group methods

9.1 INTRODUCTION TO RENORMALIZATION GROUP THEORY

The concepts of scaling and universality presented in the second chapter of this book can be given concrete foundation through the use of renormalization group (RG) theory. The fundamental physical ideas underlying RG theory were introduced by Kadanoff (Kadanoff, 1971) in terms of a simple coarse-graining approach, and a mathematical basis for this viewpoint was completed by Wilson (Wilson, 1971). Kadanoff divided the system up into cells of characteristic size ba where a is the nearest neighbor spacing and $ba < \xi$ where ξ is the correlation length of the system (see Fig. 9.1). The singular part of the free energy of the system can then be expressed in terms of cell variables instead of the original site variables, i.e.

$$F_{\text{cell}}(\tilde{\varepsilon}, \tilde{H}) = b^d F_{\text{site}}(\varepsilon, H), \tag{9.1}$$

where $\varepsilon = |1 - T/T_c|$, $\tilde{\varepsilon}$ and $\tilde{H}$ are cell variables, and d is the lattice dimensionality. This is merely a statement of the homogeneity of the free energy and yields the scaling expression

$$F(\lambda^{a_T}\varepsilon, \lambda^{a_T}H) = \lambda F(\varepsilon, H). \tag{9.2}$$

According to formal RG (renormalization group) theory the initial Hamiltonian is transformed, or *renormalized* to produce a new Hamiltonian; this process may be repeated many times and the resultant Hamiltonians, which may be given a characteristic index n to describe the number of times the transformation has been applied, are related by

$$\mathcal{H}^{(n+1)} = R_b \mathcal{H}^{(n)}. \tag{9.3}$$

The renormalization group operator R_b acts to reduce the number of degrees of freedom by b^d where b is the spatial rescaling factor and d the spatial dimensionality. (It is perhaps worthwhile pointing out that this generally does not constitute a true group theory since R_b typically has no inverse.) Note that the renormalized Hamiltonian may contain terms (i.e. additional couplings) which were not originally present and which appear only as a result of the renormalization transformation. Of course, the partition function Z must not be changed by this process since it is only being expressed in

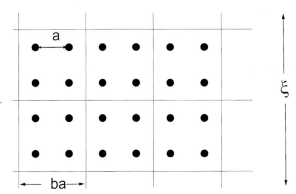

Fig. 9.1 Schematic subdivision of the lattice into cells. The lattice constant is a, the rescaling factor is b, and the correlation length is denoted as ξ.

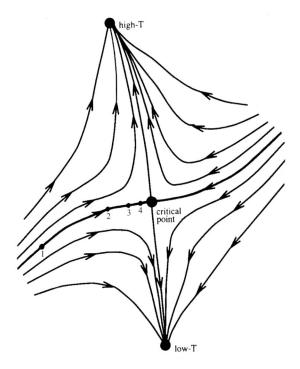

Fig. 9.2 Schematic RG flow diagram in a two-dimensional parameter space. The heavy curve represents the critical hypersurface. Point 1 is the critical value and the other points labeled show the flow towards the fixed point (heavy filled circle).

terms of new variables. After the transformation has been applied many times the Hamiltonian has reached an invariant or 'fixed point' form $\mathcal{H}^*$ and no longer changes, i.e.

$$\mathcal{H}^* = \mathbf{R}_b \mathcal{H}^*. \tag{9.4}$$

This means that the Hamiltonian of a system at its critical point will move, or 'flow', *towards* the fixed point Hamiltonian upon successive applications of the RG transformation until the form no longer changes. Conversely, if the system is not initially at a critical point, upon renormalization the Hamiltonian will 'flow' *away* from the fixed point of interest (see Fig. 9.2). In a study of an Ising-type Hamiltonian for $T > T_c$ one ultimately reaches a trivial fixed point corresponding to the ideal paramagnet at $T \to \infty$. (After a few rescalings the block size ab^n exceeds ξ and the different blocks are then uncorrelated.) For $T < T_c$ the flow is to a different, zero temperature fixed point. The Hamiltonian is written in the same general framework at each application of the transformation, e.g. an Ising-type Hamiltonian

$$\mathcal{H}/k_B T = K_1 \sum_i S_i + K_2 \sum_{\langle i,j \rangle} S_i S_j + K_3 \sum_{\langle i,j,k \rangle} S_i S_j S_k$$
$$+ K_4 \sum_{\langle i,j,k,l \rangle} S_i S_j S_k S_l + \cdots . \tag{9.5}$$

The space of coupling constants $\{K_1, K_2, \ldots\}$ is then the space in which the flow is considered. A model Hamiltonian can generally be extended to include other interactions such that an entire hypersurface of critical points is produced; in all cases in which we begin on the critical hypersurface, the system Hamiltonian should move, or 'flow', towards the fixed point of interest. When a system is at a multicritical point, it will flow towards a new 'fixed point' instead of towards the critical fixed point. Close to the multicritical point there may be complex crossover behavior and the system may at first appear to flow towards one fixed point, but upon further application of the RG transformation it begins to flow towards a different fixed point. Thus, RG theory very nicely illuminates the universality principle of critical phenomena: each type of criticality is controlled by a particular fixed point of the RG transformation that is considered (Fisher, 1974a).

Near the fixed point one can generally linearize the problem so that the Hamiltonian $\mathcal{H}'$ is related to the fixed point form by

$$\mathcal{H}' = R_b[\mathcal{H}^*] + hLQ = \mathcal{H}^* + hLQ + \cdots , \tag{9.6}$$

where the linear operator L has the eigenvalue equation

$$LQ_j = \lambda_j Q_j \tag{9.7}$$

with λ_j being the eigenvalue and Q_j the eigenvector. In terms of the spatial rescaling factor b

$$\lambda_j = b^{y_j}, \tag{9.8}$$

where y_j is termed an 'exponent eigenvalue' which can be related to the usual critical exponents, as we shall see later. We can then write an expression for the transformed Hamiltonian in terms of these eigenvalues

$$\mathcal{H}' = \mathcal{H}^* + \sum h_j \lambda_j Q_j + \cdots . \tag{9.9}$$

From this equation we can immediately write down recursion relations for the h_j

$$h_j^{(k+1)} \approx \lambda_j h_j^{(k)} \tag{9.10}$$

which may be solved to give values for the eigenvalues. The free energy in terms of the original and renormalized variables is again unchanged:

$$f(h_1, h_2, h_3, \ldots) \approx b^{-d} f(b^{\lambda_1} h_1, b^{\lambda_2} h_2, \ldots) \qquad (9.11)$$

where we may identify $h_1 = k_1 t$, $h_2 = k_2 H$, etc. Choosing b so that $b^{\lambda_1} t = 1$, we can rewrite this equation with k_1, k_2 constants

$$f(t, H, h_3) \approx t^{d/\lambda_1} f(k_1, k_2, H/t^{\lambda_1/\lambda_2}, \ldots). \qquad (9.12)$$

Thus, if we identify $d/\lambda_1 = 2 - \alpha$ and $\lambda_2/\lambda_1 = \Delta$, we have 'derived' scaling.

For completeness, we briefly mention the momentum space approach to renormalization group theory. In this case the coarse-graining and rescaling which occurs as part of the RG process is defined in k-space (momentum space instead of real space). In terms of a Landau-like Hamiltonian, the Fourier space form is

$$\mathcal{H}(m) = 1/2 \int dk (k^2 + r_0) |m(k)|^2 + \cdots. \qquad (9.13)$$

A cutoff momentum Λ is then introduced, the k values which lie between Λ and Λ/b are integrated out, and then the variable of integration is rescaled by $k' = bk$. The order parameter is then renormalized and one subsequently repeats the same steps. A perturbation expansion is then realized which leads to recursion relations for the effective interaction parameters. The solution to these equations gives rise to the 'fixed points' in the problem. Perturbation parameters may include the difference in lattice dimensionality from the upper critical dimension $\varepsilon = (d_u - d)$ or the inverse of the number of components of the order parameter n. For simple magnetic systems with isotropic, short range couplings the upper critical dimension is $d_u = 4$ and the leading order estimates for critical exponents are (Wilson and Fisher, 1972):

$$\alpha = \frac{4 - n}{2(n + 8)} \varepsilon + \cdots \qquad \text{where } \varepsilon = 4 - d, \qquad (9.14a)$$

$$\beta = \frac{1}{2} - \frac{3}{2(n + 8)} \varepsilon + \cdots, \qquad (9.14b)$$

$$\gamma = 1 + \frac{(n + 2)}{2(n + 8)} \varepsilon + \cdots. \qquad (9.14c)$$

Of course, for simple models of statistical mechanics higher order expressions have been derived with the consequence that rather accurate estimates for critical exponents have been extracted, see e.g. Brezin et al. (1973) and Zinn-Justin and Le Guillou (1980). A rather sophisticated analysis of the expansions is required in general. Renormalization group theory was used to successfully understand the behavior of the tricritical point by Wegner and Riedel (1973) who showed that Landau theory exponents are indeed correct in three dimensions but that the critical behavior is modified by the presence of logarithmic corrections. Further, a renormalization group analysis of bicritical and related tetracritical points has been carried out by Nelson et al. (1974). While the momentum space RG has yielded fairly accurate results for the critical exponents of the n-vector model, the accuracy that was reached for other problems is far more modest, e.g. universal scaling functions describing the equation of state, or describing the crossover from one universality class to another, typically are available in low-order ε-expansion only, and hence describe real systems qualitatively but not quantitatively. Moreover, the momentum space RG in principle yields information on universal properties only, but neither information on the critical coupling constants (T_c, etc.) nor on critical amplitudes (Chapter 2) is provided. The real space RG can yield this information, and hence we turn to this approach now. This work has been augmented by Monte Carlo simulations which have examined tricritical behavior in the three-dimensional Ising model and explored the four-dimensional phase diagram, i.e. in $H_\parallel, H_\perp, H_\parallel^+, T$ space, of the anisotropic Heisenberg model.

Of course, RG theory is a huge subject with many subtle aspects which can fill volumes (e.g. Domb and Green, 1976). Here we only wish to convey the flavor of the approach to the reader and emphasize those aspects which are absolutely indispensible for understanding the literature which uses Monte Carlo renormalization group methods.

9.2 REAL SPACE RENORMALIZATION GROUP

A number of simple RG transformations have been used with generally good success. By 'simple' we mean that the space of coupling constants that is allowed for is kept low-dimensional: this is an arbitrary and uncontrolled approximation, but it allows us to carry out the calculations needed for the renormalization transformation in a fast and convenient way. One approach is the 'blockspin' transformation in which a $b \times b$ block of spins is replaced by a 'superspin' whose state is determined by the state of the majority of spins in the block. If the number of spins in a block is even, one site in each block is designated as a 'tie-breaker'. Another alternative is the 'decimation' process in which the lattice is renormalized by taking every bth spin in all directions. In a nearest neighbor antiferromagnet a simple majority rule over a (2×2) blockspin would give zero for all blockspins when the system was in the groundstate. Thus a more natural and useful choice is to have the 'blockspins' composed of more complex structures where each block resides on a single sublattice. Examples of several blockspin choices are shown in Fig. 9.3. Note that the $\sqrt{5} \times \sqrt{5}$ transformation rotates the lattice through an angle $\varphi = \pi/4$ (this rotation effect is shown more clearly for the $\sqrt{7} \times \sqrt{7}$ transformation on the right in Fig. 9.3) but preserves the square lattice symmetry. If a second transformation is applied but chosen to rotate the lattice through angle $-\varphi$, the original orientation is recovered. The underlying ideas of RG theory are demonstrated in Fig. 9.4 where we have taken Monte Carlo generated configurations in a spin-$\frac{1}{2}$ Ising model on a 512×512 square lattice with periodic boundaries at three different temperatures near T_c. A $b = 2$ blockspin transformation is applied and then the lattice is rescaled to the original size. At $0.95T_c$ the system rapidly becomes almost completely

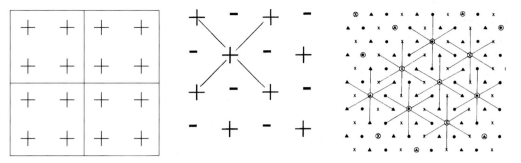

Fig. 9.3 Examples of simple blockspins: (left) (2×2) blockspin arrangement for a ferromagnet; (center) $\sqrt{5} \times \sqrt{5}$ blockspin for a nearest neighbor antiferromagnet in which each spin in a blockspin is on the same sublattice; (right) $\sqrt{7} \times \sqrt{7}$ blockspin for a nearest neighbor antiferromagnet on a triangular lattice in which each spin in a blockspin is on the same sublattice.

ordered under application of the RG transformation. At T_c the system is virtually invariant with successive application of the transformation. Since the initial lattice was finite there is still a finite size effect and the total magnetization is not zero for this particular configuration. At $1.05T_c$ the system is disordered and the renormalized magnetization becomes even smaller. As for the rescaling transformation in Eqn. (9.3), if one could carry this out exactly an increasing number of couplings $\{K_i\}$ in a Hamiltonian like Eqn. (9.5) would be generated. However, in practice, as the rescaling is iterated the space of coupling constants has to be truncated dramatically, and in an analytic approach other uncontrolled approximations may be necessary to relate the new couplings to the old couplings. These latter problems can be avoided with the help of Monte Carlo renormalization group methods which we wish to describe here.

9.3 MONTE CARLO RENORMALIZATION GROUP

9.3.1 Large cell renormalization

The large cell renormalization group transformation was used to study both spin systems (Friedman and Felsteiner, 1977; Lewis, 1977) and the percolation problem (Reynolds *et al.*, 1980). In this discussion we shall consider the method in the context of the two–dimensional Ising model with nearest neighbor coupling only. A system of size $L \times 2L$ is considered and two blockspins, σ_1 and σ_2, are created from application of the majority rule to 'large' cells of size $L \times L$. The blockspins interact with Hamiltonian

$$\mathcal{H} = K'\sigma_1'\sigma_2', \tag{9.15}$$

where the magnitude of the new effective coupling constant K' is determined from

$$\langle \sigma_1'\sigma_2' \rangle = \tanh(qK'). \tag{9.16}$$

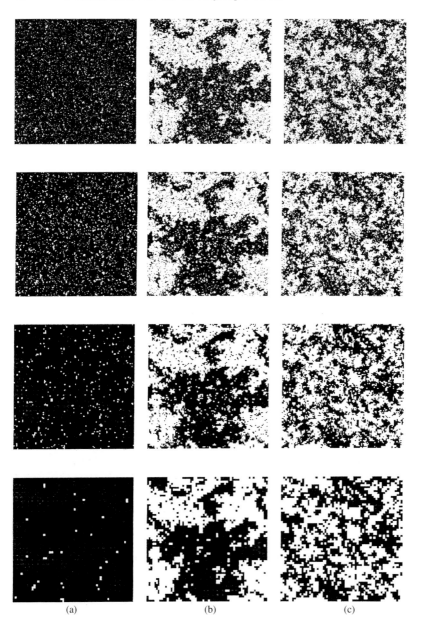

Fig. 9.4 'Snapshots' of the two–dimensional Ising model at: (a) $T = 0.95\,T_c$; (b) $T = T_c$; (c) $T = 1.05\,T_c$. The upper row shows Monte Carlo generated configurations on a 512×512 lattice with periodic boundaries. Successive rows show the configurations after 2×2 blockspin transformations have been applied and the lattices rescaled to their original size.

(a)   (b)   (c)

Note that this corresponds to a transformation with scale factor $b = L$. The thermal eigenvalue y_T is then determined from the expression

$$\frac{dK'}{dK} = L^{y_T}, \qquad (9.17)$$

where the derivative can be calculated via Monte Carlo simulation from averages, i.e.

$$\frac{dK'}{dK} = \langle \sigma_1' \sigma_2' S_1 \rangle - \langle \sigma_1' \sigma_2' \rangle \langle S_1 \rangle. \qquad (9.18)$$

If L is increased with the system held at the critical coupling the estimates for y_T should converge to the correct value of $1/\nu$.

Problem 9.1 Simulate a 16 × 32 Ising square lattice at T_c and use the large cell Monte Carlo renormalization method to estimate the value of the thermal exponent y_T.

9.3.2 Ma's method: finding critical exponents and the fixed point Hamiltonian

The Monte Carlo method was first used within the framework of renormalization group theory by Ma (1976) who applied it to the study of critical exponents in a simple Ising model system. The basic idea of this approach is to determine the behavior of the Hamiltonian upon renormalization and by following the 'flow' towards the fixed point Hamiltonian to study critical exponents. By measuring effective interaction parameters between coarse-grained blocks of spins, one can extract exponent estimates from this information. The method begins by generating a sequence of states. 'Probes' of different sizes are then used to measure interactions by observing how a spin behaves in a given environment. The length of time it takes for a spin to flip in a given environment is a reflection of the interaction parameters as well as the 'local' structure, and by examining different local environments one can produce a set of linear equations that may be solved for the individual interaction constants. This process may be repeated by examining the behavior of 'blockspins', i.e. of a 2 × 2 set of spins whose 'blockspin' value is chosen to be $\tilde{S} = 1$ if a majority of the spins in the block are 1s and $\tilde{S} = -1$ if the majority are −1s. Applying the same procedure outlined above provides a set of interaction parameters at a scale which is twice as large as that defined by the small probe.

The actual implementation demonstrated by Ma was for the Ising model with a set of interaction parameters $\mu = (\mathcal{J}, K, L)$ which represent nearest neighbor, next-nearest neighbor, and four-spin interaction parameters and has much of the flavor of the N-fold way algorithm. The rate of flipping for each spin is determined in the following way. The probability that no spin flips during the time period t' (the 'lifetime' of the state) is $\exp(-\Omega t')$ where Ω is the total transition rate for the entire system. The probability that no spin flips in the initial interval but then flips in the following dt' interval is $\exp(-\Omega t') \times dt'$. The lifetime for a given spin is determined by generating one random number to select a spin and then a second random number to determine the lifetime through $t' = -(\ln x)/\Omega$. The small probe looks at 3 × 3 blocks of spins and determines τ_+ and τ_-, i.e. the lifetimes of the states where the spin is +1 and −1 respectively. The ratio $\tau_+/\tau_- = \exp(H - H')$ gives an equation for $\mathcal{J}, K$, and L. (For example, if all the spins in the probe are $+1$, then $(H - H') = 4(\mathcal{J} + K + L)$.) If the ratio of lifetimes is measured for three different neighbor environments, a set of linear equations is obtained which can be solved to extract each individual interaction parameter. To determine the critical exponent we want to repeat this procedure with the large probe and then construct the matrix $(\partial \mathcal{J}_i'/\partial \mathcal{J}_i)$, the largest eigenvalue of which is $\lambda_T = 2^{1/\nu}$. Unfortunately, in actual practice it proves quite difficult to determine the fixed point Hamiltonian with significant accuracy.

9.3.3 Swendsen's method

Ma's method proved difficult to implement with high accuracy because it was very difficult to calculate the renormalized Hamiltonian accurately enough. A very different approach, which is outlined below, proved to be more effective in finding exponent estimates because it is never necessary to calculate the renormalized couplings. For simplicity, in the discussion in this subsection we shall express the Hamiltonian in the form

$$\mathcal{H} = \sum_\alpha K_\alpha S_\alpha, \tag{9.19}$$

where the S_α are sums of products of spin operators and the K_α are the corresponding dimensionless coupling constants with factors of $-1/kT$ absorbed. Examples of spin products are:

$$S_1 = \sum \sigma_i, \tag{9.20a}$$

$$S_2 = \sum \sigma_i \sigma_j, \tag{9.20b}$$

$$S_3 = \sum \sigma_i \sigma_j \sigma_k. \tag{9.20c}$$

Near the fixed point Hamiltonian $\mathcal{H}^*(K^*)$ the linearized transformation takes the form

$$K_\alpha^{(n+1)} - K_\alpha^* = \sum_\beta T_{\alpha\beta}^*(K_\beta^{(n)} - K_\beta^*), \tag{9.21}$$

where the sum is over all possible couplings. The eigenvalues λ_i of $T_{\alpha\beta}^*$ are related to eigenvalue exponents by

$$\lambda = b^y, \tag{9.22}$$

where the y are in turn related to the usual critical exponents, e.g. $y_T = \nu^{-1}$.

Equations (9.21) and (9.22) are still common to all real space RG methods, and the challenge becomes how to find the matrix elements $T_{\alpha\beta}^*$ at the fixed point in practice. Perhaps the most accurate implementation of real space RG methods has been through the use of Monte Carlo renormalization group (MCRG) methods (Swendsen, 1982). In this approach the elements of the linearized transformation matrix are written in terms of expectation values of correlation functions at different levels of renormalization. Thus,

Table 9.1 *Variation of the thermal eigenvalue exponent for the Ising square lattice with the number of couplings N_c, the number of iterations N_r, and for different lattice sizes. From Swendsen (1982).*

N_r	N_c	$L = 64$	$L = 32$	$L = 16$
1	1	0.912(2)	0.904(1)	0.897(3)
1	2	0.967(3)	0.966(2)	0.964(3)
1	3	0.968(2)	0.968(2)	0.966(3)
1	4	0.969(4)	0.968(2)	0.966(3)
2	1	0.963(4)	0.953(2)	0.937(3)
2	2	0.999(4)	0.998(2)	0.993(3)
2	3	1.001(4)	1.000(2)	0.994(3)
2	4	1.002(5)	0.998(2)	0.984(4)
3	1	0.957(2)	0.936(3)	0.921(5)
3	2	0.998(2)	0.991(3)	1.013(4)
3	3	0.999(2)	0.993(3)	1.020(3)
3	4	0.997(2)	0.987(4)	. . .

$$T_{\alpha\beta} = \frac{\partial K_\alpha^{(n+1)}}{\partial K_\beta^{(n)}}, \tag{9.23}$$

where the elements can be extracted from solution of the chain rule equation

$$\partial\langle S_\gamma^{(n+1)}\rangle/\partial K_{\beta}{}^{(n)} = \sum \left\{ \partial K_\alpha^{(n+1)}/\partial K_\beta^{(n)} \right\} \left\{ \partial\langle S_\gamma^{(n+1)}\rangle/\partial K_\alpha^{(n+1)} \right\}. \tag{9.24}$$

The derivatives can be obtained from correlation functions which can be evaluated by Monte Carlo simulation, i.e.

$$\partial\langle S_\gamma^{(n+1)}\rangle/\partial K_\beta^{(n)} = \langle S_\gamma^{(n+1)} S_\beta^{(n)}\rangle - \langle S_\gamma^{(n+1)}\rangle\langle S_\beta^{(n)}\rangle \tag{9.25}$$

and

$$\partial\langle S_\gamma^{(n)}\rangle/\partial K_\alpha^{(n)} = \langle S_\gamma^{(n)} S_\alpha^{(n)}\rangle - \langle S_\gamma^{(n)}\rangle\langle S_\alpha^{(n)}\rangle. \tag{9.26}$$

The $T_{\alpha\beta}$ matrix is truncated in actual calculations and the number of renormalizations is, of course, limited as well. Results for the estimates for eigenvalues are then examined as a function of the number of couplings N_c used in the analysis and the number of iterations N_r. Exact results are expected only for $N_r \to \infty$ and $N_c \to \infty$, but in practice the convergence to this limit is rather fast. By performing the calculations on different size lattices one can also determine if finite lattice size is playing a role. As a simple example, in Table 9.1 we show data for the thermal eigenvalue exponent for $L \times L$ square lattice Ising models. As the number of iterations increases the exponent rapidly converges to the exact value $y_T = 1.0$, but this is true only as long as at least one additional coupling is generated. Finite size effects also begin to appear slowly and become increasingly important as the iteration number increases.

Experience with other models has shown that in general the convergence is not as rapid as for the two-dimensional Ising model and great care must be used to insure that a sufficient number of couplings and renormalizations have been used. This also means, of course, that often rather large lattices must be used to avoid finite size effects in the renormalized systems.

Problem 9.2 Simulate a 27×27 Ising square lattice ferromagnet at T_c and use Swendsen's method with $b = 3$ to estimate the thermal exponent y_T.

9.3.4 Location of phase boundaries

9.3.4.1 Critical points

How do we determine the location of the critical point using the ideas of MCRG? This may be accomplished by matching correlation functions for transformed and untransformed systems: only at the critical point will they be the same. Finite size effects can be subtle, however, so the preferred procedure is to start with two different lattices which differ in size by the scale factor b to be used in the transformation (see Fig. 9.5). In the vicinity of the critical point we can use a linear approximation to relate the difference between the original and renormalized correlation functions to the distance to the critical point, i.e.

$$\langle S_\alpha^{(n)} \rangle_L - \langle S_\alpha^{(n-1)} \rangle_S = \sum_\beta \left[\frac{\partial \langle S_\alpha^{(n)} \rangle_L}{\partial K_\beta^{(0)}} - \frac{\partial \langle S_\alpha^{(n-1)} \rangle_S}{\partial K_\beta^{(0)}} \right] \delta K_\beta^{(0)}. \qquad (9.27)$$

The predicted 'distance' from the critical coupling $\delta K_\beta^{(0)}$ can be extracted by inverting Eqn. (9.27) for different values of n. Thus, an initial estimate for the critical coupling is chosen and the above process is carried out. The simulation is then repeated at the updated estimate and a check is made to see if this is, in fact, a good value.

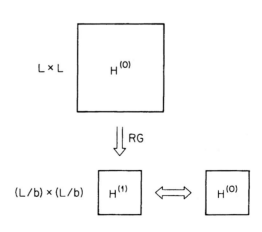

$L \times L$ $H^{(0)}$

RG

$(L/b) \times (L/b)$ $H^{(1)}$ $H^{(0)}$

Fig. 9.5 Schematic view of the two lattice comparison for the determination of the critical temperature.

9.3.4.2 Multicritical points

The methods described above can also be used to investigate multicritical behavior (Griffiths, 1970; Fisher, 1974a). Such studies are usually complicated by the fact that the multicritical point must be located in a two-dimensional parameter space, and this process often involves an iterative procedure. In addition there are usually additional critical eigenvalue exponents due to the presence of additional scaling fields for the multicritical point. This process has been carried out quite carefully by Landau and Swendsen (1986) for the two-dimensional Blume–Capel ferromagnet and for the two-dimensional Ising antiferromagnet with next-nearest neighbor interactions in a magnetic field. Mean field predicts that for certain values of the interactions there is a tricritical point on the phase boundary whereas beyond a certain value the tricritical point is decomposed into a double critical point and a critical endpoint. The MCRG study showed that for quite a wide range of couplings below the predicted critical value there was only an ordinary tricritical point with no indication of the predicted change. The numerical estimates obtained for both the dominant and sub-dominant eigenvalue exponents also remained unchanged with modifications in the couplings and were in good agreement with the predicted values for an ordinary tricritical point. This study strongly suggests that the fluctuations in the two-dimensional model destroy the mean-field behavior and retain the normal tricritical behavior.

9.3.5 Dynamic problems: matching time-dependent correlation functions

The ideas described above can be extended to the consideration of time-dependent properties. The general idea behind this approach is to generate a sequence of states which have been blocked at different levels and compute the correlation functions as functions of time. Then attempt to 'match' these correlation functions at different blocking levels at different times. The relationship between the blocking level and the time at which they match gives the dynamic exponent z. Mathematically this can be expressed by

$$C(N, m, T_2, t) = C(Nb^d, m + 1, T_1, b^z t), \qquad (9.28)$$

where the critical temperature is given by $T_1 = T_2 = T_c$. It is necessary to use two different size lattices for the comparison so that there are the same number of spins in the large lattice after the blocking as in the smaller lattice with one less blocking. Of course, we expect that the matching can be carried out successfully only for some sufficiently large value of m for which the effect of irrelevant variables has become small. This approach was first implemented by Tobochnik et al. (1981) for simple one- and two-dimensional Ising models. For best results multiple lattice sizes should be used so that finite size effects can be determined and the procedure should be repeated for different times to insure that the asymptotic, long time behavior is really being probed (Katz et al., 1982).

REFERENCES

Brezin, E., LeGuillou, J. C., Zinn-Justin, J., and Nickel, B. G. (1973), Phys. Lett. **44A**, 227.

Domb, C. and Green, M. S. (1976), (eds.) *Phase Transitions and Critical Phenomena* Vol 6 (Academic Press, London).

Fisher, M.E. (1974a), Rev. Mod. Phys. **46**, 597.

Fisher, M. E. (1974b), Phys. Rev. Lett. **32**, 1350.

Friedman, Z. and Felsteiner, J. (1977), Phys. Rev. B **15**, 5317.

Griffiths, R. B. (1970), Phys. Rev. Lett. **24**, 715.

Kadanoff, L. P. (1971), in *Critical Phenomena*, ed. M. S. Green (Academic Press, London).

Katz, S. L., Gunton, J. D., and Liu, C. P. (1982), Phys. Rev. B **25**, 6008.

Landau, D. P. and Swendsen, R. H. (1986), Phys. Rev. B **33**, 7700.

Lewis, A. L. (1977), Phys. Rev. B **16**, 1249.

Ma, S.-K. (1976), Phys. Rev. Lett. **37**, 461.

Nelson, D. R., Kosterlitz, J. M. and Fisher, M. E. (1974), Phys. Rev.Lett. **33**, 813.

Reynolds, P. J., Stanley, H. E., and Klein, W. (1980), Phys. Rev. B **21**, 1223.

Swendsen, R. H. (1982), *Real Space Renormalization*, eds. T. W. Burkhardt and J. M. J. van Leeuwen (Springer Verlag, Heidelberg).

Tobochnik, J., Sarker, S., and Cordery, R. (1981), Phys. Rev. Lett. **46**, 1417.

Wegner, F. J. and Riedel, E. K. (1973), Phys. Rev. B **7**, 248.

Wilson, K. G. (1971), Phys. Rev. B **4**, 3174, 3184.

Wilson, K. G. and Fisher, M. E. (1972), Phys. Rev. Lett. **28**, 248.

Zinn-Justin, J. and Le Guillou, J. C. (1980), Phys. Rev. B **27**, 3976.

10 Non-equilibrium and irreversible processes

10.1 INTRODUCTION AND PERSPECTIVE

In the preceding chapters of this book we have dealt extensively with equilibrium properties of a wide variety of models and materials. We have emphasized the importance of insuring that equilibrium has been reached, and we have discussed the manner in which the system may approach the correct distribution of states, i.e. behavior before it comes to equilibrium. This latter topic has been treated from the perspective of helping us understand the difficulties of achieving equilibrium. The theory of equilibrium behavior is well developed and in many cases there is extensive, reliable experimental information available.

In this chapter, however, we shall consider models which are inherently non-equilibrium! This tends to be rather uncharted territory. For some cases theory exists, but it has not been fully tested. In other situations there is essentially no theory to rely upon. In some instances the simulation has preceded the experiment and has really led the way in the development of the field. As in the earlier chapters, for pedagogical reasons we shall concentrate on relatively simple models, but the presentation can be generalized to more complex systems.

10.2 DRIVEN DIFFUSIVE SYSTEMS (DRIVEN LATTICE GASES)

Over a decade ago a deceptively simple modification of the Ising-lattice gas model was introduced (Katz *et al.*, 1984) as part of an attempt to understand the behavior of superionic conductors. In this 'standard model' a simple Ising-lattice gas Hamiltonian describes the equilibrium behavior of a system, i.e.

$$\mathcal{H} = -\mathcal{J} \sum_{\langle i,j \rangle} n_i n_j, \qquad n_i = 0, 1. \qquad (10.1)$$

In equilibrium the transition rate from state $\mathcal{N}$ to state $\mathcal{N}'$, $W(\mathcal{N} \to \mathcal{N}') = w(\beta \Delta \mathcal{H})$, is some function which satisfies detailed balance (see Section 4.2). A simple, uniform driving field $\mathbf{E}$ is applied in one direction of the lattice and 'spins' (or particle-hole pairs) are exchanged with a probability which is biased by this driving field. This process drives the system away from equilibrium regardless of which kinetic rule is used for the exchange, and the transition rate then becomes

$$W(\mathcal{N} \rightarrow \mathcal{N}') = w[(\Delta\mathcal{H} + lE)/k_B T], \qquad (10.2)$$

where $l = +1, 0,$ or -1 is the distance the particle moved along $\mathbf{E}$, and w is the same function used for the transition in the absence of the driving field. Periodic boundary conditions are applied and the system eventually reaches a non-equilibrium steady state in which a current then flows in the direction parallel to the driving field. These driven lattice gases are perhaps the simplest examples of NESS (non-equilibrium steady state) in which the Hamiltonian alone is *not* the governing feature of the resultant behavior. Since the number of particles (in lattice gas language) is held fixed, the procedure is carried out at constant magnetization (in Ising model language) and spins are exchanged instead of flipped.

Patterns form and produce regions which are relatively free of particles and other regions which are quite densely occupied. As an example, in Fig. 10.1 we show the development of a pattern in a simple Ising model at fixed magnetization with a screw periodic boundary condition in the direction parallel to the driving field. Depending upon the magnitude of the shift in the boundary, different numbers of stripes appear in the steady state. Not only are 'snapshots' of the system generated, but the usual bulk properties are calculated as well. These may show indications of phase transitions just as they would in the case of equilibrium behavior. In addition to the bulk properties, the structure factor $S(\mathbf{k}, L)$ provides important information about the correlations. Indeed, phase transitions can be observed in these systems and peaks in the structure factor offer convincing evidence of the transitions. Because the driving field distinguishes one direction from all others, the behavior is strongly anisotropic with the consequence that the usual scaling relations must be modified, so that for an infinite system

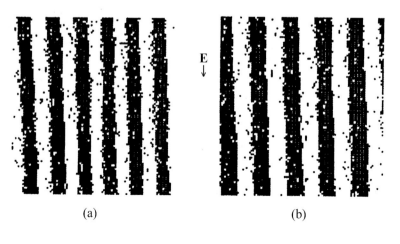

Fig. 10.1 Typical configurations for the 'standard model' driven diffusive lattice gas on a 100×100 lattice with a periodic boundary condition in the horizontal direction and a shifted periodic boundary condition in the vertical direction. The shift is given by h, where: (a) $h = 12$; (b) $h = 16$. From Schmittmann and Zia (1995).

(a)   (b)

$$S(k_\perp, k_\parallel) = k_\perp^{-2+\eta} \mathcal{S}_\perp(k_\parallel/k_\perp^{1+\Delta}), \tag{10.3}$$

where Δ characterizes the anomalous dimension of the longitudinal momenta $k_\parallel$. Of course, modifications may be made in the nature of the interactions, the lattice size, and the aspect ratio of the system. At this time there is still some controversy about the values of the critical exponents in different models, and it is likely that the question of anisotropy will prove to be essential to the understanding of the behavior. In fact, a good framework for the understanding of recent Monte Carlo data (Wang, 1996) has been provided by an extension of finite size scaling which takes into account two different correlation length exponents, $\nu_\parallel$ and $\nu_\perp$, in the directions parallel and perpendicular to the flow, respectively (Binder and Wang, 1989; Leung, 1991).

Problem 10.1 Consider a 40 × 40 Ising lattice gas with periodic boundary conditions and a field **E** in the y-direction. Calculate the structure factor $S(1, 0)$ as a function of temperature for $E/k_B = 0$ and 10.0.

10.3 CRYSTAL GROWTH

The growth of crystals from a melt or a vapor has been a topic of extensive study because of the technological implications as well as because of a desire to understand the theoretical nature of the growth phenomenon (e.g. Kashchiev *et al.*, 1997; Gilmer and Broughton, 1983). Microscopic simulations of crystal growth have long been formulated in terms of solid–on–solid Kossel models in which particles are treated as 'building blocks' which may be stacked upon each other. (Although this model neglects the expected deviations from a perfect lattice structure and the corresponding elastic energies, etc., it does provide the simplest approach to growth with the multiple processes to be outlined below.) Particles may be 'adsorbed' from the vapor or melt with some probability and may diffuse from one surface site to another using a rule which is the equivalent of the spin–exchange mechanism for spin systems. No voids or overhangs are allowed and the resultant growth is 'compact'. Three different processes are allowed: deposition, evaporation, and diffusion, and the goal is to understand what the effect of varying the respective rates for each mechanism is.

Three different kinds of 'bonds' are allowed between nearest neighbors, $-\phi_{ss}$ is the average potential energy of a solid–solid pair, $-\phi_{sf}$ is the average potential energy of a solid–fluid pair, and $-\phi_{ff}$ is the average potential energy of a fluid–fluid pair. Thus, the 'cost' of depositing an adatom on the surface can be calculated by counting the number of bonds of each kind which are created or destroyed and calculating the total energy change. From this approach we can write the effective Hamiltonian for the system as $(\varepsilon = \frac{1}{2}(\phi_{ss} + \phi_{ff}) - \phi_{sf})$

$$\mathcal{H} = -\frac{\varepsilon}{2}\sum_{\langle i,j\rangle}\sigma_i\sigma_j - \Delta\mu\sum_i\sigma_i + V(\{\sigma_i\}), \qquad (10.4)$$

where the occupation variable $\sigma_i = +1$ for an occupied site and $\sigma_i = -1$ for an unoccupied site. $\Delta\mu = (\mu_{\text{vapor}} - \mu_{\text{solid}})$. The potential V enforces the solid-on-solid approximation and is infinite for unallowed configurations. In the absence of supersaturation, the rates of deposition and evaporation are the same, but in the case of a chemical potential difference between the solid and liquid states of $\Delta\mu$ the relative rate of deposition is

$$k^+ = \nu\exp(\Delta\mu/k_{\text{B}}T), \qquad (10.5)$$

where the prefactor ν gives the 'frequency rate' and that for evaporation becomes

$$k_n^- = \nu\exp(-n\phi_{\text{ss}}/k_{\text{B}}T), \qquad (10.6)$$

where the number of bonds which must be broken is n. (Note, the chemical potential required for equilibrium is determined by setting the deposition rates and evaporation rates equal to each other for kink sites.) Diffusion of a particle from a site with energy E_A to a nearest neighbor site with energy E_B is given by

$$k_{\text{d}} = \nu_{\text{d}}\exp[(E_B - E_A)/k_{\text{B}}T], \qquad (10.7)$$

where ν_{d} is the 'frequency rate' for diffusion. As the crystal grows, the surface begins to roughen, but the morphology depends upon the competition between all three processes. Characteristic surfaces after growth has proceeded for a short time for both small supersaturation and large supersaturation are shown in Fig. 10.2.

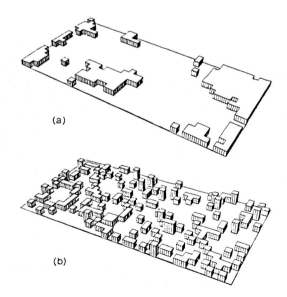

(a)

(b)

Fig. 10.2 'Snapshots' of crystal surfaces after growth of 25% of a monolayer: (a) $L/k_{\text{B}}T = 12$ (L is the binding energy of a simple cubic crystal) and $\Delta\mu/k_{\text{B}}T = 2$ (only 1.8% of the deposited atoms remained on the surface); (b) $L/k_{\text{B}}T = 12$ and $\Delta\mu/k_{\text{B}}T = 20$ (100% of the deposited atoms remained on the surface). From Gilmer et al. (1974).

Spiral crystal growth was studied in a similar fashion (Swendsen *et al.*, 1976) but used a Kossel model which contained a dislocation along one crystal edge. Under typical conditions for spiral growth, evaporation is rapid except along the dislocation (growth) edge and heterogeneous nucleation plays essentially no role in the growth. Thus, a standard Monte Carlo simulation of crystal growth used in the first part of this section would lead to extremely slow growth because very few of the deposited atoms would remain on the surface unless they encountered the spiral growth edge. Instead, in the simulation the creation of isolated particles (or holes) in the surface layer was excluded, leading to an increase in the speed of the simulation algorithm by a factor of $\exp(\varepsilon/k_B T)$. This procedure allowed rather large surfaces to be used so that the system could be followed for long enough to permit the formation of multiple spirals. Typical spiral growth is shown in Fig. 10.3. (In some earlier simulations rather small rectangular systems had been used to simulate the growth along a small strip of the surface which cut through the spirals. In these studies a number of 'steps' were placed on the strip and periodic boundary conditions were applied. The resultant 'growth' resulted in 'step train' behavior, but the spacing between steps was controlled by the number of steps and the lattice size in the direction perpendicular to the steps. Results from the spiral growth algorithm showed that the spacing between spiral arms could become quite large. A 'step train' simulation with multiple steps on a small lattice would thus probably impose an incorrect spacing between the arms and provide results for a system which was inherently non-steady state.)

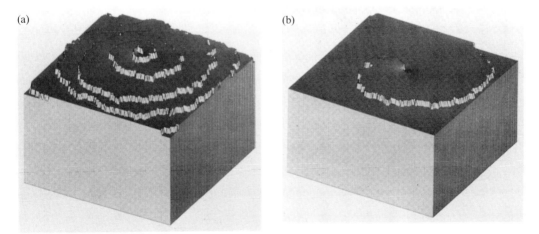

(a)   (b)

Fig. 10.3 Spiral crystal growth at high temperature for the center 200×200 sites of a simple cubic lattice surface: (a) large chemical potential difference $\Delta\mu = 0.6$; (b) small chemical potential difference $\Delta\mu = 0.1$. From Swendsen *et al.* (1976).

10.4 DOMAIN GROWTH

The general area of the temporal development of domains spans a wide range
of different physical phenomena. Background information about phase
separation was provided in Section 2.3 where we saw that at a first order
transition regions of aligned spins, i.e. 'domains', would grow as phase
separation proceeds. Simple models may be used to study the properties of
domains, and the kinetics may be due either to 'spin exchange' or 'spin flip'
mechanisms. The behavior may, in fact, be quite different for different
kinetics. For example, in an Ising model which has been quenched to
below T_c there will be many small domains formed immediately after the
quench, but if spin-flip kinetics are used, some domains will grow at the
edges and coalesce but others will shrink and simply disappear, even from
their interior. Eventually all 'large' domains except one will disappear with a
few overturned spin clusters remaining as a result of thermal excitation. With
spin-exchange kinetics the size of the domains is expected to grow with time,
but the overall magnetization remains constant; thus two equal size domains
will result in the long time limit.

The exponent which describes the domain growth is dependent upon the
kinetic mechanism, although a considerable amount of time may need to pass
before the asymptotic behavior appears. For non-conserved order parameter
models the mean domain radius $\overline{R}$ grows as

$$\overline{R} = Bt^x \tag{10.8}$$

where $x = \frac{1}{2}$. In contrast for conserved order parameter, the domain growth is
much slower and proceeds as given in Eqn. (10.8) but with $x = \frac{1}{3}$. Examples
of each kind of domain growth are shown in Fig. 10.4 for the Ising square
lattice (after Gunton et al., 1988). While in the Ising model shown in Fig.
10.4 there are just two types of domains (up and down are represented by
black and white as usual) and only one kind of domain wall exists, the
situation is more subtle when one considers generalizations to more complicated lattice model problems like domain growth in the Potts model (Grest
and Srolovitz, 1985), or Ising antiferromagnets with competing nearest and
next-nearest neighbor exchange that exhibit a four-fold degenerate groundstate (Sadiq and Binder, 1984), or Ising models with annealed or quenched
impurities (Mouritsen, 1990), etc. In many of these models the asymptotic
growth laws for the domain radius $\overline{R}(t)$ and for the dynamic structure factor
$S(\mathbf{q}, t)$ cf. Eqn. (2.97), are not yet sorted out with fully conclusive evidence
(and the situation is even worse for the analogous molecular dynamics studies
of domain growth for realistic off-lattice models of various pure fluids or fluid
mixtures, as briefly reviewed by Toxvaerd (1995)).

The reasons for these difficulties come from several sources: first of all,
neither the structure factor $S(\mathbf{q}, t)$ nor the domain size – which in the non-
conserved case can simply be found from the order parameter square $\psi^2(t)$ at
elapsed time t after the quench as $\overline{R}(t) = [\psi^2(t)/\langle\psi\rangle_{eq}^2]^{1/d}L$ in d dimensions,
where L is the linear dimension of the system – are self-averaging quantities

Fig. 10.4 Domain
growth in a 150×150
Ising square lattice
quenched from a
random configuration
to $T = 0.6T_c$: (left)
non-conserved order
parameter with $t = 2$,
15, 40, and 120 MCS/
spin; (right) conserved
order parameter with
$t = 10$, 60, 200, and
10 000 MCS/spin.
From Gunton *et al.*
(1988).

t
$\downarrow$

(Milchev *et al.*, 1986). Thus, meaningful results are only obtained if one averages the simulated 'quenching experiment' over a large number of independent runs (which should be of the order of 10^2 to 10^3 runs). Secondly, often several mechanisms of domain growth compete, such as evaporation and condensation of single atoms on domains may compete with the diffusion and coagulation of whole domains, etc., and thus there are slow transients before one growth mechanism wins. As a consequence, it is necessary to study times where $\overline{R}(t)$ is very much larger than the lattice spacing, but at the same time $\overline{R}(t)$ must be very much smaller than L, because otherwise one runs into finite size effects which invalidate the scaling behavior postulated in Eqn. (2.97). From these remarks it is already clear that the computational demands for obtaining meaningful results are huge. A further difficulty is that random numbers of high quality are needed, since the 'random' fluctuations contained in the initial disordered configuration are dramatically amplified. If there are some hidden long range correlations in this initial state – or if the random numbers used in the growth process would introduce such correlations – the growth behavior could become disturbed in a rather artificial manner. This caveat is not an academic one – in fact in their study of domain growth for the ϕ^4 model on the square lattice Milchev *et al.* (1986) ran into this problem.

Nevertheless, simulations of domain growth and of phase separation kinetics have played a very stimulating role both for the development of analytical concepts on the subject, as well as for experiments. For example, scaling concepts on the subject such as Eqn. (2.79) were postulated some time ago (Binder and Stauffer, 1974) in an attempt to interpret corresponding early simulations. This type of scaling now can be derived from rather elaborate theory (Bray, 1994) and has also been seen in experiments both on phase separation (Komura and Furukawa, 1988) and on the ordering kinetics of monolayers adsorbed on surfaces (Tringides *et al.*, 1987). Thus, the above caveats are by no means intended to prevent the reader working on such problems, but rather to make the pitfalls clear.

Problem 10.2 Consider a 40 × 40 Ising model with periodic boundary conditions. Starting with a random spin configuration, use Kawasaki dynamics to carry out a Monte Carlo simulation at $T = 1.5 J/k_B$. Measure the mean domain size.

10.5 POLYMER GROWTH

10.5.1 Linear polymers

The study of the growth of linear polymers from a solution may be easily modeled using very simple models. We begin with a lattice filled with bifunctional monomers, i.e. each monomer may form only two bonds. Each monomer is allowed to randomly atttempt to form bonds with nearest neighbors subject, of course, to the limitation in the number of bonds per monomer. A series of linear polymers will result. If bonds are also allowed to break, the model is appropriate for reversible polymerization, otherwise the polymerization is irreversible. If empty sites are included, they may play the role of solvent atoms. As a result of the growth process a distribution of chain lengths and radii of gyration will result.

10.5.2 Gelation

The formation of cross-linked polymers such as gels, is an extremely important problem which is of particular interest for those who are developing new 'designer materials'. The study of addition polymerization and the subsequent formation of gels is a problem which is well suited for simulation (Family and Landau, 1984). We describe the kinetic gelation model for irreversible, addition polymerization, see Manneville and de Seze (1981) and Herrmann *et al.* (1983), in which we begin with a lattice which contains a mixture of bi-functional and four-functional monomers. In addition, there are a few randomly placed radicals (with concentration c_I) which serve as initiators for the growth process. When a bond is formed between a monomer and an active site (initially an initiator site), the unpaired electron is transferred to the newly bonded site and it becomes 'active'. In addition polymerization, growth may only proceed from these active sites. Bi-functional monomers can only participate in a self-avoiding walk process, whereas the four-functional monomers may be involved in loop formation and crosslinking between growing chains. Initially the solution of unconnected monomers is called a 'sol', but as the growing chains link up they may form an infinite cluster called a gel. This process may involve a phase transition known as the sol–gel transition in which a finite fraction of the system is in the largest cluster. This is analogous to the percolation transition discussed in Chapter 4. Figure 10.5a shows a schematic view of a portion of a three-dimensional system in which gelation is occuring (see Chhabra *et al.* (1986)). The gel fraction G plays the role of the size of the largest cluster in percolation, and its behavior can be analyzed using finite size scaling (see Fig. 10.5b) just as in the case of percolation. Unlike percolation, however, the cluster size distribution n_s is *not* monotonically decreasing. As shown in Fig. 10.5c there are distinct peaks in the distribution at characteristic values of s. These peaks result from the approximately uniform growth of each cluster until two clusters of size s_0 combine to form a single cluster of size $2s_0 + 1$. Since the characteristic size of the smallest 'unit' of the system as it approaches the

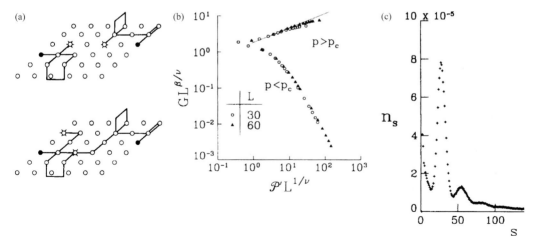

Fig. 10.5 Kinetic gelation model: (a) schematic view of growth within a single layer of a three-dimensional model just before and just after two growing clusters link up, the solid dots show the initial positions of the initiators and data for the cluster size distribution; (b) finite size scaling plot for the gel fraction for $c_I = 3 \times 10^{-2}$; (c) cluster size distribution for $c_I = 3 \times 10^{-4}$ and $p = 0.16$. From Chhabra *et al.* (1986).

sol–gel transition becomes a cluster, rather than a monomer, very large lattices are needed for the simulations.

10.6 GROWTH OF STRUCTURES AND PATTERNS

The formation of structures due to diverse growth mechanisms offers a rich and rapidly growing area of investigation (Herrmann, 1986a) which we can only briefly treat here.

10.6.1 Eden model of cluster growth

First designed as a simple model for cancer growth, the Eden model (Eden, 1961) allows the study of growing compact clusters. Growth begins with a seed particle, one neighboring site of which is then randomly occupied. Then, one neighboring site of the enlarged cluster is occupied, and the process continues in the same fashion. Perhaps the most interesting question about the growth process is the nature of the surface after growth has proceeded for a long time, i.e. how does the width of the surface depend upon the total number of particles which have been added?

In the actual implementation, one may construct a list of the 'growth sites', i.e. a list of perimeter sites which are adjacent to the cluster and at which new particles may be added. A separate array is used to keep track of those sites which have never been touched. At each step of the growth process a site is randomly chosen from the perimeter list. (The alternative approach, of searching for nearest neighbors of 'surface sites' has the danger that some sites may be chosen with too high a probability, i.e. a site may be the nearest neighbor of two different surface sites.) This site is removed from the perimeter list and one must then check to see if any of its neighboring sites have not been touched. If so, they are added to the perimeter list before the next particle is added.

10.6.2 Diffusion limited aggregation

Diffusion limited aggregation (DLA) was first proposed as a simple model for the description of the formation of soot (Witten and Sander, 1981). It has played an extraordinary role, not only in the development of the examination of fractal matter, but also in the use of color coding to effectively portray a third dimension, time, in the development of the system. The fundamental idea of DLA growth is quite simple. A 'seed' particle is placed in the center of the system and another particle is turned loose from a randomly chosen point on a large 'launch circle' which surrounds the seed. This new particle executes a random walk until it encounters the seed particle and then sticks to it. At this point another particle is turned loose from the launch circle and the process is repeated. A beautiful, fractal object results from this procedure and we find that the outer arms of the growth object shield the inner 'fjords' from the particles which are released at later times. Particles may be color coded according to the time at which they were released, and the distribution of adsorbed particles of different colors provides information about the effective 'shielding' of different portions of the cluster. The fractal dimension d_f of the DLA cluster can be determined by measuring the mass M of the cluster within a radius R of the seed and using the relation

$$M \propto R^{d_f} \tag{10.9}$$

to extract an estimate. The effective fractal dimension as a function of cluster size and dimension has been the object of extensive study (Barabási and Stanley, 1995); in two dimensions, DLA clusters with more than 10^7 particles have been grown and the fractal dimension has been estimated at $d_f = 1.71 \pm 0.01$. It was realized fairly quickly that for large systems on a lattice, effects of the anisotropy imposed by the lattice structure began to affect the properties of the cluster. Thus, DLA clusters have been grown in continuous space ('off-lattice') as well as on a variety of lattices.

10.6.2.1 On-lattice DLA

As is often the case, the restriction of a model to a lattice simplifies the situation and enables the use of time saving tricks. In the most straightforward implementation of the DLA algorithm, the particles execute a simple random walk on the lattice with each step being of unit length in a random direction. Each particle is started from a random position on a circle which has the seed at its center. (As the DLA cluster grows, the radius of this 'launch circle' is increased so that it remains larger than the greatest extent of the cluster.) The random walk process is very slow in reaching the growing cluster and can be accelerated in a very simple fashion. The lattice sites surrounding the growing cluster are each assigned an integer which is large far away from the cluster and becomes smaller as the distance to the cluster decreases. This integer specifies the size of the random step that the particle will take when it moves from that site. In the immediate vicinity of the growing DLA cluster the movement reverts to a simple nearest neighbor random walk. An example of the structure which results from this procedure is shown in Fig. 10.6a. For comparison, in Fig. 10.6b we show a pattern which was produced in a Hele–Shaw cell by pumping air into liquid epoxy which filled the spaces between a monolayer of glass balls, all between two parallel glass plates. As the size of the cluster increases, the shape of the cluster begins to reflect the underlying lattice. This effect can be made even more pronounced by using the technique of 'noise smoothing': a particle is finally absorbed only after it has experienced N-collisions, where the integer N becomes a parameter of the simulation and may be varied. The result is a structure which is much more anisotropic than for a simple DLA.

(a)

Fig. 10.6 (a) DLA
cluster of 50 000 atoms
grown on a square
lattice (Feder, 1988);
(b) Hele–Shaw cell
pattern resulting from
air displacing liquid
epoxy in a monolayer
of glass spheres
(Måløy *et al.*, 1985).

(b)

10.6.2.2 Off-lattice DLA

Growth on a lattice is intrinsically affected by the presence of the underlying
lattice structure. Any such effects can be removed simply by avoiding the use
of a lattice. Eliminating the use of a lattice complicates the simulation and, in
particular, the determination of when a particle actually encounters the cluster becomes non-trivial, but it does also remove any effects attributable to any

underlying anisotropy. It becomes necessary to compute a trajectory for each step of the random walk and check to see if the particle touches the cluster at some point along its path. If so, the particle is attached to the cluster at that point and a new particle is released from the launch circle so that the growth process proceeds just as for the on-lattice case.

Problem 10.3 Grow a DLA cluster on a square lattice with 10 000 part-icles. Then grow a DLA cluster of the same size on a triangular lattice. Comment on the similarities and the differences between the two clusters.

10.6.3 Cluster–cluster aggregation

An alternative growth mechanism involves the simultaneous activity of many 'seeds' through the consideration of an initial state which consists of many small clusters (Jullien *et al.*, 1984). Each cluster is allowed to diffuse ran-domly, but if two clusters touch at any point, they stick and begin to move as a single cluster. This model is expected to be well suited to the study of colloid formation and the coagulation processes in for example aerosols. In the simplest case, the clusters all move at the same speed. A more realistic approach is to allow the speed of a cluster to depend upon the inverse of the mass of the cluster, i.e. $\sim m^{\alpha}$. The choice of the exponent α does not affect the fractal dimension of the resulting aggregates except at very low concen-trations but it does enter the distribution function and the dynamical beha-vior.

10.6.4 Cellular automata

Cellular automata are simple lattice or 'cell' models with deterministic time dependence. The time development can, however, be applied to many of the same systems as Monte Carlo processes, and methods of analysis of cellular automata have impacted stochastic simulations. For completeness, we shall thus say a few words about cellular automata. A more complete treatment of this topic is available in Herrmann (1992). These models are defined by a collection of 'spins' or 'cells' on a d–dimensional lattice where each cell contains either a '0' or a '1'. Time is discretized and the value of a cell, σ_i, at time $(t+1)$ is determined by a simple 'rule' which involves the local environment of the ith cell at time t. A simple example is the XOR (exclu-sive-or) rule in which $\sigma_i(t+1) = \sigma_{i-1}(t).\text{XOR}.\sigma_{i+1}(t)$. Different rules result in quite different dynamic features; some produce patterns which are simple and others produce quite complex structures in time. An example of the 'growth' of a one-dimensional cellular automaton, i.e. the time development, with an XOR-rule is shown in Fig. 10.7. The application of the rule to a single site is shown along with the full configurations at times t and $(t+1)$. The major question to be answered is 'what is the nature of the behavior after a long time has elapsed?' One very simple approach is to study the 'damage spreading' (Stauffer, 1987). Consider two cellular automata which follow the

$$0\ 1\ 1\ 0\ 1\ 1\ 0\ 0\ 1\ 0 \qquad t$$
$$\lfloor_0\rfloor \qquad \downarrow$$

$$1\ 1\ 1\ 0\ 1\ 1\ 1\ 1\ 0\ 1 \qquad (t+1)$$

Fig. 10.7. Example of the time development of a simple cellular automaton using a nearest neighbor XOR rule.

same rule. Choose initial states which are identical except for some small region which is different, i.e. 'damaged' in one system. Allow both systems to propagate forward in time and then see what happens to the damage. The damage may disappear completely with the passage of time, may remain localized or may spread throughout the system. This latter behavior is indicative of the onset of chaos as is only observed for a small fraction of the rules. An equivalent approach can be taken in Monte Carlo simulations by considering two systems with almost identical initial states. The same random number sequence is then used in a simulation of each system, and the differences in the configurations for the two systems are then followed as a function of time. The critical dynamics of a cellular automata rule called Q2R in two dimensions appears to be consistent with model A Ising behavior (or possibly model C), but in three dimensions the behavior appears to be quite different (Stauffer, 1997).

Using a random initial configuration, one can model the Ising model by a Q2R cellular automaton in which a spin is flipped only if it involves no change in energy (Herrmann, 1986b). This can be carried out quite efficiently if the checkerboard decomposition is used. Unfortunately the cellular automaton algorithm is not ergodic. A solution to this problem is to randomly flip a spin occasionally while maintaining the energy within a narrow band of energies.

Probabilistic rules, e.g. the Hamiltonian formulation of the Kauffman model, may also be used.

Problem 10.3 Use the nearest neighbor XOR rule described in Fig. 10.7 to follow a 32-bit cellular automaton with p.b.c. in time with the following initial conditions: (a) a single bit is 1 and all other bits are 0; (b) 16 of the bits (randomly chosen) are 1 and the other bits are 0.

10.7 MODELS FOR FILM GROWTH

10.7.1 Background

The growth of films and the characterization of the resultant surface has formed a topic of great experimental, theoretical and simulational interest. One standard measure of the nature of this growth surface, whose local position at time t is $h(\mathbf{r}, t)$, is given by the long-time dependence of the interfacial or surface width W,

$$W^2(t) = \langle h^2 \rangle - \langle h \rangle^2 \qquad (10.10)$$

which diverges as $t \to \infty$. Note that the mean position of the surface $\langle h \rangle$ is given merely by the rate at which particles are deposited and is uninteresting. The manner in which the surface width diverges can be described by a 'critical' or growth exponent which places the systems into 'universality classes' which are analogous to the classes which have been identified for static critical behavior. Thus, the temporal variation of the surface width after growth has proceeded for a long time may be given by

$$W(t \to \infty) = Bt^\beta, \tag{10.11}$$

where the prefactor B is relatively unimportant but the growth exponent β defines the nature of the growth. In a finite system the surface width saturates at long times and instead it is the size dependence of the saturated width which is of interest:

$$W(L \to \infty) = AL^\alpha \tag{10.12}$$

where α is termed the 'roughening' exponent. The ratio of the exponents defines a dynamic exponent z , i.e.

$$z = \alpha/\beta. \tag{10.13}$$

The time-dependent and size-dependent behavior can be condensed into a dynamic scaling relation (Vicsek and Family, 1985)

$$W = L^\alpha \mathcal{F}(tL^z) \tag{10.14}$$

which should be valid in a general case. Since both relations, Eqns. (10.11) and (10.12), hold only in the asymptotic limit of large substrate size and long times, the extraction of accurate estimates for these exponents is non-trivial. These relations are expected to be generally valid, so we may attempt to analyze the behavior of many growth models using this formalism.

10.7.2 Ballistic deposition

Growth models such as ballistic deposition (see Barabási and Stanley, 1995) are relatively easy to study and the results can be displayed and interpreted graphically. In the simplest case particles are dropped from random positions above a surface and fall in a straight line until they either land on the surface or encounter a particle which has already been deposited. In the latter case, the new particle sticks to the old one either on the top or on the side. Particles are dropped sequentially and a very perforated structure grows. From a computational perspective ballistic deposition is very easy to simulate. For deposition onto a line, we randomly choose a horizontal position x_n and check to find the height of the uppermost occupied site y_n in the column above x_n and that of its two neighbor columns, i.e. y_{n-1}, y_{n+1}. If y_n is the largest of these numbers, the particle is deposited at height $(y_n + 1)$; if one of the neighboring columns is higher, the particle is deposited at a height which is the highest of $(y_{n-1} + 1)$ or $(y_{n+1} + 1)$. For deposition using a point seed, the process proceeds exactly the same as for the line 'substrate', but most of the particles never strike the seed, at least at early times. As an example, in Fig. 10.8 we show a ballistic deposition cluster which has resulted from growth with a point seed.

10.7.3 Sedimentation

In an effort to describe growing surfaces which are more compact than those described by ballistic deposition, Edwards and Wilkinson (1982) introduced a simple model which could be solved exactly. In this EW model, a particle is dropped from a random position above a growing surface. The particle lands on top of the column below the point from which it is dropped and then diffuses once to the neighboring site which is lowest lying. Another particle is then dropped and the process is repeated. Edwards and Wilkinson (1982) map this model onto a simple differential growth equation in which the variable h_i is the height of the growing surface above the mean position and

$$\frac{\partial h}{\partial t} = \nu \nabla^2 h + \zeta(r, t) \qquad (10.15)$$

where $\zeta(r, t)$ is δ-correlated noise in both space and time. The solution to this differential equation yields a dynamic exponent $z = 2.0$. However, in the simulation of the atomistic model an interesting question arises: what does one do when there is more than one neighboring site of the same 'lowest' depth? While it might seem intuitive to make a choice between the different possibilities by generating a random number, this procedure in fact leads to an additional source of (correlated) noise and changes the value of z! If a particle with multiple choices does not diffuse at all, diffusion becomes deterministic and $z = 2$ is recovered. This finding points out the subtleties involved in obtaining a complete understanding of film growth (Pal and Landau, 1999).

There are variations of this model, e.g. by Wolf and Villain (1990), which use different rules for hopping and which result in different behavior. (For example, in the WV model, particles hop to the nearest neighbor site in which

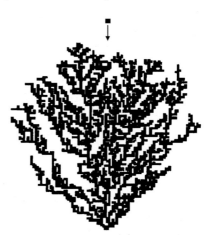

Fig. 10.8 Pattern formed by ballistic deposition simulation using a point substrate.

they will have the greatest number of bonds rather than the lowest height.)
All of these models may be compared with the KPZ model (Kardar *et al.*,
1986) which is defined by a differential equation which includes the tilt of the
surface and the surface curvature. One issue that remains to be resolved is the
delineation of the criteria which determine non–equilibrium universality
classes.

Problem 10.4 Grow a 1 + 1 dimensional Edwards–Wilkinson film for substrates of size $L = 20, 40$, and 80. Measure the interfacial widths and plot
them as a function of time. Estimate α, β, and z.

10.7.4 Kinetic Monte Carlo and MBE growth

More recently, attention has turned to the simulation of thin films grown by
molecular beam epitaxy (MBE). The growth of films by molecular beam
epitaxy (MBE) requires the inclusion of both deposition and diffusion processes. Some efforts have been directed at fully understanding the behavior of
relatively realistic models for small films using empirical potentials for short
times, and other studies have been directed at the scaling behavior of simpler
models. In this section we shall concentrate on the simplest, lattice models for
MBE growth. This approach is also in terms of solid on solid models with
nearest neighbor interactions. Particles are deposited with some fixed flux **F**.
Any of the particles may then undergo activated diffusion with probability

$$p = \exp(-E_A/kT). \tag{10.16}$$

For simple models with nearest neighbor coupling, the activation energy may
be simply dependent upon the number of occupied nearest neighbors, i.e.
$E_A = J \Sigma n_j$. An atom which has been activated may then hop to a nearest
neighbor site either randomly or with a probability which depends upon the
energy that the atom will have in that site. Thus, the rate of hopping does not
depend merely upon the relative energies of the configuration before and
after hopping as it would in a simple 'spin–exchange' Monte Carlo process
but rather the barrier plays an essential role. Diffusion thus proceeds via a
two-step process and the simulation technique which matches this process is
called kinetic Monte Carlo. Kinetic Monte Carlo methods also find widespread application for the study of surface diffusion in adsorbed monolayers
(see e.g. Uebing and Gomer, 1991, 1994). The differences between the two
processes are shown schematically in Fig. 10.9. The nature of the growth
depends upon the magnitude of the flux as well as the temperature. At very
low temperatures there is little diffusion and the surface width grows monotonically as shown in Fig. 10.10. As the temperature is raised oscillations in
the data indicate layer-by-layer like growth, i.e. atoms which land on a
'plateau' diffuse off the edge and nucleation of a new layer begins only
after the layer below is filled. (Calculations of the RHEED intensity from
the surface configuration generated show that even at the very lowest temperature studied there are small oscillations remaining, and at sufficiently

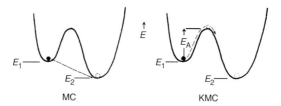

Fig. 10.9 Schematic comparison between Monte Carlo and kinetic Monte Carlo methods for diffusion of surface adatoms between two sites with energy E_1 and E_2, respectively. E_A is the activation energy for KMC.

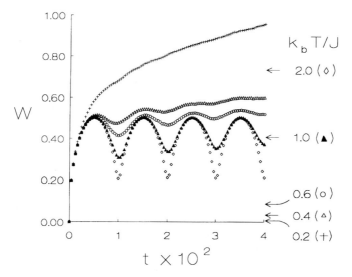

Fig. 10.10 Time dependence of the surface width for MBE models on $L \times L$ substrates with p.b.c. Values of the surface width for equilibrium are shown by the arrows to the right. After Pal and Landau (1994).

long times the width diverges for the higher temperatures shown in Fig. 10.10 for preferential hopping. Thus, there is no true transition between layer-by-layer growth and rough growth.) Note that Fig. 10.10 compares the equilibrium surface width with that obtained for the MBE growth model: the trends for the variation of the mean surface width are exactly reversed because the equilibrium surface width is quite small at low temperatures. The growth process may be repeated multiple times with different random number sequences. Each of the resultant 'growth histories' is independent, so that statistical accuracy can be improved by simply taking the average over many runs and the error bars are then straightforward to calculate. Of course, data for successive times for a given simulation will be correlated, so care must be exercised in analyzing 'structure' which is seen in a single run or a small number of runs. The long time behavior can be difficult to ascertain, because the 'asymptotic region' appears for quite different times for different values of the relevant parameters. Extensive simulations have shown that it is possible to find quite different 'effective' growth exponents for different fluxes, and we recommend that a particular exponent be observed to describe the data over *at least* two decades in time before being deemed acceptable. Finite size effects also become important at long time and

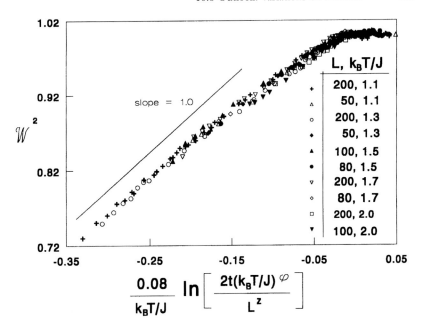

Fig. 10.11 Dynamic finite size scaling of the surface width for $2 + 1$ dimensional MBE models. The growth exponent $\beta = 0$ and the dynamic exponent $z = 1.63$ for this plot. From Pal and Landau (1999).

dynamic finite size scaling, Eqn. (10.14), can be used to analyze the data and extract exponent estimates. A typical finite size scaling plot for the surface width of a $2 + 1$ dimensional MBE growth model is shown in Fig. 10.11. Note that scaling of the surface width can be made to include the temperature dependence.

Problem 10.5 Grow a $1 + 1$ dimensional MBE film using a KMC method with a deposition rate of 1 layer/sec and a prefactor for activation of 0.1. Plot the interfacial width, averaged over multiple runs, as a function of time for $L = 20, 40$, and 80. How does the time at which finite size effects become obvious vary with L?

Problem 10.6 Grow a $1 + 1$ dimensional MBE film using 'spin exchange' Monte Carlo with a deposition rate of 1 layer/sec and a diffusion rate constant of 0.1. Plot the interfacial width as a function of time for $L = 20$. Compare your result with that obtained by kinetic Monte Carlo in Problem 10.5.

10.8 OUTLOOK: VARIATIONS ON A THEME

In this chapter we have only mentioned a small fraction of the problems that have been considered in the literature. There are many related problems of non-equilibrium growth phenomena for which Monte Carlo simulation is an extremely useful tool. In this regard, we wish to cite just one more example, that of random sequential adsorption (e.g. Evans, 1993): consider the growth of coverage of a monolayer formed by dimers (or n-mers) which are randomly adsorbed but which obey excluded volume constraints. A special 'jamming coverage' then appears where further adsorption becomes impossible. Near this jamming coverage, slow dynamics is observed. This simple model and its extensions form another rich area for investigation that we have not really examined here.

REFERENCES

Barabási, A.-L. and Stanley, H. E. (1995), *Fractal Concepts in Surface Growth* (Cambridge University Press, Cambridge).

Binder, K. and Stauffer, D. (1974), Phys. Rev. Lett. **33**, 1006.

Binder, K. and Wang, J. S. (1989), J. Stat. Phys. **55**, 87.

Bray, A. (1994), Adv. Phys. **43**, 357.

Chhabra, A., Matthews-Morgan, D., and Landau, D. P. (1986), Phys. Rev. B **34**, 4796.

Eden, M. (1961), in *Proc. 4th Berkeley Symposium on Mathematical Statistics and Probability, Vol. IV*, ed. J. Neyman (University of California, Berkeley) p. 223.

Edwards, S. F. and Wilkinson, D. R. (1982), Proc. R. Soc. A **381**, 17.

Evans, J. W. (1993), Rev. Mod. Phys. **65**, 1281.

Family, F. and Landau, D. P. (1984), *Kinetics of Aggregation and Gelation* (North Holland, Amsterdam).

Feder, J. (1988), *Fractals* (Plenum Press, NY).

Gilmer, G. H. and Broughton, J. Q. (1983), J. Vac. Sci. Technol. B **1**, 298.

Gilmer, G. H., Leamy, H. J., and Jackson, K. A. (1974), J. Cryst. Growth **24/25**, 495.

Grest, G. S. and Srolovitz, D. J. (1985), Phys. Rev. B **32**, 3014.

Gunton, J. D., Gawlinski, E., and Kaski, K. (1988), *Dynamics of Ordering Processes in Condensed Matter*, eds. S. Komura and H. Furukawa (Plenum, New York) p. 101.

Herrmann, H. J. (1986a), Physics Reports **136**, 153.

Herrmann, H. J. (1986b), J. Stat. Phys. **45**, 145.

Herrmann, H. J. (1992), in *The Monte Carlo Method in Condensed Matter Physics*, ed. K. Binder (Springer, Berlin).

Herrmann, H. J. , Stauffer, D., and Landau, D. P. (1983), J. Phys. A **16**, 1221.

Jullien, R., Kolb, M., and Botet, R. (1984), in *Kinetics of Aggregation and Gelation*, eds. F. Family and D. P. Landau (North Holland, Amsterdam).

Kardar, M., Parisi, G., and Zhang, Y.-C. (1986), Phys. Rev. Lett. **56**, 889.

Kashchiev, D., van der Eerden, J. P., and van Leeuwen, C. (1977), J. Cryst. Growth **40**, 47.

Katz, S., Lebowitz, J. L., and Spohn, H. (1984), Phys. Rev. B **28**, 1655.

Komura, S. and Furukawa, H. (1988), *Dynamics of Ordering Processes in Condensed Matter Theory* (Plenum, New York).

Leung, K.-T. (1991), Phys. Rev. Lett. **66**, 453.

Måløy, K. J., Feder, J., and Jøssang, T. (1985), Phys. Rev. Lett. **55**, 2688.

Manneville, P. and de Seze, L. (1981), in *Numerical Methods in the Study of Critical Phenomena*, eds. I. Della Dora, J. Demongeot and B. Lacolle (Springer, Berlin).

Milchev, A., Binder, K., and Herrmann, H. J. (1986), Z. Phys. B **63**, 521.

Mouritsen, O. G. (1990), in *Kinetics of Ordering and Growth at Surfaces*, ed. M. G. Lagally (Plenum Press, New York) p. 1.

Pal, S. and Landau, D. P. (1994), Phys. Rev. B **49**, 10,597.

Pal, S. and Landau, D. P. (1999), Physica A **267**, 406.

Sadiq, A. and Binder, K. (1984), J. Stat. Phys. **35**, 517.

Schmittmann, B. and Zia, R. K. P. (1995), in *Phase Transitions and Critical Phenomena* Vol 17, (Academic Press, London) p. 1.

Stauffer, D. (1987), Phil. Mag. B **56**, 901.

Stauffer, D. (1997), Int. J. Mod. Phys. C **8**, 1263.

Swendsen, R. H., Kortman, P. J., Landau, D. P., and Müller-Krumbhaar, H. (1976), J. Cryst. Growth **35**, 73.

Toxvaerd, S. (1995), in *25 Years of Nonequilibrium Statistical Mechanics*, eds. J. J. Brey, J. Marro, J. M. Rubi, and M. San Miguel (Springer, Berlin) p. 338.

Tringides, M. C., Wu, P. K., and Lagally, M. G. (1987), Phys. Rev. Lett. **59**, 315.

Uebing, C. and Gomer, R. (1991), J. Chem. Phys. **95**, 7626, 7636, 7641, 7648.

Uebing, C. and Gomer, R. (1994), Surf. Sci. **306**, 419.

Vicsek, T. and Family, F. (1985), J. Phys. A **18**, L75.

Wang, J.-S. (1996), J. Stat. Phys. **82**, 1409.

Witten, T. A. and Sander, L. M. (1981), Phys. Rev. Lett. **47**, 1400.

Wolf, D. E. and Villain, J. (1990), Europhys. Lett. **13**, 389.

11 Lattice gauge models: a brief introduction

11.1 INTRODUCTION: GAUGE INVARIANCE AND LATTICE GAUGE THEORY

Lattice gauge theories have played an important role in the theoretical description of phenomena in particle physics, and Monte Carlo methods have proven to be very effective in their study. In the lattice gauge approach a field theory is defined on a lattice by replacing partial derivatives in the Lagrangian by finite difference operators. For physical systems a quantum field theory on a four-dimensional space–time lattice is used, but simpler models in lower dimension have also been studied in hope of gaining some understanding of more complicated models as well as for the development of computational techniques.

We begin by describing the potential $A_\mu^\alpha(x)$ in terms of the position x in space–time. The rotation U of the frame which relates neighboring space–time points x^μ and $x^\mu + dx^\mu$ is given by

$$U = \exp\{igA_\mu^\alpha(x)\lambda_\alpha dx^\mu\}, \tag{11.1}$$

where g is the coupling constant and the λ_α are the infinitesimal generators of the gauge group. When the field is placed on a lattice, an element U_{ij} of the gauge group is assigned to each link between neighboring sites i and j of the lattice, subject to the condition that

$$U_{ij} \rightarrow U_{ji}^{-1}. \tag{11.2}$$

Gauge transformations are then defined by

$$U_{ji} \rightarrow U_{ji}' = g_i U_{ji} g_i^{-1} \tag{11.3}$$

where g_i is a group element. There will be some elementary closed path on the lattice which plays the role of the infinitesimal rectangular closed path which defines the transporter; for example, the path around an elementary square on a hypercubical lattice (or 'plaquette') is

$$U_p = U_i U_j U_k U_i, \tag{11.4}$$

where the 'action' associated with a plaquette is

$$S_P = \beta f(U_P). \tag{11.5}$$

$f(U_P)$ is commonly referred to as the (internal) energy of the plaquette, and the choice

$$f(U_P) = 1 - \tfrac{1}{2}\mathrm{Tr}\,U_P = 1 - \cos\theta_P \qquad (11.6)$$

is termed the Wilson action, although many other forms for the action have been studied.

By first making a Wick rotation to imaginary time, we can define the observables in a Euclidean four-dimensional space, i.e.

$$\langle O \rangle = \frac{1}{Z}\int \mathcal{D}A_\mu\, O(A_\mu)\exp[-S(A_\mu)] \to \frac{1}{Z}\sum_{\{U_{ij}\}} O(U_{ij})\exp\{-S(U_{ij})\}, \qquad (11.7)$$

where

$$Z = \int dA\,\exp(-S(A)) \to \sum_{\{U_{ij}\}} \exp\{-S(U_{ji})\}, \qquad (11.8)$$

where the sums are over the dynamic variables U_{ij}. Note that the above equations are equivalent, in a formal sense, to those which describe the behavior of an interacting particle system within the framework of statistical mechanics. In this view, β becomes equivalent to the inverse temperature and $f(U_P)$ plays the role of the Hamiltonian. With the analogy to statistical mechanics, one can carry out Monte Carlo simulations by updating the link variables, e.g. using a Metropolis method, and then calculating expectation values of quantities of interest. Thus, all of the tools needed for the study of lattice gauge models are already in place. In order to recover a non-trivial continuum field theory, the lattice constant must be allowed to go to zero, but the product $a\lambda(g)$ must remain constant. The critical point g_{cr} for which this occurs must then have scaling properties, and in the language of statistical mechanics this means that a phase transition must occur. For any 'interesting' behavior to remain, this means that the equivalent of the correlation length must diverge, i.e. a second order phase transition appears. Thus, one important goal is to determine the phase diagram of the theory. As a consequence, many of the methods of analysis of the Monte Carlo data are identical to those of the systems discussed in earlier chapters, although the interpretation of the various quantities is completely different.

Note that the same problems with finite size effects, boundary conditions, etc. which we encountered in Chapters 4 and 5 in the study of spin systems apply here and we refer the reader back to these earlier chapters for a detailed discussion. Indeed, the problems are even more severe for the four-dimensional lattice gauge theories of real interest since there is a much higher percentage of 'spins' on the boundary than for lower dimensional magnetic systems. Furthermore, the determination of new link values may be very complicated, particularly for groups such as SU(2) and SU(3), so special sampling methods have been devised.

11.2 SOME TECHNICAL MATTERS

Various specialized techniques have been devised to try to make Monte Carlo sampling more efficient for lattice gauge theories The special problem which one encounters in lattice gauge studies is that the determination of the new configuration and its energy are often extremely time consuming. As a result the 'standard' importance sampling methods often become inefficient. Among the techniques that are used are:

(1) The heatbath method. Here a new link U'_{ji} is chosen with probability $\exp\{-S(U'_{ji})\}$ regardless of the previous value of the link.

(2) Multi-hit methods. Here the Metropolis algorithm is used, but the entire process is repeated on a single link n-times before another link is chosen for consideration. This is efficient because the complexity of the interaction makes the computation of the possible new states considerably more complex than for spin models.

(3) Mixed initial states. To overcome problems with metastability, one can begin with a state in which half of the system is in a 'cold' state and half in a disordered state. The time development is followed for different values of β to see towards which state the entire system evolves.

Another simplification which has also been used is to use a discrete subgroup as an approximation to the full group; in such cases the computation of the action is simplified although the model is obviously being modified and the consequences of these changes must be carefully examined.

11.3 RESULTS FOR $Z(N)$ LATTICE GAUGE MODELS

Perhaps the simplest lattice gauge theories are those in which the variables of interest are 'spins' which assume a finite number N of values distributed on a unit circle. While such models are not expected to be relevant to the description of physical systems, they play a useful role in the study of the phase structure of lattice gauge models since their relative simplicity allows them to be simulated rather straightforwardly. For the discrete $Z(N)$ group the special case of $N = 2$ corresponds to a gauge invariant version of the Ising model. (The U(1) theory, which will be discussed in the next section, corresponds to the $N = \infty$ limit of $Z(N)$.) Creutz *et al.* (1979) examined the four-dimensional $Z(2)$ gauge model and found evidence for a first order transition. In particular, sweeps in β exhibit strong hysteresis, and starts from either ordered or disordered states at the transition coupling show very different metastable states (see Fig. 11.1).

The critical behavior for the $(2 + 1)$-dimensional $Z(2)$ lattice gauge model at finite temperatures (Wansleben and Zittartz, 1987) was calculated by looking at the block size dependence of the fourth order cumulant. $128 \times 128 \times N_T$ lattices were examined where the number of lattice points in the tem-

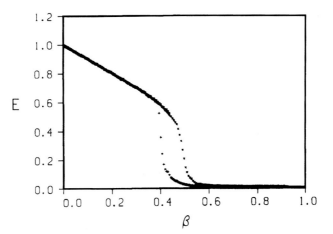

Fig. 11.1 The average energy per plaquette as a function of β for the four-dimensional $Z(2)$ lattice gauge theory. A hypercubic lattice with $L_x = L_y = L_z = 8$ and $L_t = 20$ with periodic boundary conditions was used. The 'temperature' was swept up and then back down. From Creutz *et al.* (1979).

perature direction, N_T, was varied. The value of ν is apparently unity, but the estimate for β/ν depended on N_T.

Problem 11.1 Write a Monte Carlo program for the $Z(2)$ lattice gauge model in four dimensions. Determine the behavior of the energy as a function of β for $L = 3$. Estimate the value of β at which the transition occurs. Compare your results with the data given in Fig.11.1 and comment.

11.4 COMPACT U(1) GAUGE THEORY

The U(1) model has also been extensively studied and is a prime example of the difficulties associated with obtaining clear answers for lattice gauge models. Initial Monte Carlo examinations of the simple action

$$S = -\sum_P [\beta \cos \theta_P] \tag{11.9a}$$

could not determine if the transition was first or second order. The reason for the uncertainty became clear when an adjoint coupling was added so that the total action became

$$S = -\sum_P [\beta \cos \theta_P + \gamma \cos(2\theta_P)], \tag{11.9b}$$

where θ_P is the plaquette angle, i.e. the argument of the product of U(1) variables around a plaquette P. The phase diagram in this expanded parameter space then showed that the transition actually changed order for a value of the adjoint coupling γ which was close to zero, and crossover phenomena make the interpretation for the pure U(1) model problematic. The most detailed study of this model (Jersák *et al.*, 1996a,b) simulated spherical lattices and used reweighting techniques together with finite size scaling to conclude that for $\gamma \leq 0$ the transition is indeed second order and belongs to the universality class of a non-Gaussian fixed point with the exponent ν in the range 0.35–0.40 (the best estimate is $\nu = 0.365(8)$).

Problem 11.2 Perform a Monte Carlo simulation for the simple U(1) gauge model (i.e. $\gamma = 0$) in four dimensions. Determine the variation of the energy as a function of β for $L = 3$. Estimate the location of the phase transition.

11.5 SU(2) **LATTICE GAUGE THEORY**

The transition between the weak coupling and strong coupling regimes for SU(2) lattice gauge theories at finite temperature has also been a topic of extensive study.

The Glashow–Weinberg–Salam (GWS) theory of electroweak interactions assumes the existence of a Higgs mechanism. This can be studied in the context of an SU(2) lattice gauge theory in which 'spins' are added to the lattice site and the Hamiltonian includes both gauge field and Higgs field variables:

$$
S = -\frac{\beta}{4}\sum_P \mathrm{Tr}(U_P + U_P^t) - \kappa \sum_x \sum_{\mu=1}^{4} \mathrm{Re}(\mathrm{Tr}\Phi_x^t U_{x,\mu}\Phi_{x+\mu})
$$
$$
+ \lambda \sum_x \frac{1}{2}\mathrm{Tr}(\Phi_x^t\Phi_x - 1)^2 + \sum_x \mathrm{Tr}\Phi_x^t\Phi_x.
$$
(11.10)

For fixed λ there is a confinement region for $\kappa < \kappa_c$ and a Higgs region for $\kappa > \kappa_c$. Even if λ is fixed at a physically reasonable value, the resultant phase diagram is in a two-dimensional parameter space and the nature of the transition appears to change order (Bock *et al.*, 1990). This can be seen in Fig. 11.2 where two equal peaks in the distribution develop with a very deep

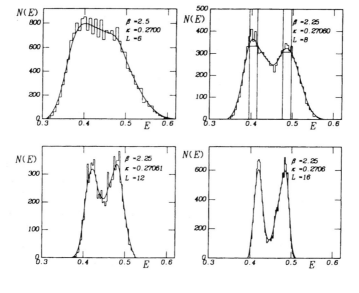

Fig. 11.2 A sequence of distribution functions $N(E)$ near the transition at $\beta = 2.25$ for different lattice sizes in the SU(2) model with Higgs fields. From Bock *et al.* (1990).

well between them as the lattice size is increased. The use of histograms and finite size scaling aids in the analysis, but the location of a tricritical point was not possible with data for lattices up to 16^4 in size.

11.6 INTRODUCTION: QUANTUM CHROMODYNAMICS (QCD) AND PHASE TRANSITIONS OF NUCLEAR MATTER

According to our current understanding of high energy physics the basic constituents of elementary particles are quarks and gluons. Quantum chromodynamics (QCD) is the relativistically invariant quantum field theory, formulated in four-dimensional space $(\mathbf{x}, \tau = it)$; note that we choose here the standard units of elementary particle physics, $\hbar = c = 1$. Since for this problem of strong interactions perturbation theory is of limited value, nonperturbative theoretical approaches must be sought. A formulation in terms of path integrals is the method of choice (Creutz et al., 1983; Kogut, 1983; Montvay and Münster, 1994). In this approach, the vacuum expectation value of a quantum observable $\mathcal{O}$ is written as (Meyer-Ortmanns, 1996)

$$\langle \mathcal{O} \rangle = \frac{1}{Z} \int \mathcal{D}A_\mu \mathcal{D}\overline{\psi} \mathcal{D}\psi \mathcal{O}(A_\mu, \overline{\psi}, \psi) \exp[-S(A_\mu, \overline{\psi}, \psi; g, m_i)], \quad (11.11)$$

where A_μ denotes the gauge fields, $\overline{\psi}$, ψ stand for the particle fields (indices $f = 1, \ldots, N_f$ for the 'flavors' and $c = 1, \ldots, N_c$ for the 'colors' classifying these quarks we suppressed, to simplify the notation). The action functional S also contains the gauge coupling and the quark masses m_i as parameters, and is the space–time integral of the Lagrange density of QCD,

$$S = \int d\tau \int d\mathbf{x} \mathcal{L}_{\text{QCD}}(A_\mu, \overline{\psi}, \psi; g, m_i); \quad (11.12)$$

the explicit form of $\mathcal{L}_{\text{QCD}}$ in full generality is rather complicated, but will not be needed here. Finally, the normalizing factor Z in Eqn. (11.11), the vacuum-to-vacuum amplitude, is

$$Z = \int \mathcal{D}A_\mu \mathcal{D}\psi \mathcal{D}\overline{\psi} \exp[-S]. \quad (11.13)$$

The formal analogy of Eqns. (11.11–11.13) with problems in statistical mechanics is rather obvious: If we interpret $\mathcal{L}_{\text{QCD}}$ as a density of an effective free energy functional, multiplied by inverse temperature β, the action can be interpreted as effective Hamiltonian $\beta\mathcal{H}$, and Z is analogous to a partition function. Now it is already well known for the path integral formulation of simple non-relativistic quantum mechanics (Feynman and Hibbs, 1965) that a precise mathematical meaning must be given to all these functional integrals over gauge and matter fields. One very attractive way to do this is the lattice formulation in which the $(3 + 1)$-dimensional space–time continuum is discretized on a hypercubic lattice. A gauge-invariant lattice action must be chosen, which then provides a gauge-invariant scheme to regularize the path integral: in the limit where the lattice linear dimensions become large, the continuum limit is recovered.

In practice such a lattice action can be chosen following Wilson (1974) associating matter variables $\psi_\mathbf{x}, \overline{\psi}_\mathbf{x}$ with the sites of the lattice and gauge variables with the links, $U_\mathbf{x}^\mu$ being associated with a link leaving a site $\mathbf{x}$ in direction $\hat{\mu}$. These link variables are elements of the gauge group $SU(N)$ and replace the continuum gauge fields A_μ. One can then show that a gauge action that produces the correct continuum limit (namely $(4g^2)^{-1} \int dt \int d\mathbf{x} \mathrm{Tr} F_{\mu\nu}^2$ where $F_{\mu\nu}$ is the Yang–Mills field strength) can be expressed in terms of products of these link variables over closed elementary plaquettes of the hypercubic lattice,

$$S = \frac{2N}{g^2} \sum_{\substack{\mathbf{x} \\ \mu<\nu}} p_\mathbf{x}^{\mu\nu}, \qquad p_\mathbf{x}^{\mu\nu} = 1 - \frac{1}{N} \mathrm{Tr} U_\mathbf{x}^\mu U_{\mathbf{x}+\hat{\mu}}^\nu U_{\mathbf{x}+\hat{\mu}}^{\mu+} U_\mathbf{x}^{\nu+}, \qquad (11.14)$$

Tr denoting the trace in color space (normally $N = 3$, quarks exist in three colors, but corresponding studies using the $SU(2)$ group are also made).

If one treats pure gauge fields, the problem closely resembles the treatment of spin problems in the lattice as encountered in previous chapters – the only difference being that β then correponds to g^{-2}, and rather than a bilinear Hamiltonian in terms of spins on lattice sites one has to deal with a Hamiltonian containing those products of link variables around elementary plaquettes.

The problem becomes far more involved if the matter fields $\psi(\mathbf{x}), \overline{\psi}(\mathbf{x})$ describing the quarks are included: after all, quarks are fermions, and hence these fields really are operators obeying anticommutation rules (so-called Grassmann variables). There is no practical way to deal with such fermionic fields explicitly in the context of Monte Carlo simulations!

Fortunately, this aspect of QCD is somewhat simpler than the many-fermion problems encountered in condensed matter physics (such as the Hubbard Hamiltonian, etc., see Chapter 8): the Lagrangian of QCD contains $\overline{\psi}$ and ψ only in bilinear form, and thus one can integrate out the matter fields exactly! The price that has to be paid is that a complicated determinant appears, which is very cumbersome to handle and requires special methods, which are beyond consideration here (Herrmann and Karsch, 1991). Thus, sometimes this determinant is simply ignored (i.e. set equal to unity), but this so-called 'quenched approximation' is clearly uncontrolled, although there is hope that the errors are relatively small.

What do we wish to achieve with this lattice formulation of QCD? One very fundamental problem that the theory should master is the prediction of the masses of the hadrons, using the quark mass as an input. Very promising results for the mass of the nucleon, the pion, the delta baryon, etc., have indeed been obtained (Butler et al., 1993), although the results are still to be considered somewhat preliminary due to the use of the 'quenched approximation' mentioned above.

There are many more problems in QCD where the analogy with problems encountered in condensed matter physics is even closer, namely phase transitions occurring in nuclear matter of very high energy (or in other words, at very high 'temperature': 100 MeV corresponds to 1.16×10^{12} K)! While the phase transitions in condensed matter physics occur at the scale from 1 K to 10^3 K, at $T_c \approx (2.32 \pm 0.6) \times 10^{12}$ K one expects a 'melting' of nuclear matter – quarks and gluons cease to be confined inside hadrons and begin to move freely (Meyer-Ortmanns, 1996). According to the big bang theory of the early universe, this deconfinement transition should have happened at about 10^{-6} sec after the big bang.

We now turn to some special aspects of the average in Eqn. (11.11). Due to the Wick rotation $(it \to \tau)$ inverse temperature appears as an integration limit of the τ integration,

$$S = \int_0^\beta d\tau \int d\mathbf{x} \mathcal{L}(A_\mu, \overline{\psi}, \psi, g, m_i) \tag{11.15}$$

and in addition boundary conditions have to be obeyed,

$$A_\mu(\mathbf{x}, 0) = A_\mu(\mathbf{x}, \beta), \qquad \psi(\mathbf{x}, 0) = -\psi(\mathbf{x}, \beta), \qquad \overline{\psi}(\mathbf{x}, 0) = -\psi(\mathbf{x}, \beta). \tag{11.16}$$

Thus, while one has periodic boundary conditions in pseudo-time direction for the gauge fields, as is familiar from condensed matter physics problems, the particle fields require antiperiodic boundary conditions. As we shall discuss in the next section, there is intense interest in understanding the order of this deconfinement transition, and the problems in its analysis have many parallels with studies of the Potts model in statistical mechanics.

11.7 THE DECONFINEMENT TRANSITION OF QCD

The deconfinement transition of a pure gauge model employing the SU(3) symmetry can be considered as the limit of QCD in which all quark masses tend to infinity. Real physics, of course, occurs at finite quark masses (remember that there exist two light quarks, called 'up' and 'down', and one heavier one, the so-called 'strange quark'). This case is difficult to treat, and therefore another simplified limit of QCD has been considered, where the quark masses are put all equal to zero. This limit is called the 'chiral limit', because a Lagrange density applies which exhibits the so-called 'chiral symmetry' and is reminiscent of Landau theory, Eqn. (2.36), the central distinction being that the scalar order parameter field $m(\mathbf{x})$ is now replaced by a $N_f \times N_f$ matrix field ϕ (Pisarski and Wilczek, 1984)

$$\mathcal{L} = \frac{1}{2}\mathrm{Tr}\left(\frac{\partial\phi^+}{\partial x_\mu}\right)\left(\frac{\partial\phi}{\partial x_\mu}\right) - \frac{f}{2}\mathrm{Tr}(\phi^+\phi) - \frac{\pi^2}{3}\left[f_1(\mathrm{Tr}\phi^+\phi)^2 + f_2\mathrm{Tr}(\phi^+\phi)^2\right]$$
$$+ g(\det\phi + \det\phi^+).$$

$$(11.17)$$

Here f, f_1, f_2 and g are constants. At zero temperature there is a symmetry-broken state, i.e. the vacuum expectation value $\langle\phi\rangle$ (which is also called the 'quark condensate') is now zero but exhibits $\mathrm{SU}(N_f)$ symmetry. This spontaneous breaking of chiral symmetry is associated with the occurrence of a multiplet of Goldstone bosons (i.e. massless excitations, loosely analogous to spin wave excitations in a Heisenberg ferromagnet).

At finite temperature this model is believed to undergo a phase transition to a phase where the chiral symmetry is restored. One believes that for g of order unity this transition is of second order for $N_f = 2$ but of first order for $N_f = 3$. The obvious problem is that QCD leads to rather different phase transitions in the limit of quark masses $m \to \infty$ and $m \to 0$: Note that the order parameter for the deconfinement transition is rather subtle, namely the expectation value of a Wilson loop, $\langle L(\mathbf{x})\rangle$, where $L(\mathbf{x})$ is defined by

$$L(\mathbf{x}) \equiv \mathrm{Tr}\hat{T}\exp\left(\int_0^\beta dt A_0(\mathbf{x}, t)\right),$$

$$(11.18)$$

where $\hat{T}$ is the time-ordering operator. One can interpret $\langle L(\mathbf{x})\rangle$ in terms of the free energy $F(\mathbf{x})$ of a free test quark inserted into the system at $\mathbf{x}$, $\langle L(\mathbf{x})\rangle = \exp[-\beta F(\mathbf{x})] = 0$ in the phase exhibiting quark confinement, while $\langle L(\mathbf{x})\rangle$ is non-zero if we have deconfinement. This behavior qualifies $\langle L(\mathbf{x})\rangle$ as an order parameter of the deconfinement transition.

The question now is what happens when we consider intermediate quark masses: are the deconfinement transition at T_d and the chiral transition at T_ch simply limits of the same transition within QCD which smoothly changes its character when the quark masses are varied, or are these transitions unrelated to each other (and then ending at critical points somewhere in the (T, m) plane), Fig. 11.3? If scenario (b) applies and if the physically relevant quark masses lie in the range in between $m_\mathrm{ch}^\mathrm{crit} < m < m_\mathrm{d}^\mathrm{crit}$, no phase transition

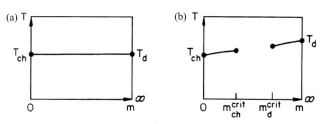

Fig. 11.3 Hypothetical phase diagrams of QCD in the (m, T) plane, where T is the temperature, and m stands for generic quark masses. (a) The transitions persist for finite non-zero m and coincide. (b) Both transitions terminate at critical points for intermediate mass values. After Meyer-Ortmanns (1996).

occurs but rather the change of nuclear matter to the quark–gluon plasma is a gradual, smooth crossover (as the change of a gas of neutral atoms in a plasma of ions and electrons when the temperature of the gas is raised).

From Fig. 11.3 we recognize that a crucial problem of QCD is the clarification of a phase diagram (whether or not a sharp phase transition occurs, and if the answer is yes, what is the order of the transition). If there were a first order transition, this should have experimentally observable consequences for heavy-ion collisions. Also the abundance of light elements in the universe has been attributed to consequences of the first order scenario, but one must consider this idea rather as an unproven speculation.

Before one can address the behavior of QCD for intermediate quark masses, it clearly is of central importance to clarify the phase transitions in the two limiting cases of Fig. 11.3, $m \to 0$ and $m \to \infty$. Even this problem has led to longstanding controversies, e.g. the order of the deconfinement transition ($m \to \infty$) has been under debate for some time, but now the controversy seems to be settled (Meyer-Ortmanns, 1996) by the finding of a (relatively weak) first order transition. The equation of state $\Delta = (\varepsilon - 3p)/T^4$ of a pure SU(3) gauge model is plotted in Fig. 11.4 (Karsch, 1995). Here ε is the energy density $\{\varepsilon = -(1/V)\partial(\ln Z)/\partial(1/T)\}$ and p is the pressure $\{p = T(\partial/\partial V)\ln Z\}$ of nuclear matter. These definitions are just the usual ones in the continuum limit, of course. In order to evaluate such derivatives in the framework of lattice gauge theory one has to introduce the lattice spacing for the 'temporal' direction (a_τ) and spatial directions (a_σ) as explicit variables (the volume then is $V = a_\sigma^3 N_\sigma^3 a_\tau N_\tau$ for a lattice of linear size N_σ in the spatial directions and N_τ in the 'time' direction). Treating a_τ and a_σ as continuous variables, one can write $\partial/\partial T = N_\tau^{-1}\partial/\partial a_\tau$, and $\partial/\partial V = (3a_\sigma^2 N_\sigma^3)\partial/\partial a_\sigma$. After performing the appropriate lattice derivatives of $\ln Z$, one can set the lattice spacings equal again, $a_\sigma = a_\tau = a$, and use a as the unit of length. However, when one wishes to extrapolate towards the continuum limit, one needs to let $N_\tau \to \infty$, $a_\tau \to 0$ keeping the temperature $(N_\tau a_\tau)^{-1} = T$ fixed at the physical scale of interest (MeV units). Therefore one needs to study the dependence of the data on N_τ carefully, as shown in Fig. 11.4. A detailed analysis of the steep rise of $\varepsilon - 3p$ at T_c shows that there indeed occurs a first order phase transition, with a latent heat of $\Delta\varepsilon/T_c^4 = 2.44 \pm 0.24\,(N_\tau = 4)$ or $\Delta\varepsilon/T_c^4 = 1.80 \pm 0.18\,(N_\tau = 6)$, respectively. An important conclusion from the equation of state as shown in Fig. 11.4 also is the fact that interaction effects are still present at temperatures far above T_c (for a non-interacting ideal gas one would have $\varepsilon = 3p$, of course).

Another quantity which has found much attention is the interface tension between low temperature and high temperature phases at the deconfinement transition, since this quantity plays a role in some of the scenarios that describe the evolution of the early universe. This interface tension was measured by Iwasaki et al. (1994) by an extension of the finite size analysis of distribution functions originally proposed for the Ising model (Binder, 1982). The result is $\sigma/T_c^3 = 0.0292 \pm 0.0022$ and for $N_\tau = 4$ and $\sigma/T_c^3 = 0.0218 \pm 0.0033$ for $N_\tau = 6$. Note that all these calculations are extremely time-con-

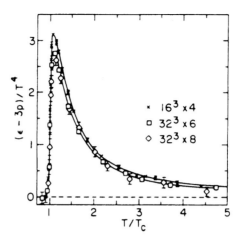

Fig. 11.4 Interaction measure $\varepsilon - 3p$ normalized to T^4 (dimensionless units) plotted vs. T/T_c for a pure SU(3) gauge theory for different lattice sizes. From Karsch (1995).

suming and difficult – early estimates for σ/T_c^3 applying different methods ended up with estimates that were nearly an order of magnitude too large. For a description of the dynamics of the early universe, this interface tension controls the extent to which the quark–gluon plasma at the deconfinement transition could be supercooled, before hadrons are nucleated. For the estimates of σ/T_c^3 quoted above, one ends up finally with the result that the average distance between hadronic bubbles should have been 22 ± 5 mm (Meyer-Ortmanns, 1996).

Of course, this brief introduction was not intended to give a representative coverage of the extensive literature on Monte Carlo applications in lattice gauge theory; we only want to give the reader a feeling for the ideas underlying the approach and to make the connections with Monte Carlo applications in the statistical mechanics of condensed matter transparent.

REFERENCES

Binder, K. (1982), Phys. Rev. A **25**, 1699.

Bock, W., Evertz, H. G., Jersak, J., Landau, D. P., Neuhaus, T., Xu, J. L. (1990), Phys. Rev. D **41**, 2573.

Butler, F., Chen, H., Sexton, J., Vaccarino, A., and Weingarten, D. (1993), Phys. Rev. Lett. **70**, 7849.

Creutz, M., Jacobs, L., and Rebbi, C. (1979), Phys. Rev. Lett. **42**, 1390.

Creutz, M., Jacobs, L., and Rebbi, C. (1983), Phys. Rep. **93**, 207.

Feynman, R. P. and Hibbs, A. R. (1965), *Quantum Mechanics and Path Integrals* (McGraw-Hill, New York).

Herrmann, H. J. and Karsch, F. (1991), *Fermion Algorithms* (World Scientific, Singapore).

Iwasaki, Y., Kanaya, K., Karkkainen, L., Rummukainen, K., and Yoshie, T. (1994), Phys. Rev. D **49**, 3540.

Jersák, J., Lang, C. B., and Neuhaus, T. (1996a), Phys. Rev. Lett. **77**, 1933.

Jersák, J., Lang, C. B., and Neuhaus, T. (1996b), Phys. Rev. D **54**, 6909.

Karsch, F. (1995), Nucl. Phys. A **590**, 367.

Kogut, J. B. (1983), Rev. Mod. Phys. **55**, 775.

Meyer-Ortmanns, H. (1996), Rev. Mod. Phys. **68**, 473.

Montvay, I. and Münster, G. (1994),
   Quantum Fields on the Lattice
   (Cambridge University Press,
   Cambridge).
Pisarski, R. D. and Wilczek, F. (1984),
   Phys. Rev. D **29**, 338.

Wansleben, S. and Zittartz, J. (1987),
   Nuclear Phys. B **280**, 108.
Wilson, K. (1974), Phys. Rev. D **10**,
   2455.

12 A brief review of other methods of computer simulation

12.1 INTRODUCTION

In the previous chapters of this text we have examined a wide variety of Monte Carlo methods in depth. Although these are exceedingly useful for many different problems in statistical physics, there are some circumstances in which the systems of interest are not well suited to Monte Carlo study. Indeed there are some problems which may not be treatable by stochastic methods at all, since the time-dependent properties as constrained by deterministic equations of motion are the subject of the study. The purpose of this chapter is thus to provide a very brief overview of some of the other important simulation techniques in statistical physics. Our goal is not to present a complete list of other methods or even a thorough discussion of these methods which are included but rather to offer sufficient background to enable the reader to compare some of the different approaches and better understand the strengths and limitations of Monte Carlo simulations.

12.2 MOLECULAR DYNAMICS

12.2.1 Integration methods (microcanonical ensemble)

Molecular dynamics methods are those techniques which are used to numerically integrate coupled equations of motion for a system which may be derived, e.g. in the simplest case from Lagrange's equations or Hamilton's equations. Thus, the approach chosen is to deal with many interacting atoms or molecules within the framework of classical mechanics. We begin this discussion with consideration of systems in which the number of particles N, the system volume V, and the total energy of the system E are held constant This is known as the NVE ensemble. In the first approach, Lagrange's equations for N particles produce a set of $3N$ equations to be solved:

$$m_i \ddot{\mathbf{r}}_i = \mathbf{F}_i = -\nabla_{r_i} \mathcal{V}, \tag{12.1}$$

where m_i is the particle mass and $\mathbf{F}_i$ the total net force acting on each particle ($\mathcal{V}$ is the appropriate potential). For N particles in three spatial dimensions ($d = 3$) this entails the solution of $3N$ second order equations. (The reader will recognize Eqn. (12.1) as Newton's second law.) If instead, Hamilton's equations are used to derive the system dynamics, a set of $6N$ first order equations will result:

$$\dot{\mathbf{r}}_i = \mathbf{p}_i/m_i, \tag{12.2a}$$

$$\dot{\mathbf{p}}_i = \mathbf{F}_i. \tag{12.2b}$$

where $\mathbf{p}_i$ is the momentum of the particle. Either set of equations can be solved by simple finite difference methods using a time interval Δ which must be made sufficiently small to maintain accuracy. It is clear from the Hamilton's equation approach that the energy of the system is invariant with time so that solution of these equations produces states in the micro-canonical ensemble. The simplest numerical solution is obtained by making a Taylor expansion of the position and velocity about the current time t, i.e.

$$\mathbf{r}_i(t + \Delta) = \mathbf{r}_i(t) + \mathbf{v}(t)\Delta + \tfrac{1}{2}\mathbf{a}(t)\Delta^2 + \cdots \tag{12.3a}$$

$$\mathbf{v}_i(t + \Delta) = \mathbf{v}_i(t) + \mathbf{a}_i(t)\Delta + \cdots. \tag{12.3b}$$

These equations are truncated after a small number of terms so that the calculation of the properties of each particle at the next time is straightfor-ward, but errors tend to build up rather quickly after many time steps have passed. In order to minimize truncation errors two–step predictor–corrector methods may be implemented. In these approaches a prediction is made for the new positions, velocities, etc. using the current and previous values of these quantities, and then the predicted acceleration is used to calculate improved (or corrected) positions, velocities, etc. A number of different predictor–corrector methods have been considered and the comparison has been made elsewhere, see e.g. Berendsen and van Gunsteren (1986).

No discussion of molecular dynamics methods, not even an introductory one, would be complete without some presentation of the Verlet algorithm (Verlet, 1967). The position $\mathbf{r}_i$ is expanded using increments $+\Delta$ and $-\Delta$ and the resultant equations are then added to yield

$$\mathbf{r}_i(t + \Delta) = \mathbf{r}_i(t) - \mathbf{r}_i(t - \Delta) + \mathbf{a}_i(t)\Delta^2 + \cdots. \tag{12.4}$$

The velocities are then determined by taking numerical time derivatives of the position coordinates

$$\mathbf{v}_i(t) = \frac{\mathbf{r}_i(t + \Delta) - \mathbf{r}_i(t - \Delta)}{2\Delta}. \tag{12.5}$$

Note that the error in Eqn. (12.4) has been reduced to order Δ^4 but the error in the velocity is of order Δ^2. There are a number of other schemes for carrying out the integration over time that have been developed and these are discussed by Allen and Tildesley(1987) and Rapaport (1995). Molecular dynamics studies have played an extremely important role in the develop-ment of computer simulations, and indeed the discovery of long time tails (algebraic decay) of the velocity autocorrelation function in a simple hard sphere model was a seminal work that provided important insights into liquid behavior (Alder and Wainwright, 1970).

In these microcanonical simulations both the kinetic energy and the potential energy will vary, but in such a way as to keep the total energy fixed. Since the temperature is proportional to the mean kinetic energy, i.e.

$$\frac{1}{2}\sum_i m_i \dot{r}_i^2 = \frac{3}{2}Nk_\mathrm{B}T, \tag{12.6}$$

it will fluctuate during the course of the simulation on a finite system. Similarly, the potential energy will vary as the particles move, but these variations can be determined by direct measurement. Obviously the use of such techniques for obtaining averages in thermal equilibrium relies on the ergodicity property of the system. Typical time steps are in the sub-picosecond range and molecular dynamics simulations can generally follow a system for only tens or hundreds of nanoseconds. Therefore, it is only possible to study problems where equilibrium is reached on such a short time scale. Characteristic of the kinds of studies that can be performed using molecular dynamics are investigations of classical fluid models in which the particles interact via a Lennard-Jones potential (see Eqn. (6.4)). Figure 12.1 shows the equilibrium correlations obtained for a dense fluid of 864 particles (Verlet, 1968).

More recently there have been improvements made in the use of higher order decompositions, which are based on the Trotter formula, for the integration of coupled equations of motion which describe different kinds of motions with very different time scales (Tuckerman et al., 1992). In this approach the 'slow' degrees of freedom are frozen while the others are updated using a rather fine time scale; the 'slow' degrees of freedom are then updated using a coarse time scale.

Some time integration methods are better at conserving energy, or other 'constants of the motion' while some methods are capable of determining other physical properties with greater accuracy or speed even though the exact preservation of conservation properties is lost. One important consideration is the conservation of phase space volume. Only integration methods which have time reversible symmetry will conserve a given volume in phase space, and algorithms which are time reversible generally have less long term drift of conserved quantities than those which are not time reversal invariant. Molecular dynamics methods have been well suited to vectorization and,

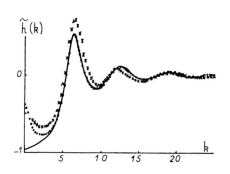

Fig. 12.1 Pair correlation function $\tilde{h}(k)$ for a classical fluid: (dots) molecular dynamics data for a Lennard-Jones potential with $T = 1.326$, $\rho = 0.5426$; (solid curve) hard-sphere model; (crosses) x-ray experiment on argon. From Verlet (1968).

more recently, efficient parallel algorithms have been constructed that allow the study of quite large systems. For example, in Fig. 12.2 we show recent results of fracture in a system of about 2×10^6 particles interacting with a modified Lennard-Jones potential. Historically, the choice of algorithm was often determined in large part by the amount of computer memory needed, i.e. the number of variables that needed to be kept track of. Given the large memories available today, this concern has been largely ameliorated. Two features that we do want to mention here which were introduced to make molecular dynamics simulations faster are potential 'cutoffs' and 'neighbor lists'. (These labor saving devices can also be used for Monte Carlo simulations of systems with continuous symmetry.) As the particles move, the forces acting on them change and need to be continuously recomputed. A way to speed up the calculation with only a modest reduction in accuracy is to cut off the interaction at some suitable range and then make a list of all neighbors which are within some slightly larger radius. As time progresses, only the forces caused by neighbors within the 'cutoff radius' need to be recomputed, and for large systems the reduction in effort can be substantial. (The list includes neighbors which are initially beyond the cutoff but which are near enough that they might enter the 'interacting region' within the number of time steps, typically 10–20 which elapse before the list is updated.) With the advent of parallel computers, molecular dynamics algorithms have been devised that will distribute the system over multiple processors and allow treatment of quite large numbers of particles. One major constraint which remains is the limitation in maximum integration time and algorithmic improvement in this area is an important challenge for the future. There

Fig. 12.2 Results of a molecular dynamics study of the time evolution of crack propagation in a model with modified Lennard-Jones interactions. The top row shows time sequences for initial motion in the stiff direction, and in the bottom row the initial motion is in the soft direction. From Abraham (1996).

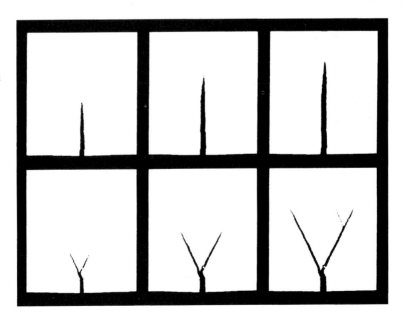

are a number of important details and we refer the reader elsewhere (Allen and Tildesley, 1987; Rapaport, 1995) for the entire story.

Problem 12.1 Consider a cubic box of fixed volume V and containing $N = 256$ particles which interact with a Lennard-Jones potential suitable for argon: $\sigma = 0.3405$ nm, $\epsilon/k_B = 119.8$ K, $m = 6.63382 \times 10^{-26}$ kg ($T^* = kT/\epsilon$, $\rho^* = \rho\sigma^3$). Use a simple Verlet algorithm with a cutoff of $r = 2.5\sigma$ to carry out a molecular dynamics simulation with a density of $\rho^* = 0.636$ and a total (reduced) energy E^* of 101.79. Please answer the following questions.

a. What is the average temperature T^* for the system?
b. What is the time dependence of the kinetic energy for the system?
c. What is the time dependence of the potential energy for the system?

12.2.2 Other ensembles (constant temperature, constant pressure, etc.)

Often the properties of the system being studied are desired for a different set of constraints. For example, it is often preferable to have information at constant temperature rather than at constant energy. This can be accomplished in several different ways. The crudest approach is to periodically simply rescale all of the velocities so that the total kinetic energy of the systems remains constant. This basic approach can also be implemented in a stochastic manner in which the velocity of a randomly chosen particle is reset using a Maxwell–Boltzmann distribution. A very popular method is that of 'thermostats' in which an additional degree of freedom is added to play the role of a reservoir (Nosé, 1984; Hoover, 1985). The time integration is then carried out for this extended system and energy is extracted from the reservoir or inputted to it from the system so as to maintain a constant system temperature. The equations of motion which must then be solved are different from the original expressions; if we denote the particle position by $\mathbf{r}$ and the 'new' degree of freedom by s, the equations to be solved for a particle of mass m become

$$\ddot{\mathbf{r}}_i = \mathbf{F}_i/m_i s^2 - 2\dot{s}\dot{\mathbf{r}}_i/s, \tag{12.7a}$$

$$Q\ddot{s} = \sum_i m_i \dot{r}_i^2 s - (f+1)k_B T/s, \tag{12.7b}$$

where f is the number of degrees of freedom, T is the desired temperature, and Q represents the size of the 'thermal ballast'. There will, of course, be some thermal lag and/or overshoot if this process is not carried out carefully, i.e. if Q is not chosen wisely, but when care is exercised the net result is usually quite good.

Molecular dynamics simulations can also be carried at constant pressure using several different techniques including 'pressurestat' methods which are the equivalent of the thermostats described above (Andersen, 1980). Constant pressure may also be maintained by changing the box size, and more sophisticated algorithms even allow for a change in the shape of the simulation box. This latter capability may be important for the study of solids which exhibit structural phase changes which may be masked or inhibited by a fixed shape for the simulation box. Obviously it is possible to include both thermostats and pressurestats to work in the *NPT* ensemble.

A rather different approach to molecular dynamics may be taken by considering a system of perfectly 'hard' particles which only interact when they actually collide. The purpose of this simplification is to enable rather large numbers of particles in relatively low density systems to be simulated with relatively modest resources. For studies of hard particles the algorithms must be modified rather substantially. The (straight line) trajectories of each of the particles are calculated and the time and location of the next collision are determined. The new velocities of the colliding particles are calculated using conservation of energy and momentum for elastic collisions and the process is resumed. Thus, instead of being a time step-driven process hard particle molecular dynamics becomes an event-driven method. Such simulations have been quite successful in producing macroscopic phenomena such as the Rayleigh–Bénard instability, shown in Fig. 12.3, in a two-dimensional system (Rapaport, 1988) confined between two horizontal plates held at different temperatures. The data show that the formation of the final, steady-state roll pattern takes quite some time to develop.

Problem 12.2 Take the system which you used in Problem 12.1 and carry out a constant temperature MD simulation at the temperature which you found from Problem 12.1. Determine:
a. the average kinetic energy for the system;
b. the average potential energy for the system;
c. the average total energy for the system. Compare with the value of E^* in Problem 12.1.

In the example shown in Fig. 12.1 we have used the pair correlation function, i.e. a static quantity in thermal equilibrium, which could have been evaluated with Monte Carlo methods as well (see Chapter 6). In fact, molecular dynamics often is used to address static equilibrium properties only, ignoring the additional bonus that dynamical properties could be obtained as well. This approach makes sense in cases where molecular dynamics actually produces statistically independent equilibrium configurations faster than corresponding Monte Carlo simulations. Such situations have been reported, e.g. in the simulation of molten SiO_2 (due to strong covalent bonds Monte Carlo moves where the random movement of single atoms to new positions has a low acceptance rate), models of polymer melts near their glass transition, etc. For problems of this type, the decision whether Monte Carlo or molecular dynamics algorithms should be used is non-trivial, because the judgment of efficiency is subtle. Sometimes Monte Carlo is superior due to non-local moves, such as pivot rotations of large parts of long polymer chains (see Chapter 6).

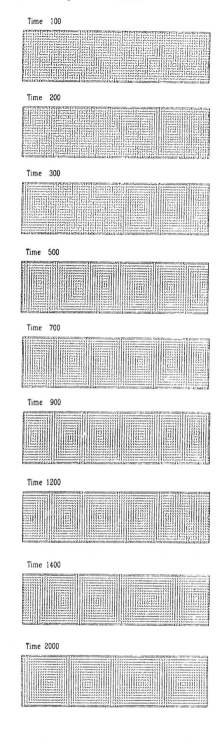

Fig. 12.3 Development of coarse-grained flow lines for the Rayleigh–Bénard instability as determined from hard particle molecular dynamics simulations. From Rapaport (1988).

12.2.3 Non-equilibrium molecular dynamics

In the entire discussion given above, the goal was to produce and study the behavior of an interacting system of particles in equilibrium. For systems which are not in equilibrium, e.g. systems subject to a large perturbation, the techniques used must be altered. In methods of non-equilibrium molecular dynamics a large perturbation is introduced and transport coefficients are then measured directly. Either the perturbation may be applied at time $t = 0$ and the correlation functions are measured and integrated to give transport coefficients, or an oscillating perturbation is applied and the real and imaginary responses are measured by Laplace transform of the correlation functions.

12.2.4 Hybrid methods (MD + MC)

For some complex systems Monte Carlo simulations have very low acceptance rates except for very small trial moves and hence become quite inefficient. Molecular dynamics simulations may not allow the system to develop sufficiently in time to be useful, however, molecular dynamics methods may actually improve a Monte Carlo investigation of the system. A trial move is produced by allowing the molecular dynamics equations of motion to progress the system through a rather large time step. Although such a development may no longer be accurate as a molecular dynamics step, it will produce a Monte Carlo trial move which will have a much higher chance of success than a randomly chosen trial move. In the actual implementation of this method some testing is generally advisable to determine an effective value of the time step (Duane *et al.*, 1987).

12.2.5 *Ab initio* molecular dynamics

No discussion of molecular dynamics would be complete without at least a brief mention of the approach pioneered by Car and Parrinello (1985) which combines electronic structure methods with classical molecular dynamics. In this hybrid scheme a fictitious dynamical system is simulated in which the potential energy is a functional of both electronic and ionic degrees of freedom. This energy functional is minimized with respect to the electronic degrees of freedom to obtain the Born–Oppenheimer potential energy surface to be used in solving for the trajectories of the nuclei. This approach has proven to be quite fruitful with the use of density functional theory for the solution of the electronic structure part of the problem and appropriately chosen pseudopotentials.

The Lagrangian for the system is

$$L = 2 \sum_i^{occ} \int d\mathbf{r} \mu_i |\dot{\psi}_i(\mathbf{r})|^2 + \frac{1}{2} \sum_I M_I \dot{R}_I^2 - E[\{\psi_a\}\mathbf{R}_I]$$
$$+ 2 \sum_{id} \Lambda_{ij} \left(\int d\mathbf{r} \psi_i^*(\mathbf{r})\psi_j(\mathbf{r}) - \delta_{ij} \right), \qquad (12.8)$$

where E is the energy functional, ψ_i the single particle wave function, M_I and R_I the ionic masses and positions respectively. μ_i is the fictitious electronic mass and the fictitious dynamics is given by

$$\dot{\psi}_i(\mathbf{r}, t) = -\frac{1}{2} \frac{\delta E}{\delta \psi_i^*(\mathbf{r}, t)}. \qquad (12.9)$$

(Note that the single particle wave functions play the role of fictitious classical dynamic variables.) The Λ_{ij} are Lagrangian multipliers that are used to maintain the orthonormality of the single particle wave functions. The resultant equations of motion are

$$\mu_i \ddot{\psi}_i(\mathbf{r}, t) = -\frac{1}{2} \frac{\delta E}{\delta \psi_i^*(\mathbf{r}, t)} + \sum_j \Lambda_{ij} \psi_j(\mathbf{r}, t), \qquad (12.10a)$$

$$M_I \ddot{\mathbf{R}}_I = -\frac{\partial E}{\partial \ddot{\mathbf{R}}_I(t)}. \qquad (12.10b)$$

These equations of motion can then be solved by the usual numerical methods, e.g. the Verlet algorithm, and constant temperature simulations can be performed by introducing thermostats or velocity rescaling. This *ab initio* method is efficient in exploring complicated energy landscapes in which both the ionic positions and electronic structure are determined simultaneously (Parrinello, 1997).

12.3 QUASI-CLASSICAL SPIN DYNAMICS

Although the static properties of a large number of magnetic systems have been well studied experimentally, theoretically and via simulation, the study of the dynamic properties of magnetic systems is far less mature. The Monte Carlo method is fundamentally stochastic in nature and in general there is no correlation between the development of a system in Monte Carlo time and in real time, although the static averages are the same (by construction). An approach to the investigation of true time-dependent properties is to generate initial states, drawn from a canonical ensemble using Monte Carlo methods, and to use these as starting points for the integration of the coupled equations of motion. For example, consider a system of N spins which interact with the general Hamiltonian

$$\mathcal{H} = -J \sum_{\langle i,j \rangle} (S_{ix} S_{jx} + S_{iy} S_{jy} + \lambda S_{iz} S_{jz}) + D \sum_i S_{iz}^2 + H \sum_i S_{iz}, \quad (12.11)$$

where the first sum is over all nearest neighbor pairs, λ represents exchange anisotropy, D is the single ion anisotropy, and H is the external magnetic field. There are a number of physical systems which are well approximated by Eqn. (12.8), although for different systems one or more of the parameters may vanish. For $\lambda = 1$ and $D = 0$ this represents the isotropic Heisenberg ferromagnet or the corresponding antiferromagnet for $\mathcal{J} > 0$ or $\mathcal{J} < 0$, respectively.

For models with continuous degrees of freedom, real equations of motion can be derived from the quantum mechanical commutator,

$$\frac{\partial \hat{S}_i}{\partial t} = -\frac{i}{\hbar}[\hat{S}_i, \mathcal{H}], \tag{12.12a}$$

by allowing the spin value to go to infinity and normalizing the length to unity to yield

$$\frac{d\mathbf{S}_i}{dt} = \frac{\partial \mathcal{H}}{\partial \mathbf{S}_i} \times \mathbf{S}_i = -\mathbf{S}_i \times \mathbf{H}_{\text{eff}}, \tag{12.12b}$$

where $\mathbf{H}_{\text{eff}}$ is an 'effective' interaction field. For the isotropic Heisenberg ferromagnet $\mathbf{H}_{\text{eff}} = -\mathcal{J} \sum_{nn} \mathbf{S}_j$ and the time dependence of each spin, $\mathbf{S}_r(t)$, can be determined from integration of these equations. These coupled equations of motion can be viewed as describing the precession of each spin about an effective interaction field; the complexity arises from the fact that since all spins are moving, the effective field is not static but rather itself constantly changing direction and magnitude.

A number of algorithms are available for the integrations of the coupled equations of motion which were derived in the previous sub-section. The simplest approach is to expand about the current spin value using the time step Δ as the expansion variable;

$$S_i^\alpha(t + \Delta) = S_i^\alpha(t) + \Delta \dot{S}_i^\alpha(t) + \frac{1}{2}\Delta^2 \ddot{S}_i^\alpha(t) + \frac{1}{3!}\Delta^3 \dddot{S}_i^\alpha(t) + \cdots \tag{12.13}$$

where the α denotes the spin component. (Compare this equation with Eqn. (12.3) for molecular dynamics.) The 'new' estimate may be made by simply evaluating as many terms as possible in the sum, although this procedure must obviously be truncated at some point. Typical values of Δ which deliver reliable results to a reasonable maximum integration time t_{max} are in the range of $\Delta = 0.005$. If the equation is truncated at the point shown in Eqn. (12.13), the errors will be of order Δ^4. A very simple improvement can be made by implementing a 'leapfrog' procedure (in the spirit of Eqn. (12.4)) to yield (Gerling and Landau, 1984)

$$S_i^\alpha(t + \Delta) = S_i^\alpha(t - \Delta) + 2\Delta \dot{S}_i^\alpha(t) + \frac{2}{3!}\Delta^3 \dddot{S}_i^\alpha(t) + \cdots. \tag{12.14}$$

The error in this integration is $O(\Delta^5)$ and allows not only larger values of Δ to be used but also allows us to extend the maximum integration time to $t_{\text{max}} \approx 100\mathcal{J}^{-1}$. Several standard numerical methods can also be applied. One excellent approach is to use a predictor–corrector method; fourth order predictor–corrector methods have proven to be quite effective for spin dynamics simulations. An example is the explicit four-step Adams–Bashforth method (Burden *et al.*, 1981) followed by an implicit Adams–Moulton corrector step, a combination which also has a local truncation error of Δ^5 and which has proven to be quite successful. The first application of this method requires that at least three time steps have already been taken; these can initially be provided using the fourth order Runge–Kutta method, starting with the initial state. Of course, this predictor–corrector method requires that the spin configuration at four time steps must be kept in memory. Note that the conservation laws discussed earlier will only be observed within the accuracy set by the truncation error of the method. In practice, this limits the time step to typically $\Delta = 0.01 \mathcal{J}^{-1}$ in $d = 3$ (Chen and Landau, 1994) for the isotropic model ($D = 0$), where $t_{max} \leq 200 \mathcal{J}^{-1}$. The same method was used in $d = 2$; with $\Delta = 0.01 \mathcal{J}^{-1}$, $t_{max} = 400 \mathcal{J}^{-1}$ (Evertz and Landau, 1996) could be achieved, and this was sufficient to provide an excellent description of the dynamic structure factor for the two-dimensional XY-model at the Kosterlitz–Thouless transition as shown in Fig. 12.4. This result presents a real theoretical challenge, since none of the existing theoretical predictions (labeled NF (Nelson and Fisher, 1977) and Villain (1974) in the figure) can explain either the central peak or the shape of the spin wave peak. Note that the high frequency intensity falls off as a power law, in agreement with the NF theory.

For a typical spin dynamics study the major part of the cpu time needed is consumed by the numerical time integration. The biggest possible time step is thus most desirable, however, 'standard' methods impose a severe restriction on the size of Δ for which the conservation laws of the dynamics are

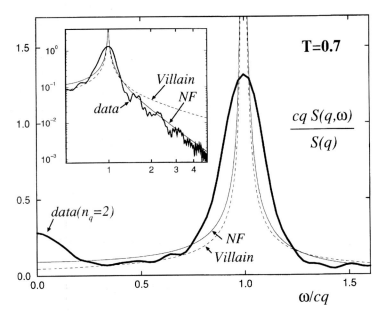

Fig. 12.4 Dynamic structure factor for the two-dimensional XY-model at T_{KT}. The heavy curve shows data obtained from spin dynamics simulations, and the light lines are theoretical predictions. From Evertz and Landau (1996).

obeyed. It is evident from Eqn. (12.11) that $|\mathbf{S}_i|$ for each lattice site i and the total energy are conserved. Symmetries of the Hamiltonian impose additional conservation laws, so, for example, for $D = 0$ and $\lambda = 1$ (isotropic Heisenberg model) the magnetization $\mathbf{m}$ is conserved. For an anisotropic Heisenberg model, i.e. $\lambda \neq 1$ or $D \neq 0$, only the z-component m_z of the magnetization is conserved. Conservation of spin length and energy is particularly crucial, and it would therefore also be desirable to devise an algorithm which conserves these two quantities exactly. In this spirit, a new, large time step integration procedure, which is based on Trotter–Suzuki decompositions of exponential operators and conserves both spin length and energy *exactly* for $D = 0$, has been devised (Krech *et al.*, 1998). Variants of this method for more general models allow very large time steps but do not necessarily conserve all quantities exactly. The conservation is nonetheless good enough for practical application.

12.4 LANGEVIN EQUATIONS AND VARIATIONS (CELL DYNAMICS)

An alternative approach to the study of a system in the canonical ensemble is to allow the particles to undergo collisions with much lighter particles, the collection of which plays the role of a heat bath. In the same way, if a system has fluctuations on both very short and relatively long time scales, it is possible to use a rather large time step and allow the effect of rapid fluctuations to be described by a random noise plus a damping term. The relevant equations to be solved are then a set of Langevin equations:

$$m_i \ddot{\mathbf{r}}_i(t) = -\Gamma \dot{\mathbf{r}}_i(t) + \mathbf{F}_i(t) + \eta(t) \tag{12.15}$$

where $\mathbf{F}_i(t)$ is the net force acting on the ith particle, Γ is the friction (damping) constant, and $\eta(t)$ is a random, uncorrelated noise with zero mean. If the damping constant is chosen carefully the system will reach equilibrium and the resultant dynamic properties will not be affected by the choice of Γ. Such Langevin simulations were quite successful in the study of distortive phase transitions (Schneider and Stoll, 1978). In a different context, Grest and Kremer (1986) used Langevin dynamics methods to study polymers in a heat bath for different values of the friction. They found that this method not only reproduced the Rouse model but remained effective at high densities and allowed differentiation between interchain couplings and the solvent.

Langevin equations often result when one is not describing the system in full atomistic detail but rather on a more coarse-grained level, e.g. binary mixtures are described by a local concentration variable $c(\mathbf{r}, t)$, fluctuating in space ($\mathbf{r}$) and time (t). For a binary solid alloy as considered in Fig. 2.9, this variable $c(\mathbf{r}, t)$ arises by averaging over the concentrations of lattice sites contained in a cell of volume L^d (in d dimensions) centered at site $\mathbf{r}$. It is then possible to derive a non-linear differential equation for $c(\mathbf{r}, t)$, supplemented by a random force. The resulting Langevin equation is used to describe spinodal decomposition (see Chapter 2) and has been studied by simulations. An efficient discretized version of this approach is known as 'cell dynamics' technique (Oono and Puri, 1988).

12.5 LATTICE GAS CELLULAR AUTOMATA

An inventive approach to the use of cellular automata to study fluid flow (Frisch *et al.*, 1986) incorporates the use of point masses on a regular lattice for simulations in which space, time, and velocity are discretized. In two dimensions, particles move on a triangular lattice, and particle number and momentum are conserved when they collide. Each particle has a vector associated with it which points along one of the lattice directions. On the triangular lattice each point has six nearest neighbors, and thus only six different values of velocity are allowed. The system progresses in time as a cellular automaton in which each particle may move one nearest neighbor distance in one time step. The system is updated by allowing particles which collide to scatter according to Newton's laws, i.e. obeying conservation of momentum. Examples of collision rules are shown schematically in Fig. 12.5. This 'lattice gas cellular automata' approach to fluid flow has been shown, at least in the limit of low velocity, to be equivalent to a discrete form of the Navier–Stokes equation, and represents a potentially very fast method to study fluid flow from a microscopic perspective. In the case of collisions which involve non-zero momentum this procedure is always used. If the total momentum of colliding particles is zero, there is a degeneracy in the resulting outcome (see Fig. 12.5) and the choice can be made by a predetermined 'tie-breaker' or through the use of a random number generator. Lattice gas models have now been used extensively to examine a number of different physical situations including flow in complex geometries, phase separation, interface properties, etc. As a demonstration of the nature of the results that one may obtain, we show in Fig. 12.6 a typical flow pattern obtained when a flat plate is inserted in front of the moving fluid.

A more complete description of lattice gas cellular automata, as well as more extensive sample results, can be found elsewhere (Rothman and Zaleski, 1994).

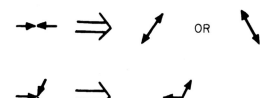

Fig. 12.5 Lattice gas cellular automata collision rules for particle movement on a triangular lattice.

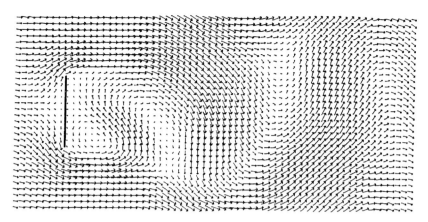

Fig. 12.6 Two-dimensional flow past a flat plate (flow from left to right) as obtained from a cellular automata lattice gas simulation. From d'Humières *et al.* (1985).

REFERENCES

Abraham, F. F. (1996), Phys. Rev. Lett. **77**, 869.

Alder, B. J. and Wainwright, T. E. (1970), Phys. Rev. A **1**, 18.

Allen, M. P. and Tildesley, D. J. (1987), *Computer Simulations of Liquids* (Clarendon Press, Oxford).

Andersen, H. C. (1980), J. Chem. Phys. **72**, 2384.

Berendsen, H. J. C. and Van Gunsteren, W. F. (1986), in *Molecular Dynamics Simulation of Statistical Mechanical Systems*, Proceedings of the Enrico Fermi Summer School, Varenna (Soc. Italiana di Fisica, Bologna).

Burden, R. L., Faires, J. D., and Reynolds, A. C. (1981), *Numerical Analysis* (Prindle, Weber, and Schmidt, Boston).

Car, R. and Parrinello, M. (1985), Phys. Rev. Lett. **55**, 2471.

Chen, K. and Landau, D. P. (1994), Phys. Rev. B **49**, 3266.

d'Humières, D., Pomeau, Y. and Lallemand, P. (1985), C. R. Acad. Sci. II **301**, 1391.

Duane, S., Kennedy, A. D., Pendleton, B. J., and Roweth, D. (1987), Phys. Lett. B **195**, 216.

Evertz, H. G. and Landau, D. P. (1996), Phys. Rev. B **54**, 12302.

Frisch, U., Hasslacher, B., and Pomeau, Y. (1986), Phys. Rev. Lett. **56**, 1505.

Gerling, R. W. and Landau, D. P. (1984), J. Magn. Mag. Mat. **45**, 267.

Grest, G. S. and Kremer, K. (1986), Phys. Rev. A **33**, 3628.

Hoover, W. G. (1985), Phys. Rev. A **31**, 1695.

Krech, M., Bunker, A., and Landau, D. P. (1998), Comput. Phys. Commun. **111**, 1.

Nelson, D. R. and Fisher, D. S. (1977), Phys. Rev. B **16**, 4945.

Nosé, S. (1984), Mol. Phys. **52**, 255.

Oono, Y. and Puri, S. (1988), Phys. Rev. A **38**, 434.

Parrinello, M. (1997), Solid State Commun. **102**, 107.

Rapaport, D. C. (1995), *The Art of Molecular Dynamics Simulation* (Cambridge University Press).

Rapaport, D. C. (1988), Phys. Rev. Lett. **60**, 2480.

Rothman, D. H. and Zaleski, S. (1994), Rev. Mod. Phys. **66**, 1417.

Schneider, T. and Stoll, E. (1978), Phys. Rev. **17**, 1302.

Tuckerman, M., Martyna, G. J., and Berne, B. J. (1992), J. Chem. Phys. **97**, 1990.

Verlet, L. (1967), Phys. Rev. **159**, 98.

Verlet, L. (1968), Phys. Rev. **165**, 201.

Villain, J. (1974), J. Phys. (Paris) **35**, 27.

13 Outlook

Within the contents of this book we have attempted to elucidate the essential features of Monte Carlo simulations and their application to problems in statistical physics. We have attempted to give the reader practical advice as well as to present theoretically based background for the methodology of the simulations as well as the tools of analysis. New Monte Carlo methods will be devised and will be used with more powerful computers, but we believe that the advice given to the reader in Section 4.8 will remain valid.

In general terms we can expect that progress in Monte Carlo studies in the future will take place along two different routes. First, there will be a continued advancement towards ultra high resolution studies of relatively simple models in which critical temperatures and exponents, phase boundaries, etc. will be examined with increasing precision and accuracy. As a consequence, high numerical resolution as well as the physical interpretation of simulational results may well provide hints to the theorist who is interested in analytic investigation. On the other hand, we expect that there will be a tendency to increase the examination of much more complicated models which provide a better approximation to physical materials. As the general area of materials science blossoms, we anticipate that Monte Carlo methods will be used to probe the often complex behavior of real materials. This is a challenge indeed, since there are usually phenomena which are occurring at different length and time scales. As a result, it will not be surprising if multiscale methods are developed and Monte Carlo methods will be used within multiple regions of length and time scales. We encourage the reader to think of new problems which are amenable to Monte Carlo simulation but which have not yet been approached with this method.

Lastly, it is likely that an enhanced understanding of the significance of numerical results can be obtained using techniques of scientific visualization. The general trend in Monte Carlo simulations is to ever larger systems studied for longer and longer times. The mere interpretation of the data is becoming a problem of increasing magnitude, and visual techniques for probing the system (again over different scales of time and length) must be developed. Coarse-graining techniques can be used to clarify features of the results which are not immediately obvious from inspection of columns of numbers. 'Windows' of various size can be used to scan the system looking for patterns which develop in both space and time; and the development of such methods may well profit from interaction with computer science.

Clearly improved computer performance is moving swiftly in the direction of parallel computing. Because of the inherent complexity of message passing, it is likely that we shall see the development of hybrid computers in which large arrays of symmetric (shared memory) multiprocessors appear. (Until much higher speeds are achieved on the Internet, it is unlikely that non-local assemblies of machines will prove useful for the majority of Monte Carlo simulations.) We must continue to examine the algorithms and codes which are used for Monte Carlo simulations to insure that they remain well suited to the available computational resources.

We strongly believe that the utility of Monte Carlo simulations will continue to grow quite rapidly, but the directions may not always be predictable. We hope that the material in this book will prove useful to the reader who wanders into unfamiliar scientific territory and must be able to create new tools instead of merely copying those that can be found in many places in the literature. If so, our efforts in developing this textbook will have been worthwhile.

Appendix

Since the thrust of the homework problems is for the student to write, debug, and run 'homemade' programs, we will not provide a compendium of simulational software. Nonetheless, to provide some aid to the student in the learning process, we will offer a few programs that demonstrate some of the basic steps in a Monte Carlo simulation. We do wish to make the reader aware, however, that these program do not have all of the 'bells and whistles' which one might wish to introduce in a serious study but are merely simple programs that can be used to test the students' approach.

Program 1 *Test a random number generator*

Note, as an exercise the student may wish to insert other random number generators or add tests to this simple program.

```
c*****************************************************************
c This program is used to perform a few very simple tests of a random
c number generator. A congruential generator is being tested
c*****************************************************************
      Real*8 Rnum(100000),Rave,R2Ave,Correl,SDev
      Integer Iseed,num
      open(Unit=1,file='result_testrng_02')
      PMod = 2147483647.0D0
      DMax = 1.0D0/PMod
c*******
c Input
c*******
      write(*,800)
800   format('enter the random number generator seed ')
      read(*,921) Iseed
921   format(i5)
      write(*,801) Iseed
      write(1,801) Iseed
801   format(' The random number seed is ', I8)
      write(*,802)
802   format('enter the number of random numbers to be generated')
      read(*,921) num
      write(*,803) num
      write(1,803)num
803   format ('number of random numbers to be generated = ',i8)
c*****************************
c Initialize variables, vectors
c*****************************
```

```fortran
      do 1 i=2,10000
1     Rnum(i)=0.0D0
      Rave=0.D0
      Correl=0.0D0
      R2Ave=0.0D0
      SDev=0.0D0
c************************
c Calculate random numbers
c************************
      Rnum(1)=Iseed*DMax
      Write(*,931) Rnum(1),Iseed
      Do 10 i=2,num
         Rnum(i)=cong16807(Iseed)
         if (num.le.100) write(*,931) Rnum(i),Iseed
931   format(f10.5,i15)
10    continue
      Rave=Rnum(1)
      R2Ave=Rnum(1)**2
      Do 20 i=2,num
         Correl=Correl+Rnum(i)*Rnum(i-1)
         Rave=Rave+Rnum(i)
20    R2Ave=R2Ave+Rnum(i)**2
      Rave=Rave/num
      SDev=Sqrt((R2Ave/num-Rave**2)/(num-1))
      Correl=Correl/(num-1)-Rave*RAve
c*******
c Output
c*******
      write(*,932) Rave,SDev,Correl
932   format('Ave. random number = ',F10.6, ' +/-', F10.6,
     1       / ' ''nn''-correlation = ' F10.6)
      write(1,932) Rave,SDev,Correl
999   format(f12.8)
      close (1)
      stop
      end


      FUNCTION Cong16807(ISeed)
c****************************************************
c This is a simple congruential random number generator
c****************************************************
      INTEGER ISeed,IMod
      REAL*8 RMod,PMod,DMax
      RMod = DBLE(ISeed)
      PMod = 2147483647.0D0
      DMax = 1.0D0/PMod
      RMod = RMod*16807.0D0
      IMod = RMod*DMax
      RMod = RMod - PMod*IMod
      cong16807=rmod*DMax
      Iseed=Rmod
      RETURN
      END
```

Program 2 *A good routine for generating a table of random numbers*

```
C***************************************************************
C This program uses the R250/R521 combined generator described in:
C A. Heuer, B. Duenweg and A.M. Ferrenberg, Comp. Phys. Comm. 103, 1
C 1997). It generates a vector, RanVec, of length RanSize 31-bit random
C integers. Multiply by RMaxI to get normalized random numbers. You
C will need to test whether RanCnt will exceed RanSize. If so, call
C GenRan again to generate a new block of RanSize numbers. Always
C remember to increment RanCnt when you use a number from the table.
C***************************************************************
      IMPLICIT NONE
      INTEGER RanSize,Seed,I,RanCnt,RanMax
      PARAMETER(RanSize = 10000)
      PARAMETER( RanMax = 2147483647 )
      INTEGER RanVec(RanSize),Z1(250+RanSize),Z2(521+RanSize)
      REAL*8 RMaxI
      PARAMETER ( RMaxI = 1.0D0/(1.0D0*RanMax) )
      COMMON/MyRan/RanVec,Z1,Z2,RanCnt
      SAVE
      Seed = 432987111
C****************************************
C Initialize the random number generator.
C****************************************
      CALL InitRan(Seed)*
C***************************************************************
c If the 10 numbers we need pushes us past the end of the RanVec vector,
C call GenRan. Since we just called InitRan, RanCnt = RanSize we must
c call it here.
C***************************************************************
      IF ((RanCnt + 10) .GT. RanSize) THEN
C**   Generate RanSize numbers and reset the RanCnt counter to 1
         Call GenRan
      END IF
      Do I = 1,10
         WRITE(*,*) RanVec(RanCnt + I - 1),RMaxI*RanVec(RanCnt + I - 1)
      End Do
      RanCnt = RanCnt + 10
C***************************************************************
C Check to see if the 10 numbers we need will push us past the end
C of the RanVec vector. If so, call GenRan.
C***************************************************************
      IF ((RanCnt + 10) .GT. RanSize) THEN
C**   Generate RanSize numbers and reset the RanCnt counter to 1
         Call GenRan
      END IF
      Do I = 1,10
         WRITE(*,*) RanVec(RanCnt + I - 1),RMaxI*RanVec(RanCnt + I - 1)
      End Do
      RanCnt = RanCnt + 10
      END
```

```
      SUBROUTINE InitRan(Seed)
C*****************************************************************
C Initialize the R250 and R521 generators using a congruential generator
C to set the individual bits in the 250/521 numbers in the table. The
C R250 and R521 are then warmed-up by generating 1000 numbers.
C*****************************************************************
      IMPLICIT NONE
      REAL*8 RMaxI,RMod,PMod
      INTEGER RanMax,RanSize
      PARAMETER ( RanMax = 2147483647 )
      PARAMETER (RanSize = 100000)
      PARAMETER ( RMaxI = 1.0D0/(1.0D0*RanMax) )
      INTEGER Seed,I,J,K,IMod,IBit
      INTEGER RanVec(RanSize),Z1(250+RanSize),Z2(521+RanSize)
      INTEGER RanCnt
      COMMON/MyRan/RanVec,Z1,Z2,RanCnt
      SAVE
      RMod = DBLE(Seed)
      PMod = DBLE(RanMax)
C**********************************
C Warm up a congruential generator
C**********************************
      Do I = 1,1000
         RMod = RMod*16807.0D0
         IMod = RMod/PMod
         RMod = RMod - PMod*IMod
      End Do
C*****************************************************************
C Now fill up the tables for the R250 & R521 generators: This
C requires random integers in the range 0-> 2*31 1. Iterate a
C strange number of times to improve randomness.
C*****************************************************************
      Do I = 1,250
         Z1(I) = 0
         IBit = 1
         Do J = 0,30
            Do K = 1,37
               RMod = RMod*16807.0D0
               IMod = RMod/PMod
               RMod = RMod - PMod*IMod
            End Do
C**   Now use this random number to set bit J of X(I).
            IF (RMod .GT. 0.5D0*PMod) Z1(I) = IEOR(Z1(I),IBit)
            IBit = IBit*2
         End Do
      End Do
      Do I = 1,521
         Z2(I) = 0
         IBit = 1
         Do J = 0,30
            Do K = 1,37
               RMod = RMod*16807.0D0
               IMod = RMod/PMod
               RMod = RMod - PMod*IMod
            End Do
C**   Now use this random number to set bit J of X(I).
            IF (RMod .GT. 0.5D0*PMod) Z2(I) = IEOR(Z2(I),IBit)
```

```
      IBit = IBit*2
      End Do
      End Do
C*********************************************************************
C Perform a few iterations of the R250 and R521 random number generators
C to eliminate any effects due to 'poor' initialization.
C*********************************************************************
      Do I = 1,1000
         Z1(I+250) = IEOR(Z1(I),Z1(I+147))
         Z2(I+521) = IEOR(Z2(I),Z2(I+353))
      End Do
      Do I = 1,250
         Z1(I) = Z1(I + 1000)
      End Do
      Do I = 1,521
         Z2(I) = Z2(I + 1000)
      End Do
C*********************************************************************
C Set the random number counter to RanSize so that a proper checking
C code will force a call to GenRan in the main program.
C*********************************************************************
      RanCnt = RanSize
      RETURN
      END


      SUBROUTINE GenRan
C*********************************************************************
C Generate vector RanVec (length RanSize) of pseudo-random 31-bit
C integers.
C*********************************************************************
      IMPLICIT NONE
      INTEGER RanSize,RanCnt,I
      PARAMETER(RanSize = 100000)
      INTEGER RanVec(RanSize),Z1(250+RanSize),Z2(521+RanSize)
      COMMON/MyRan/RanVec,Z1,Z2,RanCnt
      SAVE
C*********************************************************************
C Generate RanSize pseudo-random nubmers using the individual gen-
erators
C*********************************************************************
      Do I = 1,RanSize
         Z1(I+250) = IEOR(Z1(I),Z1(I+147))
         Z2(I+521) = IEOR(Z2(I),Z2(I+353))
      End Do
C*********************************************************************
C Combine the R250 and R521 numbers and put the result into RanVec
C*********************************************************************
      Do I = 1,RanSize
         RanVec(I) = IEOR(Z1(I+250),Z2(I+521))
      End Do
C*********************************************************************
C Copy the last 250 numbers generated by R250 and the last 521 numbers
C from R521 into the working vectors (Z1), (Z2) for the next pass.
C*********************************************************************
      Do I = 1,250
         Z1(I) = Z1(I + RanSize)
      End Do
```

```
      Do I = 1,521
         Z2(I) = Z2(I + RanSize)
      End Do
C***************************************
C Reset the random number counter to 1.
C***************************************
      RanCnt = 1
      RETURN
      END
```

Program 3 *The Hoshen–Kopelman cluster finding routine*

```
c*******************************************************************
c lx,ly = lattice size along x,y
c ntrymax = number of lattices to be studied for each concentration
c iclmax = number of clusters (including those of 0 elements) found
c in a lattice configuration for a given concentration
c ioclmax = number of different cluster sizes found
c ns(1,j) = cluster size, j=1,ioclmax
c ns(2,j) = number of clusters of that size, j=1,ioclmax
c ninf = number of infinite clusters
c ninf/ntrymax = probability of infinite cluster
c
c For more details on the method, see:
c J. Hoshen and R. Kopelman, Phys. Rev. B14, 3428 (1976).
c*******************************************************************
      Parameter(lxmax=500,lymax=500)
      Parameter(nnat=lxmax*lymax,nclustermax=nnat/2+1)
      Integer isiti(lxmax,lymax)
      Integer list(nnat),ncluster(nnat),nlabel(nclustermax)
      Integer ibott(lxmax),itop(lxmax),ileft(lymax),iright(lymax)
      Integer iperc(100),nsize(nclustermax),ns(2,nclustermax)
      Character*40 fout
c************
c Input data
c************
      read(5,*)lx
      read(5,*)ly
      read(5,*)fout
      if (lx.gt.lxmax) stop 'lx too big'
      if (ly.gt.lymax) stop 'ly too big'
c******************
c List of the sites
c******************
      num=0
      do j=1,lx
         do i=1,ly
            num=num+1
            isiti(i,j)=num
         enddo
      enddo
      nat=num
c**************
c Initialize
c**************
      ninf=0
```

```
      iocl=0
      ns(1,icl)=0
      ns(2,icl)=0
      do num=1,nat
         list(num)=0
         ncluster(num)=0
      enddo
      do icl=1,nclustermax
         nsize(icl)=0
      enddo
      open(unit=50,file=fout,status='unknown',form='formatted')
c******************
c Input spins
c******************
      do iy=1,ly
         read(5,*) (list(isiti(ix,iy)),ix=1,lx)
      enddo
c************************
c Analysis of the cluster
c************************
      icl=0
      if (list(1).eq.1) then
         icl=icl+1
         ncluster(1)=icl
         nlabel(icl)=icl
      endif
      do num=2,lx
         if (list(num).eq.1) then
             if (list(num-1).eq.1) then
             ivic1=ncluster(num-1)
             ilab1=nlabel(ivic1)
             ncluster(num)=ilab1
             icheck=1
           else
             icl=icl+1
             ncluster(num)=icl
             nlabel(icl)=icl
           endif
         endif
      enddo
      do jj=1,ly-1
         num=jj*lx+1
            if (list(num).eq.1) then
             if (list(num-lx).eq.1) then
             ivic2=ncluster(num-lx)
             ilab2=nlabel(ivic2)
             ncluster(num)=ilab2
             icheck=1
           else
             icl=icl+1
             ncluster(num)=icl
             nlabel(icl)=icl
           endif
      endif
      do num=jj*lx+2,(jj+1)*lx
         if (list(num).eq.1) then
             if (list(num-1).eq.1) then
```

```
          ivic1=ncluster(num-1)
          ilab1=nlabel(ivic1)
          if (list(num-lx).eq.1) then
          ivic2=ncluster(num-lx)
          ilab2=nlabel(ivic2)
          imax=max(ilab1,ilab2)
          imin=min(ilab1,ilab2)
          ncluster(num)=imin
          nlabel(imax)=nlabel(imin)
          do kj=1,icl
             if (nlabel(kj).eq.imax) nlabel(kj)=imin
          enddo
          icheck=1
        else
           ncluster(num)=ilab1
           icheck=1
        endif
      else
        if (list(num-lx).eq.1) then
           ivic2=ncluster(num-lx)
           ilab2=nlabel(ivic2)
           ncluster(num)=ilab2
           icheck=1
        else
           icl=icl+1
           ncluster(num)=icl
           nlabel(icl)=icl
        endif
      endif
    endif
    enddo
      if (icheck.eq.0) then
          write(*,*) 'no possible percolation'
          go to 2000
      endif
      icheck=0
    enddo
    iclmax=icl
c************************************************
c Determination of the number of infinite clusters
c************************************************
    io=0
    do num=1,lx
      itest=0
      if (list(num).eq.1) then
         ilab=nlabel(ncluster(num))
         call conta(num,ilab,ibott,itest,io,lx)
      endif
    enddo
    iomax=io
    in=0
    do num=(ly-1)*lx+1,nat
      itest=0
      if (list(num).eq.1) then
         ilab=nlabel(ncluster(num))
         call conta(num,ilab,itop,itest,in,lx)
      endif
```

```
      enddo
      inmax=in
      il=0
      do num=1,nat,lx
         itest=0
         if (list(num).eq.1) then
            ilab=nlabel(ncluster(num))
            call conta(num,ilab,ileft,itest,il,ly)
         endif
      enddo
      ilmax=il
      ir=0
      do num=lx,nat,lx
         itest=0
         if (list(num).eq.1) then
            ilab=nlabel(ncluster(num))
            call conta(num,ilab,iright,itest,ir,ly)
         endif
      enddo
      irmax=ir
      nperc=0
      nperc1=0
      np=0
      do ii=1,iomax
         do jj=1,inmax
            if (itop(jj).eq.ibott(ii)) then
               nperc=nperc+1
               np=np+1
               iperc(np)=nperc
            endif
         enddo
      enddo
      npmax=np
      itest2=0
      do ii=1,irmax
         do jj=1,ilmax
            if (ileft(jj).eq.iright(ii)) then
               do np=1,npmax
               if (ileft(jj).eq.iperc(np)) itest2=1
            enddo
            if (itest2.eq.0) nperc=nperc+1
            endif
         enddo
      enddo
      if (nperc.gt.0) nperc1=1
      if (nperc.gt.0) ninf=ninf+1
      call size(nat,iclmax,nsize,nlabel,ncluster,ns,iocl,
     *     nclustermax)
      ioclmax=iocl
      fl=1.0/float(nat)
      do icl=1,ioclmax
         fl1=log(float(ns(1,icl)))
         fl2=log(float(ns(2,icl))*fl)
         write (50,*) ns(1,icl),ns(2,icl),float(ns(2,icl))*fl,fl1,fl2
      enddo
      write (*,*) 'Number of cluster sizes = ',ioclmax
      write (*,*) 'Number of infinite clusters =',ninf
```

```
      SUBROUTINE monte(mcstps,Irng)
c********************************
c Perform a Monte Carlo step/site
c********************************
      Integer*2 Ispin(80)
      Integer*2 neigh(20)
      Real*4 prob(9,3),rn
      Common/spins/Ispin
      Common/sizes/n,nsq
      Common/trans/ prob
      nm1=n-1
      if(nm1.eq.0) nm1=1
      do 1 mc=1,mcstps
         jmc=0
         do 2 jj=1,n
            j=n*RAN(Irng)+1.0e-06
            jp=j+1
            if(jp.gt.n) j=1
            jm=j-1
            if(jm.lt.1) jm=n
            rn=RAN(Irng)
            jmc=jmc+1
            nc=Ispin(j)
            n4=Ispin(jm)+Ispin(jp)
            n4=nc*n4+3
            nh=nc+2
            if(rn.gt.prob(n4,nh)) goto 6
            Ispin(j)=-nc
6           continue
2        continue
1     continue
      return
      end


      SUBROUTINE carlo(new)
c**********************************************
c Calculate the table of flipping probabilities
c**********************************************
      Logical new
      Integer*2 Ispin(80)
      Real*4 prob(9,3)
      Common/spins/Ispin
      Common/sizes/n,nsq
      Common/trans/ prob
      Common/param/beta,betah
      nsq=n*n
      if((abs(betah).gt.30.0).or.(abs(beta).gt.30.0)) then
         write(*,6666)
6666     format(' Stop the simulation; the temperature is too cold!')
         stop
      endif
      do 11 j=1,5
         do 11 jh=1,3
            prob(j,jh)=exp(-2.0*beta*(j-3)-2.0*betah*(jh-2))
11    continue
      if(.not.new) return
      new=.false.
```

```
      do 2 j=1,n
         Ispin(j)=1
2     continue
      write(*,950)
950   format('initial state:')
      call picture
      write(*,960)
960   format(//)
      return
      end

      SUBROUTINE results(lll)
c***************
c Output results
c***************
      Real*8 e(99),ee(99),am(99),amm(99),am4(99),U(99)
      Real*8 dam(99),de(99),spheat(99),cor(20),wnum
      Real temper(99),fields(99)
      Common/inparm/ temp,field,Jint
      Common/sizes/n,nsq
      Common/index/l
      Common/corrs/cor
      if(lll) 1,2,3
1     continue
      e(l)=0.0d0
      ee(l)=0.0d0
      am(l)=0.0d0
      amm(l)=0.0d0
      am4(l)=0.0d0
      num=0
      return
2     continue
      num=num+1
      e(l)=e(l)+cor(1)
      ee(l)=ee(l)+cor(1)*cor(1)
      am(l)=am(l)+cor(2)
      amm(l)=amm(l)+cor(2)*cor(2)
      am4(l)= am4(l)+cor(2)**4
      return
3     continue
      if(lll.gt.1) goto 4
      write(*,99)
99    format(/t4,'T',t10,'H',t17,'U4',t25,'E',t31,'E*E',
     * t39,'dE**2',t50,'M',t58,'M*M',t66,'dM**2',t76,'C')
      wnum=1.0d0/num
      temper(l)=temp
      fields(l)=field
      e(l)=e(l)*wnum
      ee(l)=ee(l)*wnum
      am(l)=am(l)*wnum
      amm(l)=amm(l)*wnum
      am4(l)=am4(l)*wnum
      de(l)=ee(l)-e(l)*e(l)
      dam(l)=amm(l)-am(l)*am(l)
      U(l)=1.0d0-am4(l)/(3.0d0*amm(l)**2)
      fn=1.0d0*n
      spheat(l)=fn*de(l)/(temper(l)**2)
```

```
      write(*,100) temper(l),fields(l), U(l),e(l),ee(l),de(l),
     * am(l),amm(l),dam(l),spheat(l)
      return
4     continue
      write(*,900)
900   format('Summary of the results:')
      write(*,99)
      write(1,99)
      do 55 j=1,l
         write(*,100) temper(j),fields(j),U(j),e(j),ee(j),de(j),
     * am(j),amm(j),dam(j),spheat(j)
         write(1,100) temper(j),fields(j),U(j),e(j),ee(j),de(j),
     * am(j),amm(j),dam(j),spheat(j)
100   format(2f6.3, 3f8.4,f8.4,f9.5,f9.5,f9.5,f7.3)
55    continue
      return
      end
```

Program 5 *The bond fluctuation method*

Note, this program contains yet another random number generator.

```
c*****************************************************************
c This program simulates a simple 3-dim lattice model for polymers
c using the athermal bond-fluctuation method. For more details see:
c I. Carmesin and K. Kremer, Macromolecules 21, 2878 (1988).
c*****************************************************************
      Implicit none
      Integer seed, nrmeas, mcswait
      Character*50 infile,outfile,outres
      include ''model.common''
      include ''lattice.common''
      write(*,*) 'input file for the old configuration:'
      read(*,'(a50)') infile
      write(*,*) infile
      write(*,*) 'output file for the new configuration:'
      read(*,'(a50)') outfile
      write(*,*) outfile
      write(*,*) 'output file for measurements:'
      read(*,'(a50)') outres
      write(*,*) outres
      write(*,*) 'time lapse between two measurements:'
      read(*,*) mcswait
      write(*,*) mcswait
      write(*,*) 'number of measurements:'
      read(*,*) nrmeas
      write(*,*) nrmeas
      write(*,*) 'seed for the random number generator:'
      read(*,*) seed
      write(*,*) seed
c*******************************
c Initialize the bond vectors
c*******************************
      call bdibfl
c*************************************************
c Initialize the bond angles and index for the bond angles
c*************************************************
```

```
      call aninbfl
c*****************************************
c Initialize the table for the allowed moves
c*****************************************
      call inimove
c***********************************************
c read in the configuration and initialize the lattice
c***********************************************
      call bflin(infile)
c******************
c MC simulation part
c******************
      call bflsim(mcswait,nrmeas,seed,outres)
c****************************
c write out the end configuration
c****************************
      call bflout(outfile)
      end


      SUBROUTINE aninbfl
c************************************************
c This program calculates the possible bond-angles
c************************************************
      Implicit none
      Real skalp(108,108), winkel(100), pi
      Integer indx(100), index, i, j, k, double, new(88), sawtest
      Logical test
      include ''model.common''
c********************************
c Initializing the set of bond angles
c********************************
      pi = 4.0 * atan(1.0)
      index = 1
      do 410 i=1,108
        do 410 j=1,108
          winkel(index) = 5.0
          test = .false.
          sawtest =  (bonds(i,1)+bonds(j,1))**2 +
     *               (bonds(i,2)+bonds(j,2))**2 +
     *               (bonds(i,3)+bonds(j,3))**2
          if(sawtest.ge.4) then
             test = .true.
          skalp(i,j) = bonds(i,1)*bonds(j,1) +
     *                 bonds(i,2)*bonds(j,2) +
     *                 bonds(i,3)*bonds(j,3)
          skalp(i,j) = skalp(i,j) /(bl(i)*bl(j))
          skalp(i,j) = min(skalp(i,j),1.0)
          skalp(i,j) = max(skalp(i,j),-1.0)
          skalp(i,j) = pi - acos(skalp(i,j))
             do 411 k=1,index
                if(abs(skalp(i,j)-winkel(k)).le.0.001) then
                   test = .false.
                   angind(i,j) = k
                endif
411   continue
      if(test) then
         winkel(index) = skalp(i,j)
```

```
             angind(i,j) = index
             index = index + 1
             winkel(index) = 5.0
          endif
      else
          angind(i,j) = 100
      endif
410   continue
      do 417 i=1,108
          do 417 j=1,108
             if(angind(i,j).eq.100) angind(i,j) = index
417   continue
      call indexx(index,winkel,indx)
      do 412 i=1,index
          angles(i) = winkel(indx(i))
          new(indx(i)) = i
412   continue
      do 413 i=1,108
          do 413 j=1,108
             angind(i,j) = new(angind(i,j))
413   continue
      return
      end


      SUBROUTINE bdibfl
c***************************************************************
c This subroutine creates the allowed bond-set and passes it back.
c***************************************************************
      Implicit none
      Integer max, ipegel, i, j, k, index, ind
      Integer startvec(6,3), zielvec(50,3),testb(3),sumvec(3)
      Integer dumvec(50,3), bondnr, newbond(3), dummy
      Logical test, foundbond
      include ''model.common''
c*********************************
c INITIALIZING POSSIBLE BONDVECTORS
c*********************************
      startvec(1,1) = 2
      startvec(1,2) = 0
      startvec(1,3) = 0
      startvec(2,1) = 2
      startvec(2,2) = 1
      startvec(2,3) = 0
      startvec(3,1) = 2
      startvec(3,2) = 1
      startvec(3,3) = 1
      startvec(4,1) = 2
      startvec(4,2) = 2
      startvec(4,3) = 1
      startvec(5,1) = 3
      startvec(5,2) = 0
      startvec(5,3) = 0
      startvec(6,1) = 3
      startvec(6,2) = 1
      startvec(6,3) = 0
      max = 0
         do 210 i=1,6
```

```
            ind = 1
            do 211 j=1,2
               do 212 k=1,3
                  zielvec(ind,1) = startvec(i,1)
                  zielvec(ind,2) = startvec(i,2)
                  zielvec(ind,3) = startvec(i,3)
                  ind = ind + 1
                  zielvec(ind,1) = startvec(i,1)
                  zielvec(ind,2) = startvec(i,2)
                  zielvec(ind,3) = - startvec(i,3)
                  ind = ind + 1
                  zielvec(ind,1) = startvec(i,1)
                  zielvec(ind,2) = - startvec(i,2)
                  zielvec(ind,3) = startvec(i,3)
                  ind = ind + 1
                  zielvec(ind,1) = startvec(i,1)
                  zielvec(ind,2) = - startvec(i,2)
                  zielvec(ind,3) = - startvec(i,3)
                  ind = ind + 1
                  zielvec(ind,1) = - startvec(i,1)
                  zielvec(ind,2) = startvec(i,2)
                  zielvec(ind,3) = startvec(i,3)
                  ind = ind + 1
                  zielvec(ind,1) = - startvec(i,1)
                  zielvec(ind,2) = startvec(i,2)
                  zielvec(ind,3) = - startvec(i,3)
                  ind = ind + 1
                  zielvec(ind,1) = - startvec(i,1)
                  zielvec(ind,2) = - startvec(i,2)
                  zielvec(ind,3) = startvec(i,3)
                  ind = ind + 1
                  zielvec(ind,1) = - startvec(i,1)
                  zielvec(ind,2) = - startvec(i,2)
                  zielvec(ind,3) = - startvec(i,3)
                  ind = ind + 1
                  dummy = startvec(i,1)
                  startvec(i,1) = startvec(i,2)
                  startvec(i,2) = startvec(i,3)
                  startvec(i,3) = dummy
212   continue
      dummy = startvec(i,1)
      startvec(i,1) = startvec(i,2)
      startvec(i,2) = dummy
211   continue
      dumvec(1,1) = zielvec(1,1)
      dumvec(1,2) = zielvec(1,2)
      dumvec(1,3) = zielvec(1,3)
      ipegel = 2
      do 213 k=1,48
         index = 1
         test = .false.
333   if((.not.test).and.(index.lt.ipegel)) then
         test =   ((zielvec(k,1).eq.dumvec(index,1)).and.
     *            (zielvec(k,2).eq.dumvec(index,2))).and.
     *            (zielvec(k,3).eq.dumvec(index,3))
         index = index + 1
         goto 333
```

```
      endif
      if(.not.test) then
         dumvec(ipegel,1) = zielvec(k,1)
         dumvec(ipegel,2) = zielvec(k,2)
         dumvec(ipegel,3) = zielvec(k,3)
         ipegel = ipegel + 1
      endif
213   continue
      do 214 j=1,ipegel-1
         bonds(max+j,1) = dumvec(j,1)
         bonds(max+j,2) = dumvec(j,2)
         bonds(max+j,3) = dumvec(j,3)
214   continue
      max = max + ipegel - 1
210   continue
      do 220 i=1,108
         bl2(i) = bonds(i,1)**2 + bonds(i,2)**2 + bonds(i,3)**2
         bl(i) = sqrt(bl2(i))
220   continue
      return
      end


SUBROUTINE bflin(infile)
c*****************************************************************
c This subroutine reads in an old configuration. The first line of the
c configuration file contains the number of chains and degree of poly-
c merization. The chain conformations are stored in consecutive lines:
c One line contains x, y and z coordinates of the start monomer of the
c chain, and the next lines each contain 10 integers which are the
c numbers of the bonds connecting adjacent monomers. For each chain
c the last bond number is 109, indicating a chain end without a bond.
c This works only for chains with length N=k*10. The coordinates of
c monomers 2 to N are then reconstructed from this information.
c*****************************************************************
      Implicit none
      Character*50 infile
      Integer i, j, jj, k, kd, kk, xp, yp, zp, xp1, yp1, zp1, nb,base
      Include ''model.common''
      Include ''lattice.common''
      open(11,file=infile,form='formatted',status='old')
      read(11,*) nrchains,polym
      ntot = nrchains * polym
      nb = polym / 10
      do 1 j=1,nrchains
         base = polym * (j-1)
         read(11,*) monpos(base+1,1),monpos(base+1,2),monpos
            (base+1,3)
         do 2 jj = 0,nb-1
         read(11,*) (monbd(k+10*jj+base),k=1,10)
2        continue
      do 3 k=2,polym
      do 3 kd=1,3
         monpos(base+k,kd) =  monpos(base+k-1,kd) +
     *                        bonds(monbd(base+k-1),kd)
         monlatp(base+k,kd) = mod(monpos(base+k,kd),ls) + 1
         if(monlatp(base+k,kd).le.0) then
            monlatp(base+k,kd) = monlatp(base+k,kd) + ls
```

```
        endif
3       continue
1       continue
        monbd(0) = 109
        monbd(ntot+1) = 109
c*****************************************************************
c These are the arrays for the periodic boundary conditions.
c*****************************************************************
        do 10 i=1,ls
           ip(i) = i+1
              ip2(i) = i+2
              im(i) = i-1
10      continue
        ip(ls) = 1
        ip2(ls-1) = 1
        ip2(ls) = 2
        im(1) = ls
c*****************************************************************
c Now we initialize the lattice, setting all occupied vertices to unity
c*****************************************************************
        do 4 j=1,ls
           do 4 k=1,ls
              do 4 kk=1,ls
                 latt(j,k,kk) = 0
4       continue
        do 5 j=1,ntot
           xp = monlatp(j,1)
           yp = monlatp(j,2)
           zp = monlatp(j,3)
           xp1 = ip(xp)
           yp1 = ip(yp)
           zp1 = ip(zp)
           latt(xp,yp,zp) = 1
           latt(xp1,yp,zp) = 1
           latt(xp,yp1,zp) = 1
           latt(xp,yp,zp1) = 1
           latt(xp1,yp1,zp) = 1
           latt(xp1,yp,zp1) = 1
           latt(xp,yp1,zp1) = 1
           latt(xp1,yp1,zp1) = 1
5       continue
        end

        SUBROUTINE bflout(outfile)
c*****************************************************************
c Stores the final configuration of the simulation into a configura-
c tion file for use as a start configuration for a continuation run.
c*****************************************************************
        Implicit none
        Character*50 outfile
        Integer j, jj, k, nb, base
        include ''model.common''
        open(13,file=outfile,form='formatted',status='unknown')
        write(13,*) nrchains,polym
        nb = polym / 10
        do 1 j=1,nrchains
           base = polym*(j-1) + 1
```

```
         write(13,*) monpos(base,1),monpos(base,2),monpos(base,3)
         do 2 jj = 0,nb-1
            base = polym * (j-1) + 10 * jj
            write(13,'(10I4)') (monbd(k+base),k=1,10)
2        continue
1        continue
         end



   SUBROUTINE bflsim(mcswait,nrmeas,seed,outres)
c****************************************************************
c Performs the actual Monte Carlo simulation using jumps to nearest-
c neighbor sites as the only type of moves.
c****************************************************************
      Implicit none
      Double precision r2m,r4m,rg2m,rg4m,lm,l2m
      Double precision rgnorm, blnorm, accept
      Real u(97), c, cd, cm
      Integer mcswait, nrmeas, seed, dir
      Integer i97, j97, imeas, iwait, ind, mono, xp, yp, zp
      Integer xm1, xp1, xp2, ym1, yp1, yp2, zm1, zp1, zp2
      Iinteger newbl, newbr, testlat
      Logical test
      Character*50 outres
      include ''model.common''
      include ''lattice.common''
      Common/raset1/ u,c,cd,cm,i97,j97
      Common/static/ r2m,r4m,rg2m,rg4m,lm,l2m
      open(12,file=outres,form='formatted',status='unknown')
c***********************************************************
c Initialize the cumulative measurement variables.
c***********************************************************
      r2m = 0.0d0
      r4m = 0.0d0
      rg2m = 0.0d0
      rg4m = 0.0d0
      lm = 0.0d0
      l2m = 0.0d0
      accept = 0.0d0
c***************************************
c Initialize the random number generator
c***************************************
      call rmarin(seed)
c****************************************************************
c loop over the number of measurements we wish to perform.
c****************************************************************
      do 10 imeas=1,nrmeas
c****************************************************************
c loop over the number of Monte Carlo steps between two measurements
c****************************************************************
      do 20 iwait=1,mcswait
         call ranmar(rand,3*ntot)
         ind = 1
         mono = ntot * rand(ind) + 1
         dir = 6 * rand(ind+1) + 1
         newbl = move(monbd(mono-1),dir)
         newbr = move(monbd(mono),dir)
```

```
       test = (newbl.eq.0).or.(newbr.eq.0)
       if(.not.test) then
            xp = monlatp(mono,1)
            yp = monlatp(mono,2)
            zp = monlatp(mono,3)
            if(dir.eq.1) then
c*************************
c jump in +x direction
c*************************
      xp2 = ip2(xp)
      xp1 = ip(xp)
      yp1 = ip(yp)
      zp1 = ip(zp)
      testlat =  latt(xp2,yp,zp) + latt(xp2,yp1,zp) +
     *           latt(xp2,yp,zp1) + latt(xp2,yp1,zp1)
      if(testlat.eq.0) then
c*************************************
c new monomer positions and new bonds
c*************************************
      monpos(mono,1) = monpos(mono,1) + 1
      monlatp(mono,1) = xp1
      monbd(mono-1) = newbl
      monbd(mono) = newbr
c*******************************************************************
c set the newly occupied vertices to one and the old to zero.
c*******************************************************************
            latt(xp2,yp,zp) = 1
            latt(xp2,yp1,zp) = 1
            latt(xp2,yp,zp1) = 1
            latt(xp2,yp1,zp1) = 1
            latt(xp,yp,zp) = 0
            latt(xp,yp1,zp) = 0
            latt(xp,yp,zp1) = 0
            latt(xp,yp1,zp1) = 0
            accept = accept + 1.0d0
         endif
      endif
      if(dir.eq.6) then
c*************************
c jump in -x direction
c*************************
      xm1 = im(xp)
      xp1 = ip(xp)
      yp1 = ip(yp)
      zp1 = ip(zp)
      testlat =  latt(xm1,yp,zp) + latt(xm1,yp1,zp) +
     *           latt(xm1,yp,zp1) + latt(xm1,yp1,zp1)
      if(testlat.eq.0) then
c*************************************
c new monomer positions and new bonds
c*************************************
      monpos(mono,1) = monpos(mono,1) - 1
      monlatp(mono,1) = xm1
      monbd(mono-1) = newbl
      monbd(mono) = newbr
```

```
c*****************************************************************
c set the newly occupied vertices to one and the old to zero.
c*****************************************************************
            latt(xm1,yp,zp) = 1
            latt(xm1,yp1,zp) = 1
            latt(xm1,yp,zp1) = 1
            latt(xm1,yp1,zp1) = 1
            latt(xp1,yp,zp) = 0
            latt(xp1,yp1,zp) = 0
            latt(xp1,yp,zp1) = 0
            latt(xp1,yp1,zp1) = 0
            accept = accept + 1.0d0
         endif
      endif
      if(dir.eq.2) then
c************************
c jump in +y direction
c************************
      xp1 = ip(xp)
      yp1 = ip(yp)
      yp2 = ip2(yp)
      zp1 = ip(zp)
      testlat = latt(xp,yp2,zp) + latt(xp1,yp2,zp) +
     *            latt(xp,yp2,zp1) + latt(xp1,yp2,zp1)
      if(testlat.eq.0) then
c*****************************************
c new monomer positions and new bonds
c*****************************************
      monpos(mono,2) = monpos(mono,2) + 1
      monlatp(mono,2) = yp1
      monbd(mono-1) = newbl
      monbd(mono) = newbr
c*****************************************************************
c set the newly occupied vertices to one and the old to zero.
c*****************************************************************
            latt(xp,yp2,zp) = 1
            latt(xp1,yp2,zp) = 1
            latt(xp,yp2,zp1) = 1
            latt(xp1,yp2,zp1) = 1
            latt(xp,yp,zp) = 0
            latt(xp1,yp,zp) = 0
            latt(xp,yp,zp1) = 0
            latt(xp1,yp,zp1) = 0
            accept = accept + 1.0d0
         endif
      endif
      if(dir.eq.5) then
c************************
c jump in -y direction
c************************
      xp1 = ip(xp)
      yp1 = ip(yp)
      ym1 = im(yp)
      zp1 = ip(zp)
      testlat = latt(xp,ym1,zp) + latt(xp1,ym1,zp) +
     *            latt(xp,ym1,zp1) + latt(xp1,ym1,zp1)
      if(testlat.eq.0) then
```

```
c*******************************************
c new monomer positions and new bonds
c*******************************************
      monpos(mono,2) = monpos(mono,2) - 1
      monlatp(mono,2) = ym1
      monbd(mono-1) = newbl
      monbd(mono) = newbr
c***********************************************************************
c set the newly occupied vertices to one and the old to zero.
c***********************************************************************
            latt(xp,ym1,zp) = 1
            latt(xp1,ym1,zp) = 1
            latt(xp,ym1,zp1) = 1
            latt(xp1,ym1,zp1) = 1
            latt(xp,yp1,zp) = 0
            latt(xp1,yp1,zp) = 0
            latt(xp,yp1,zp1) = 0
            latt(xp1,yp1,zp1) = 0
            accept = accept + 1.0d0
         endif
      endif
      if(dir.eq.3) then
c**************************
c jump in +z direction
c**************************
      xp1 = ip(xp)
      yp1 = ip(yp)
      zp1 = ip(zp)
      zp2 = ip2(zp)
      testlat =  latt(xp,yp,zp2) + latt(xp1,yp,zp2) +
     *           latt(xp,yp1,zp2) + latt(xp1,yp1,zp2)
      if(testlat.eq.0) then
c*******************************************
c new monomer positions and new bonds
c*******************************************
      monpos(mono,3) = monpos(mono,3) + 1
      monlatp(mono,3) = zp1
      monbd(mono-1) = newbl
      monbd(mono) = newbr
c***********************************************************************
c set the newly occupied vertices to one and the old to zero.
c***********************************************************************
            latt(xp,yp,zp2) = 1
            latt(xp1,yp,zp2) = 1
            latt(xp,yp1,zp2) = 1
            latt(xp1,yp1,zp2) = 1
            latt(xp,yp,zp) = 0
            latt(xp1,yp,zp) = 0
            latt(xp,yp1,zp) = 0
            latt(xp1,yp1,zp) = 0
            accept = accept + 1.0d0
         endif
      endif
      if(dir.eq.4) then
c**************************
c jump in -z direction
c**************************
```

```fortran
      xp1 = ip(xp)
      yp1 = ip(yp)
      zp1 = ip(zp)
      zm1 = im(zp)
      testlat =  latt(xp,yp,zm1) + latt(xp1,yp,zm1) +
     *           latt(xp,yp1,zm1) + latt(xp1,yp1,zm1)
      if(testlat.eq.0) then
c*************************************
c new monomer positions and new bonds
c*************************************
      monpos(mono,3) = monpos(mono,3) - 1
      monlatp(mono,3) = zm1
      monbd(mono-1) = newbl
      monbd(mono) = newbr
c*********************************************************************
c set the newly occupied vertices to one and the old to zero.
c*********************************************************************
                latt(xp,yp,zm1) = 1
                latt(xp1,yp,zm1) = 1
                latt(xp,yp1,zm1) = 1
                latt(xp1,yp1,zm1) = 1
                latt(xp,yp,zp1) = 0
                latt(xp1,yp,zp1) = 0
                latt(xp,yp1,zp1) = 0
                latt(xp1,yp1,zp1) = 0
                accept = accept + 1.0d0
              endif
            endif
          endif
        ind = ind + 3
20    continue
c*******************************************
c calculation of equilibrium properties
c*******************************************
        call chainst
10    continue
c***********************************
c normalization of measurements
c***********************************
      rgnorm = nrchains*nrmeas
      blnorm = rgnorm*(polym-1)
      r2m = r2m / rgnorm
      r4m = r4m / rgnorm
      rg2m = rg2m / rgnorm
      rg4m = rg4m / rgnorm
      lm = lm / blnorm
      l2m = l2m / blnorm
      accept = accept/(1.0d0*ntot*mcswait*nrmeas)
c***********************************
c output of measured quantities
c***********************************
      write(12,*) 'Mean squared end-to-end distance: ',r2m
      write(12,*) 'Mean quartic end-to-end distance: ',r4m
      write(12,*) 'Mean squared radius of gyration : ',rg2m
      write(12,*) 'Mean quartic radius of gyration : ',rg4m
      write(12,*) 'Mean bond length : ',lm
      write(12,*) 'Mean squared bond length : ',l2m
```

```
      write(12,*) 'Mean acceptance rate : ',accept
      end


      SUBROUTINE chainst
c*******************************************************************
c This subroutine calculates some simple chain properties, e.g. the
c average end-to-end distance, radius of gyration and bond length.
c*******************************************************************
      Implicit none
      Double precision r2m,r4m,rg2m,rg4m,lm,l2m
      Double precision r2,r4,rg2,rg4,rcm(3),dpolym
      Integer base, mon1, mon2, i, j
      Common/static/r2m,r4m,rg2m,rg4m,lm,l2m
      include ''model.common''
      include ''lattice.common''
      dpolym = polym*1.0d0
c*******************************************************************
c Calculate 2^{nd} and 4th moment of the end-to-end vector of the chains
c*******************************************************************
      do 10 i=1,nrchains
        mon1 = polym*(i-1) + 1
        mon2 = polym*i
        r2 =  (monpos(mon2,1) - monpos(mon1,1)) ** 2 +
     *        (monpos(mon2,2) - monpos(mon1,2)) ** 2 +
     *        (monpos(mon2,3) - monpos(mon1,3)) ** 2
        r4 = r2 * r2
        r2m = r2m + r2
        r4m = r4m + r4
10    continue
c*******************************************************************
c Calculate 2nd and 4th moments of the radius of gyration of the chains
c*******************************************************************
      do 20 i=1,nrchains
        rcm(1) = 0.0d0
        rcm(2) = 0.0d0
        rcm(3) = 0.0d0
        base = polym*(i-1)
        do 21 j=1,polym
          mon1 = base + j
          rcm(1) = rcm(1) + monpos(mon1,1)
          rcm(2) = rcm(2) + monpos(mon1,2)
          rcm(3) = rcm(3) + monpos(mon1,3)
21      continue
        rcm(1) = rcm(1) / dpolym
        rcm(2) = rcm(2) / dpolym
        rcm(3) = rcm(3) / dpolym
        rg2 = 0.0d0

        do 22 j=1,polym
          mon1 = base + j
          rg2 = rg2 + (monpos(mon1,1) - rcm(1)) **2 +
     *                (monpos(mon1,2) - rcm(2)) **2 +
     *                (monpos(mon1,3) - rcm(3)) **2
22        continue
        rg2 = rg2 / dpolym
        rg4 = rg2 * rg2
        rg2m = rg2m + rg2
```

```
          rg4m = rg4m + rg4
20        continue
c***************************************************************
c Calculate the 1st and 2nd moments of the bond length
c***************************************************************
      do 30 i=1,nrchains
        base = polym*(i-1)
        do 30 j=1,polym-1
          mon1 = base + j
          lm = lm + bl(monbd(mon1))
          l2m = l2m + bl2(monbd(mon1))
30        continue
      end

      SUBROUTINE INDEXX(N,ARRIN,INDX)
      DIMENSION ARRIN(N),INDX(N)
      DO 11 J=1,N
        INDX(J)=J
11      CONTINUE
      L=N/2+1
      IR=N
10      CONTINUE
        IF(L.GT.1)THEN
          L=L-1
          INDXT=INDX(L)
          Q=ARRIN(INDXT)
        ELSE
          INDXT=INDX(IR)
          Q=ARRIN(INDXT)
          INDX(IR)=INDX(1)
          IR=IR-1
          IF(IR.EQ.1)THEN
            INDX(1)=INDXT
            RETURN
          ENDIF
        ENDIF
        I=L
        J=L+L
20      IF(J.LE.IR)THEN
          IF(J.LT.IR)THEN
            IF(ARRIN(INDX(J)).LT.ARRIN(INDX(J+1)))J=J+1
          ENDIF
          IF(Q.LT.ARRIN(INDX(J)))THEN
            INDX(I)=INDX(J)
            I=J
            J=J+J
          ELSE
            J=IR+1
          ENDIF
        GO TO 20
        ENDIF
        INDX(I)=INDXT
      GO TO 10
      END
```

```
     SUBROUTINE inimove
C**********************************
     Implicit none
     Integer i, j, k, new(6,3)
     Logical test
     include ''model.common''
     do 1 i=1,108
        new(1,1) = bonds(i,1) + 1
        new(1,2) = bonds(i,2)
        new(1,3) = bonds(i,3)
        new(2,1) = bonds(i,1)
        new(2,2) = bonds(i,2) + 1
        new(2,3) = bonds(i,3)
        new(3,1) = bonds(i,1)
        new(3,2) = bonds(i,2)
        new(3,3) = bonds(i,3) + 1
        new(4,1) = bonds(i,1)
        new(4,2) = bonds(i,2)
        new(4,3) = bonds(i,3) - 1
        new(5,1) = bonds(i,1)
        new(5,2) = bonds(i,2) - 1
        new(5,3) = bonds(i,3)
        new(6,1) = bonds(i,1) - 1
        new(6,2) = bonds(i,2)
        new(6,3) = bonds(i,3)
        do 2 j=1,6
           test = .false.
           do 3 k=1,108
           test =  (new(j,1).eq.bonds(k,1)).and.
     *              (new(j,2).eq.bonds(k,2)).and.(new(j,3).eq.-
bonds(k,3))
           if(test) then
              move(i,j) = k
           else
              move(i,j) = 0
           endif
3          continue
2          continue
1    continue
     do 4 i=1,6
        move(109,i) = 109
4    continue
     end


     SUBROUTINE RANMAR(RVEC,LEN)
C******************************************************************
C Random number generator proposed in: G. Marsaglia and A. Zaman,
C Ann. Appl. Prob. 1, 462 (1991). It generates a vector 'RVEC' of
C length 'LEN' OF pseudorandom numbers; the commonblock includes
C everything needed to specify the state of the generator.
C******************************************************************
     DIMENSION RVEC(*)
     COMMON/RASET1/U(97),C,CD,CM,I97,J97
     DO 100 IVEC=1,LEN
        UNI = U(I97) - U(J97)
        IF(UNI.LT.0.) UNI = UNI + 1.
        U(I97) = UNI
```

```
          I97 = I97 - 1
          IF(I97.EQ.0) I97 = 97
          J97 = J97 - 1
          IF(J97.EQ.0) J97 = 97
          C = C - CD
          IF(C.LT.0.) C = C + CM
          UNI = UNI - C
          IF(UNI.LT.0.) UNI = UNI + 1.
          RVEC(IVEC) = UNI
100   CONTINUE
      RETURN
      END


      SUBROUTINE RMARIN(IJKL)
C*****************************************************************
C Initializes RANMAR. The input value should be in the range:
C 0 <= IJKL <= 900 000 000 . To obtain the standard values in the
C MARSAGLIA - ZAMAN PAPER (I=12, J=34, K=56, L=78) PUT IJKL = 54217137
C*****************************************************************
      COMMON/RASET1/U(97),C,CD,CM,I97,J97
      IJ = IJKL / 30082
      KL = IJKL - IJ * 30082
      I = MOD(IJ/177,177) + 2
      J = MOD(IJ,177) + 2
      K = MOD(KL/169,178) + 1
      L = MOD(KL,169)
C     WRITE(*,*) 'RANMAR INITIALIZED: ',IJKL,I,J,K,L
      DO 2 II=1,97
        S = 0.
        T = 0.5
        Do 3 JJ=1,24
           M = MOD(MOD(I*J,179)*K,179)
           I = J
           J = K
           L = MOD(53*L+1,169)
           IF(MOD(L*M,64).GE.32) S = S + T
3          T = 0.5 * T
2       U(II) = S
      C = 362436. / 16777216.
      CD = 7654321. / 16777216.
      CM = 16777213. / 16777216.
      I97 = 97
      J97 = 33
      RETURN
      END


c lattice.common
c*****************************************************************
c ls = the linear size of the lattice in lattice constants
c nmax = the maximum number of monomers on the lattice
c maxch = the maximum number of chains.
C nmax, maxch > the requirements for the standard melt simulation: a
C volume fraction of 0.5 translates into 4000 monomers on the lattice
c Monomer positions and bonds are stored in arrays indexed by the
c number (n*k + j) for the j-th monomer in the k-th chain. Fake bonds
c lead to monomer 1 and from the last monomer so we won't have to
```

```
c distinguish between them and the other monomers (same for chain
  ends).
C****************************************************************
      Integer ls, nmax, maxch
      Parameter (ls=40, nmax=10001, maxch=500)
C************************************************
c For use with real random numbers and ranmar
C************************************************
      Real rand(3*nmax)
      Integer latt(ls,ls,ls),monbd(-1:nmax),monpos(nmax,3),
     * monlatp(nmax,3),ip(ls),ip2(ls),im(ls),
     * nrchains,polym,nrends,ntot
      Common/lattice/ rand,latt,monbd,monpos,monlatp,ip,ip2,im,
     * nrchains,polym,nrends,ntot


c     model.common
C*****************************************************
   Real angles(0:100),
   Real bl(108),bl2(108)
   Integer bonds(110,3),angind(110,110),move(109,6)
   Common/model/ angles,bl,bl2,bonds,angind,move
```

Index